Radiation Detection for Nuclear Physics

Methods and industrial applications

IOP Series in Nuclear Spectroscopy and Nuclear Structure

About the Series:

The IOP Series in Nuclear Spectroscopy and Nuclear Structure provides up-to-date coverage of theoretical and experimental topics in all areas of modern nuclear spectroscopy and structure. Books in the series range from student primers, graduate textbooks, research monographs and practical guides to meet the needs of students and scientists.

All aspects of nuclear spectroscopy and structure research are included, for example:

- Nuclear systematics and data
- Ab-initio models
- Shell model-based descriptions
- Nuclear collective models
- Nuclear symmetries and algebraic descriptions
- Many-body aspects of nuclear structure
- Quantum mechanics for nuclear structure study
- Spectroscopy with gamma-rays, charged particles and neutrons following radioactive decay and reactions
- Spectroscopy of rare isotopes
- Detectors employed for gamma-ray, neutron and charged-particle detection in nuclear spectroscopy and related societal applications.

Radiation Detection for Nuclear Physics

Methods and industrial applications

David Jenkins
Department of Physics, University of York, York, UK

IOP Publishing, Bristol, UK

ISBN 978-0-7503-1428-2 (ebook)
ISBN 978-0-7503-1429-9 (print)
ISBN 978-0-7503-1784-9 (myPrint)
ISBN 978-0-7503-1430-5 (mobi)

DOI 10.1088/978-0-7503-1428-2

Version: 20201101

IOP ebooks

British Library Cataloguing-in-Publication Data: A catalogue record for this book is available from the British Library.

Published by IOP Publishing, wholly owned by The Institute of Physics, London

IOP Publishing, Temple Circus, Temple Way, Bristol, BS1 6HG, UK

US Office: IOP Publishing, Inc., 190 North Independence Mall West, Suite 601, Philadelphia, PA 19106, USA

Contents

Preface

Nuclear Physics is a subject dating back over one hundred years to Rutherford's initial suggestion of a tiny dense nucleus at the centre of the atom to explain the results of his famous gold foil scattering experiment. Over that time period, advances in the subject have been heavily driven by advances in detector technology. The necessary technology is broad in character as it has to address the challenges of detecting different types of ionising radiation such as α and β particles, γ rays, heavy ions and neutrons. The detector technology never stops developing—driven by the demands of researchers for greater efficiency, energy resolution and sensitivity to aid their studies of the nuclear system to the limits of nuclear stability. The technology frequently draws on the state-of-the-art in materials research and advances in electronics such as digital data acquisition.

A key feature of the technology used in experimental nuclear physics is that not only does it find application in other areas of science such as astronomy and

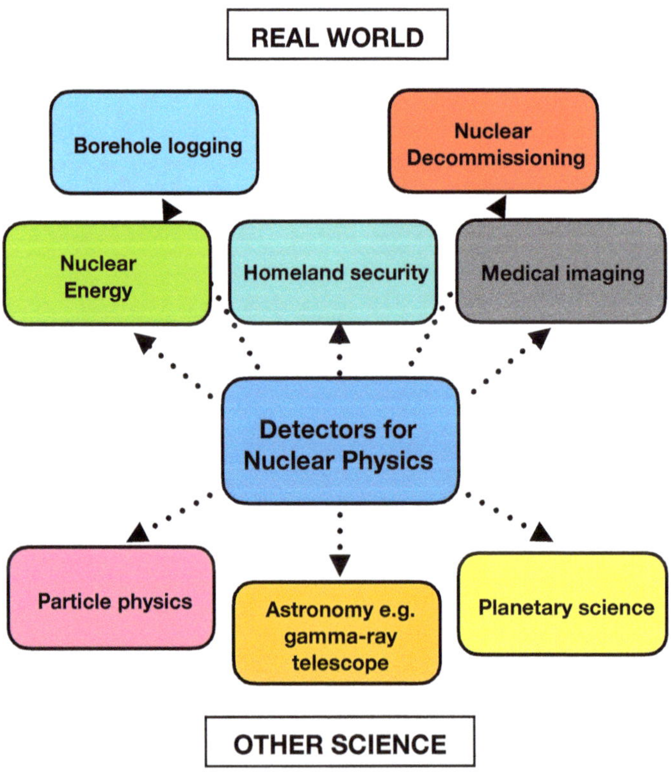

Figure 0.1. The tentacles associated with the detector technology used in experimental nuclear physics stretch far and wide. In terms of other scientific disciplines, strong commonality is found with particle physics even though the detection challenges are often very different. There is also significant interaction with the demands of gamma-ray astronomy and some aspects of planetary science. In the real world, nuclear physics detectors embrace many areas of significant societal impact from homeland security to medical imaging.

planetary science but it is also closely aligned with the demands of various industrial processes and technologies for societal benefit such as homeland security and medical imaging (see figure 0.1). It could be argued therefore that the underpinning detector technology of nuclear physics can have greater societal impact than that from related fields such as particle physics and astronomy/astrophysics.

This e-book aims to introduce readers such as PhD students in experimental nuclear physics to the state-of-the-art in nuclear detector technology and how it impacts on the wider world. It should also provide a valuable update for more experienced researchers. The e-book will also be valuable to those working in industry and medicine who wish to better understand aspects of nuclear detectors and measurements made with them which can benefit their work. The philosophy behind the book is not to give a historical overview nor to greatly focus on well-proven technology dating back to the Geiger counter, but rather to emphasise what the reader would come across and use in a well-equipped international laboratory or a modern industrial setting. Different detector technologies will be compared and contrasted and a focus will be on how to choose relevant detector technologies and designs to solve particular measurement problems.

Author biography

David Jenkins

Professor David Jenkins is Head of the Nuclear Physics Group at the University of York, UK. He is also a Fellow of the Institute of Advanced Study, University of Strasbourg (USIAS) and an Extraordinary Professor of the University of Western Cape in South Africa. Professor Jenkins' DPhil was in experimental nuclear physics at the University of York. Following postdoctoral research at Argonne National Laboratory and the University of Liverpool, he returned to the University of York as an EPSRC Advanced Fellow and subsequently joined the academic faculty there. His research in experimental nuclear physics focusses on several topics such as nuclear astrophysics (the origin of the chemical elements), clustering in nuclei and the study of proton-rich nuclei. His research is truly international and he has led experiments at accelerator facilities across Europe, the US, Canada, Japan and South Africa. In recent years, the author has developed a strong strand of applications-related research with extensive industrial collaboration. He has led the development of bespoke radiation detectors for homeland security, nuclear decommissioning, borehole logging and medical applications.

Part I

Theory and context

IOP Publishing

Chapter 1

Nuclear structure and radioactive decay

This chapter is intended as an essential introduction to the topic of radiation detection. It starts with a basic overview of nuclear structure and the essential properties of nuclei. This leads on to consideration of the concept of radioactive decay. This is illustrated with the principal decay modes of nuclear ground states (and long-lived isomeric states) by beta decay, and (generally) for heavy nuclei: alpha decay and fission. The chapter goes on to consider excited states of nuclei and the electromagnetic transitions between them; the principal modes of such electromagnetic transitions being gamma-ray emission, and in certain cases, internal conversion.

Concepts: essentials of nuclear physics; table of isotopes; fundamental nuclear models: liquid drop, shell model; concept of radioactive decay; radioactive decay modes—α, β, γ and internal conversion, and fission; excited states in nuclei; electromagnetic transitions: angular correlations and polarisation.

1.1 Introduction to basic atomic and nuclear structure

The focus of this e-book is on the detection of different types of ionising radiation whether it be in the context of experimental nuclear physics or in the many applications of the techniques and tools of nuclear physics in real-word applications. The response of detectors to different kinds of ionising radiation varies strongly in terms of the manner in which such radiation deposits energy into detectors; this issue will be reviewed in chapter 2. A stepping stone to a description of the various mechanisms by which ionising radiation interacts is to first classify it in terms of its origin, and achieve a basic understanding of how such ionising radiation relates to changes in configuration of nuclear (and atomic) states. A comprehensive understanding is well beyond the scope of the present work, and so we elect to focus only on the key features pertinent to the themes of this e-book. At the very least, however, we will be compelled to address the quantum mechanical nature of our physical world at its smallest scale, which many potential readers may find challenging.

doi:10.1088/978-0-7503-1428-2ch1

Accordingly, we summarise the key points in each section and review them in more detail for the interested reader.

1.1.1 The atomic nucleus

A paradigm shift in our understanding of the composition of the physical world occurred in 1909 when Geiger and Marsden discovered that alpha particles directed at a very thin gold foil exhibited completely unexpected behaviour, in that some small fraction of the incident alpha particles were found to be scattered through very large angles. It was Ernest Rutherford, a couple of years later, who realised that the correct interpretation of this experiment was that the mass and positive charge of the atom must be concentrated in a very small volume—the *nucleus*—at its centre [1]. It was later shown that the positive charge of the nucleus is made up from *protons* each contributing a charge, $+e$, to the nuclear system which is balanced in a neutral atom by an equal number of orbiting electrons. Chadwick [2] discovered through some of the earliest nuclear reaction studies that in addition to protons as positive charge carriers, the nucleus may also contain a neutral particle—the *neutron*. These two particles, the proton and neutron, which have almost the same mass, comprise a common set of particles, called *nucleons*.

The *atomic number* is defined by the number of protons in the nucleus, $+Z$, and it effectively corresponds to the chemical element since it also defines the number of atomic electrons in a neutral atom. The number of neutrons for a given atomic number may, however, vary leading to *isotopes*. Some isotopes are *stable* with respect to radioactive decay but there are a far larger number of isotopes, which can be produced in the laboratory or in astrophysical environments, that are *unstable* or *radioactive* to a greater or lesser degree. Such unstable isotopes are the playground of contemporary nuclear physics since they allow physicists to explore the diversity of nuclear structure and challenge the state-of-the-art in nuclear models.

Figure 1.1 illustrates the concept of isotopes using the example of the common element, Carbon. Figure 1.1 is annotated with respect to half-life (see section 1.2.1), showing the manner in which isotopes become progressively more unstable moving away from the line of stability. The relative stability of different isotopes is defined by the radioactive decay processes to be outlined below (see section 1.2). A notable isotope of carbon is ^{14}C whose unusual 5730 yr half-life makes it suitable for radiocarbon dating.

In total, there are 253 isotopes that are stable as well as a further 33 whose half-life is so long as to still be present in solar system bodies such as the Earth—described as *primordial* isotopes. However, there are many thousands more isotopes which have been synthesised in the laboratory and whose study is the province of nuclear

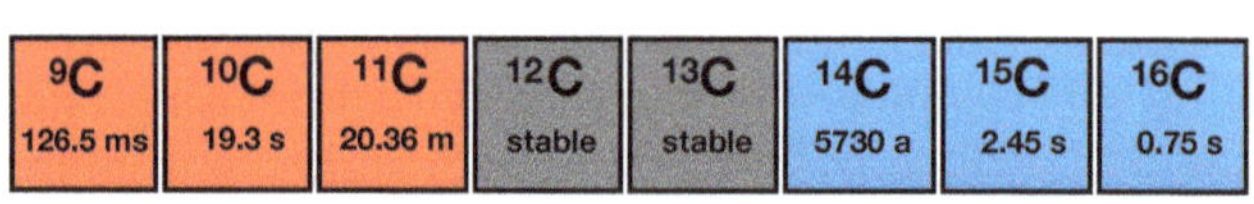

Figure 1.1. A subset of the known isotopes of carbon. Each isotope is labelled along with its respective half-life (in seconds, minutes, years). The stable isotopes are coloured grey. Those isotopes which are unstable to β^+/EC decay are indicated in red, while those which are unstable to β^- decay are shown in blue. The convention in nuclear physics is to label the isotopes with a superscript for the atomic mass on the left-hand side.

structure theory and experiment. Many further isotopes are posited to exist by nuclear models but the extreme physical conditions needed to produce them may only be manifested in exotic astrophysical scenarios such as supernova or merging neutron stars.

The extent of our knowledge of the existence of isotopes is presented in figure 1.2 where isotopes are colour-coded according to their stability and

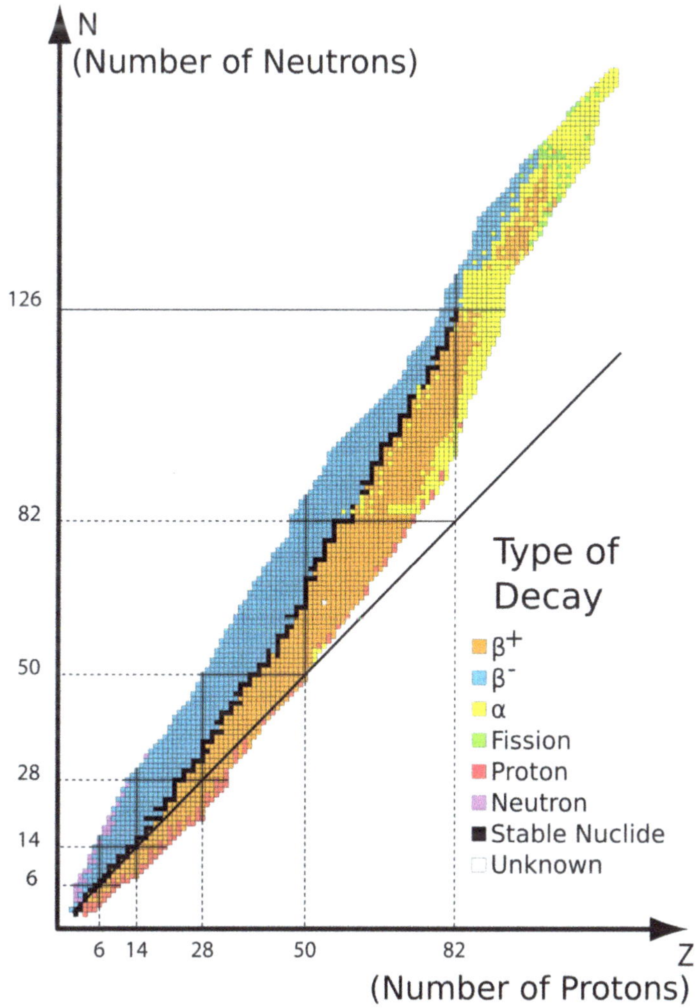

Figure 1.2. The 'table of isotopes' representing each known isotope with a square coloured box. The horizontal axis corresponds to increasing atomic number, Z, while the vertical axis represents increasing neutron number, N. The black squares are the stable isotopes. The light blue squares are isotopes which undergo β^- decay while the orange squares are isotopes which undergo β^+ or electron capture (EC) decay. Yellow squares are isotopes which decay principally by α emission, while the green squares are those isotopes which mostly decay by spontaneous fission. The purple squares are isotopes which decay by prompt neutron emission, while the red squares are those decaying by prompt proton emission. Image has been obtained by the author from the Wikimedia website where it was made available by Quibik under a CC BY-SA 3.0 licence. It is included within this article on that basis. It is attributed to Quibik.

different decay mechanisms; the decay mechanisms will be described in detail in sections 1.3–1.7.1.

1.1.2 What drives nuclear stability?

In order to understand the pattern of isotopes which exist, we need to understand something about the underlying physics which sets the limits on nuclear stability, in other words, basic nuclear physics theory. It is perhaps remarkable that over a hundred years since Rutherford's 'discovery' of the nucleus, nuclear physics theory is not something that can predict parameters to high precision in the same way that models such as quantum electrodynamics (QED) can model the interaction of light and matter. The main aspect of the challenge for nuclear theory is that it does not consider simple, point particles like the electron but rather it attempts to treat a composite object—the nucleus—which contains an inconvenient number of particles interacting principally through not one but three of the fundamental forces—electromagnetism, and the strong and weak nuclear forces. The number of particles involved in a given nucleus ranges from a single nucleon (hydrogen—^{1}H) to several hundred nucleons meaning that for all but the very lightest nuclei[1], a simple description is not possible and there are not nearly enough particles to apply a statistical model. Indeed, while the nucleons which make up the nucleus are known to be made up themselves from quarks and gluons, such information does not particularly help in describing, for example, the pattern of excited states in a heavy nucleus. The nucleus is a good example, therefore, of an *emergent phenomenon* which is difficult to understand from its fundamental constituents. The interested reader is directed to the *further reading* section in appendix A.1 where a number of e-books and physical books are suggested to learn more about nuclear structure and nuclear structure models.

Nuclear stability is determined by the binding energy of a given nuclear system. Figure 1.3 presents the binding energy per nucleon for different nuclear systems as a function of the number of nucleons making them up. Several characteristic features can be observed in this plot. First, that the binding energy per nucleon rises rapidly through the light nuclei before flattening off and reaching a maximum in the region around Fe and Ni. It then slowly decreases towards the heaviest elements such as uranium. This explains why energy is gained from the system by splitting a heavy nucleus into two parts—the energetic driver of fission observed in heavy nuclei (see section 1.5).

There are a number of excursions from the general trend in binding energy (see figure 1.3). For example, ^{4}He stands out as an exceptionally stable i.e. tightly-bound nuclear system. This stability is seen to be important in explaining the origin of alpha decay (to be discussed in section 1.3). ^{6,7}Li are located just next to ^{4}He on figure 1.3. Clearly, these two isotopes are significantly less bound than ^{4}He, with ^{6}Li less bound than ^{7}Li. This indicates the importance of pairing in nuclei where there is binding energy gained by pairing off like nucleons. ^{4}He has two protons and two neutrons

[1] There are attempts to describe light nuclei in a so-called *ab initio* approach starting with fundamental exchange forces but such techniques are so computationally intensive, they are not readily scaleable to heavier nuclei.

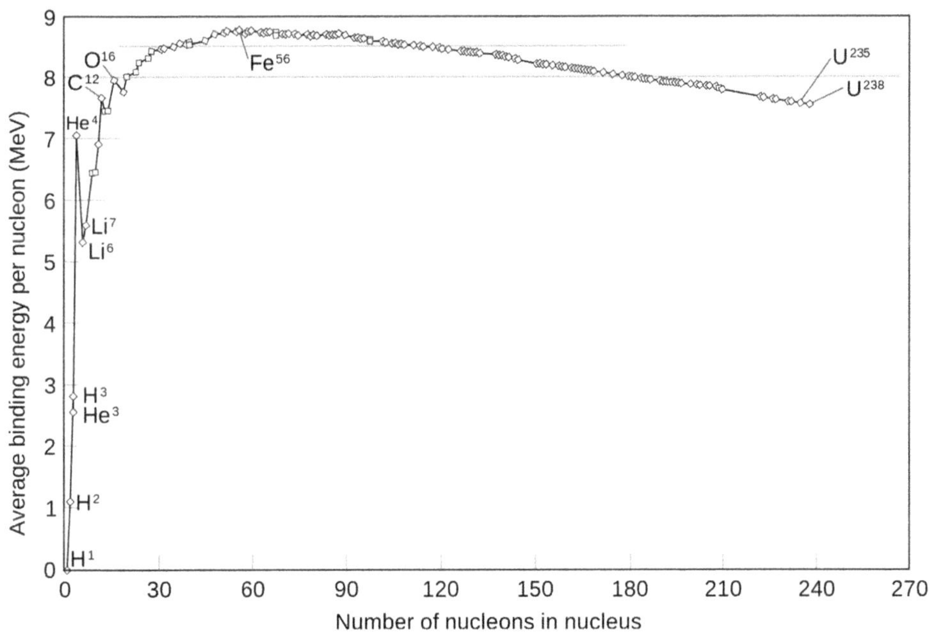

Figure 1.3. Binding energy per nucleon as a function of nucleon number. The location of a few specific nuclei are indicated on the diagram as a guide to the eye. This image has been obtained by the author from the Wikimedia website https://commons.wikimedia.org/wiki/File:Binding_energy_curve_-_common_isotopes.svg, where it is stated to have been released into the public domain. It is included within this article on that basis.

which can pair, while ^{6}Li has three protons and three neutrons leading to an unpaired neutron and proton. Such *odd–odd* nuclei are significantly less bound than their *even–even* neighbours. Indeed, there are only a handful of stable odd–odd isotopes found in nature, e.g. ^{6}Li, ^{10}B, ^{14}N[2].

1.1.3 Liquid drop model

The simplest model of the nucleus is to treat it as a liquid drop. This ignores most of the details of nuclear structure but it does a reasonably good job of describing the gross properties of the system such as the nuclear masses.

The mass of a nucleus, m_A with atomic mass A, is given by:

$$m_A = Zm_p + Nm_n - \frac{E_B}{c^2}, \tag{1.1}$$

where m_p and m_n are the masses of the proton and neutron, respectively, and E_B is the binding energy of the nucleus. E_B is expressed via the semi-empirical mass formula as:

[2] As will be discussed in more detail in section 9.2, the nuclear structure and relative instability of ^{6}Li and ^{10}B make them very suitable for components of thermal neutron detectors.

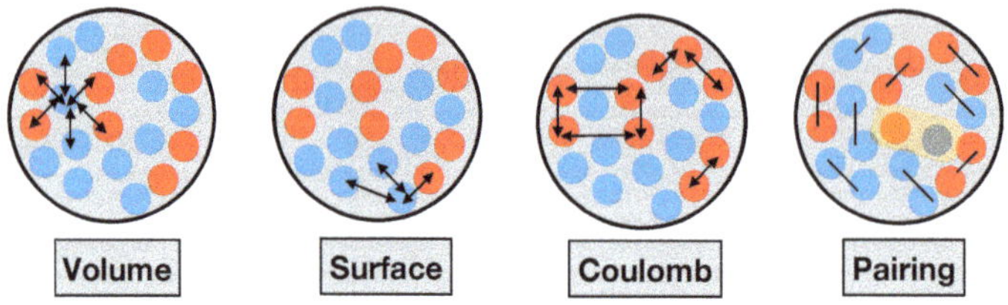

Figure 1.4. Schematic representation of four of the terms in the semi-empirical mass formula (equation (1.2)). Protons are shown in red and neutrons in blue. From left to right: the volume term which represents the fact the strong nuclear force acts most effectively on nearest neighbours; the surface term which corrects for nucleons at the surface having fewer neighbours; the Coulomb term which represents repulsion between protons (in this schematic only a few of the relevant repulsions are shown); the pairing term which reflects that like nucleons pair off (the unpaired nucleons in this example are highlighted in yellow).

$$E_B = a_V A - a_S A^{\frac{2}{3}} - a_C \frac{Z(Z-1)}{A^{\frac{1}{3}}} - a_{asym} \frac{(A-2Z)^2}{A} \pm \delta(A, Z), \qquad (1.2)$$

Let us look in some detail at the terms which make up the semi-empirical mass equation (see figure 1.4):

Volume term: $a_V A$

The strong nuclear force acts between nucleons at short range acting most effectively on nearest neighbours. The interaction scales as the number of nucleons i.e. the atomic mass, A.

Surface term: $-a_S A^{\frac{2}{3}}$

The surface term is a correction to the volume term reflecting the fact that nucleons at the surface of the nuclear system have fewer neighbours with which they can interact. The magnitude of the surface term is proportional to the surface area of the nucleus. Since the radius of the nucleus scales as $A^{\frac{1}{3}}$, the surface term scales as $A^{\frac{2}{3}}$.

Coulomb term: $-a_C \frac{Z(Z-1)}{A^{\frac{1}{3}}}$

The counterbalance to the attractive nature of the strong force at short range is the Coloumb repulsion between protons which wants to drive the system apart. This Coulomb term scales approximately as Z^2 and is inversely proportional to the radius of the system which scales as $A^{\frac{1}{3}}$. The Coulomb repulsion will become very large in high-Z nuclei and this drives the tendency towards *fission* in such systems (see section 1.5).

Asymmetry term: $-a_{\text{asym}} \frac{(A-2Z)^2}{A}$

The origin of the asymmetry term is somewhat complex but relates to the fact that nucleons are *fermions* meaning they have a half-integer spin. This means that they must observe the Pauli exclusion principle in how they fill available nuclear orbitals with no more that two fermions occupying the same quantum state. This effect serves to reduce the nuclear binding energy.

Pairing term: $\pm\delta(A, Z)$
This term reflects the fact that like nucleons prefer to pair off, increasing the binding energy of the system. This feature was noted in the discussion of the binding energy curve, figure 1.3, above.

1.1.4 Nuclear shell model

The liquid drop model solely considers bulk properties but not the details of nuclear structure. Additional observations with respect to nuclear stability need to be accounted for, namely, the increased binding of the so-called *magic nuclei* which possess a *magic number* of component nucleons, $(Z \text{ or } N) = 2, 8, 20, 28, 50, 82, \ldots$. Particularly stable examples of doubly magic nuclei include ^{40}Ca ($Z = 20$, $N = 20$) and ^{208}Pb ($Z = 82$, $N = 126$). The explanation for the occurrence of magic numbers came about through the development of the nuclear shell model in the late 1940s by Maria Goeppert Mayer and Hans Jensen[3]. The shell model posits that nuclear energy levels are quantised in a similar manner to the energy levels of atomic electrons. The path to this solution is to seek eigenstates in a three-dimensional harmonic oscillator potential. This does not immediately generate the correct sequence of magic numbers. Introducing the spin–orbit interaction ($l \cdot s$), which reflects the coupling of the intrinsic spin of the nucleon to the orbital angular momentum, serves to split the nuclear energy levels and then correctly reproduces the observed magic numbers (see figure 1.5).

1.2 Radioactive decay

Having defined the basic parameters responsible for nuclear stability, let us explore the physics of radioactive decay which governs what happens to unstable nuclei. We first define radioactive decay and concepts such as half-life before turning to the details of the decay mechanisms for the ground states of unstable nuclei.

1.2.1 Basic definitions

Radioactive decay is a *stochastic process* meaning that the decay of individual nuclei takes place randomly but the decay process for an ensemble of nuclei can be considered in statistical terms as exhibiting an exponential decay (see figure 1.6) over time governed by the decay constant, λ, in units of s^{-1}:

$$N(t) = N_0 e^{-\lambda t}, \tag{1.3}$$

where $N(t)$ is the number of radioactive nuclei at time t, and N_0 is the initial number of radioactive nuclei.

The *lifetime (mean life)*, τ, of a given decay process is given by

$$\tau = \frac{1}{\lambda}, \tag{1.4}$$

[3] Later recognised through the award of the Nobel Prize in Physics in 1963.

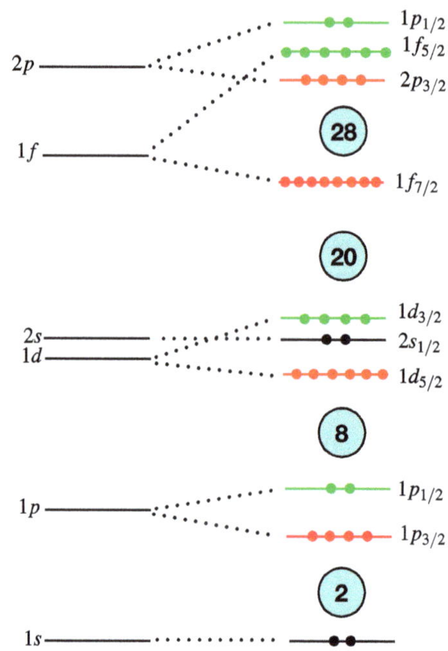

Figure 1.5. Diagram showing the nuclear shell model orbitals. The solutions for the harmonic oscillator potential are further split under the spin–orbit interaction. These orbitals are labelled with j. The $j = l + \frac{1}{2}$ orbitals are coloured red while the $j = l - \frac{1}{2}$ orbitals are coloured green. Each orbital can contain $2j + 1$ nucleons. This is shown with the coloured circles for each orbital. The position of the magic numbers is marked.

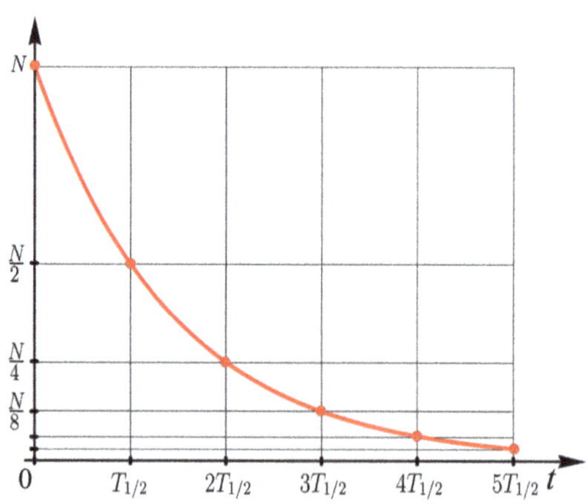

Figure 1.6. The exponential decay characteristic of radioactive decay. This image has been obtained by the author from the Wikimedia website where it was made available by Svjo under a CC BY-SA 4.0 licence. It is included within this article on that basis. It is attributed to Svjo.

which corresponds to the time for the initial sample of radioactive nuclei to decay to $\frac{1}{e}$ of the initial value. This can be converted into the more familiar *half-life*, $t_{1/2}$, the time taken for half the initial sample to decay, where

$$t_{1/2} = 0.693\tau. \tag{1.5}$$

Activity

The *activity* of a sample, $A(t)$ can be defined as

$$A(t) = \lambda N(t). \tag{1.6}$$

Activity corresponds to the number of disintegrations per second, conventionally expressed in the SI unit, the becquerel (Bq). The specific activity is the activity per unit mass

$$a = \frac{A(t)}{M} = \frac{\lambda N_A}{A}, \tag{1.7}$$

where N_A is Avogadro's number and A is the atomic mass.

1.3 Alpha decay

Alpha decay is an important decay mode for ground states and long-lived isomeric (see section 1.7.2) states in heavy nuclei. Alpha particles spontaneously emitted from heavy elements were first detected by Rutherford in 1899. It was latterly shown by Rutherford and Royds in 1907 [3] that the alpha particle corresponds to a doubly-charged ^{4}He ion (see figure 1.7). This process is, in a sense, mysterious since the alpha particle is not itself a fundamental particle and therefore must be 'pre-formed' inside the nucleus. The fact that alpha particles are emitted as a discrete unit from nuclei is closely related to ^{4}He being a very strongly bound nuclear system relative to its neighbours (see figure 1.3). This strong binding is, in turn, attributed to the fact that the alpha particle comprises two protons and two neutrons meaning closed shells for both species of nucleon and strong like-nucleon pairing.

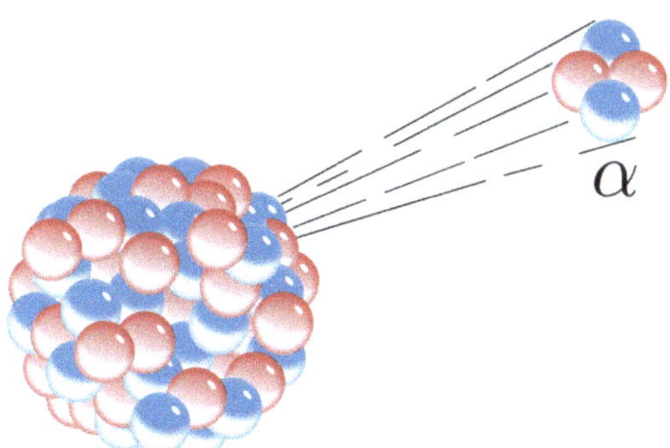

Figure 1.7. Cartoon of the alpha decay process where a doubly-charged ^{4}He nucleus is spontaneously emitted from a heavy nucleus. This image has been obtained by the author from the Wikimedia website where it was made available by Svjo under a CC BY-SA 4.0 licence. It is included within this article on that basis. It is attributed to Svjo.

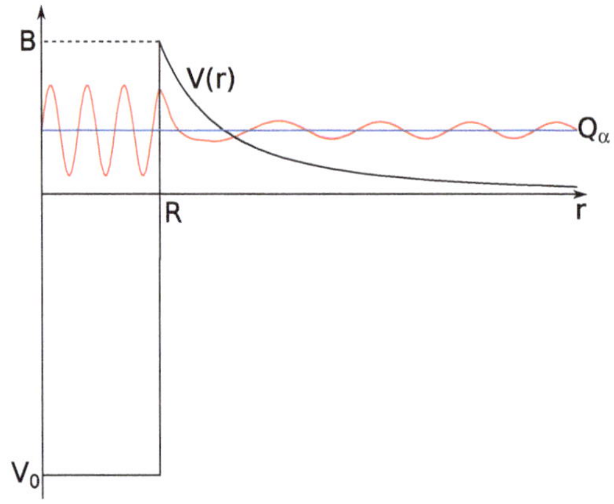

Figure 1.8. General schematic of alpha tunnelling showing the nuclear square well potential with depth, V_0 and the Coulomb barrier height, B. The potential falls off as a function of radius, $V(r)$, outside the nuclear radius, R. The decay Q value, Q_α is indicated.

The physics of alpha particle emission is governed by quantum-mechanical tunnelling[4]. This means that a pre-formed alpha particle within the nucleus does not have sufficient kinetic energy, in a classical sense, to overcome the Coulomb barrier and escape from the nucleus. Instead, the decay process takes place by tunnelling through the barrier (see figure 1.8). A recent review of some aspects of alpha decay, in particular its implications for nuclear structure measurements, may be found in reference [5].

The probability of alpha-particle emission is strongly related to the decay Q value, i.e. the amount of energy released in the decay process, which can be calculated from the masses of the nuclei involved:

$$M\left({}^{A}_{Z}X_N\right) = M\left({}^{A-4}_{Z-2}X_{N-2}\right) + M\left({}^{4}_{2}\mathrm{He}\right) + Q, \tag{1.8}$$

where Q is in mass units ($1u = 931.49$ MeV/c^2).

Alpha decay half-lives generally obey a simple relationship to the Q value given by the Geiger–Nuttall rule [6, 7]:

$$\log_{10} T_{1/2} = A(Z)Q^{-1/2} + B(Z), \tag{1.9}$$

where $A(Z)$ and $B(Z)$ are functions of the atomic number, Z. Figure 1.9 shows the general applicability of this relationship for a wide range of isotopic chains.

The Q value of the decay is shared between the kinetic energies of the alpha particle and the recoiling daughter nucleus. Taking account of the conservation of momentum, this implies that the corresponding kinetic energy of the alpha particle, T_α is given by:

$$T_\alpha = \frac{Q}{1 + \frac{m_\alpha}{m_d}}, \tag{1.10}$$

[4] In addition to alpha emission, certain heavy nuclei may undergo cluster emission, for example, ^{14}C is spontaneously emitted from ^{223}Ra [4].

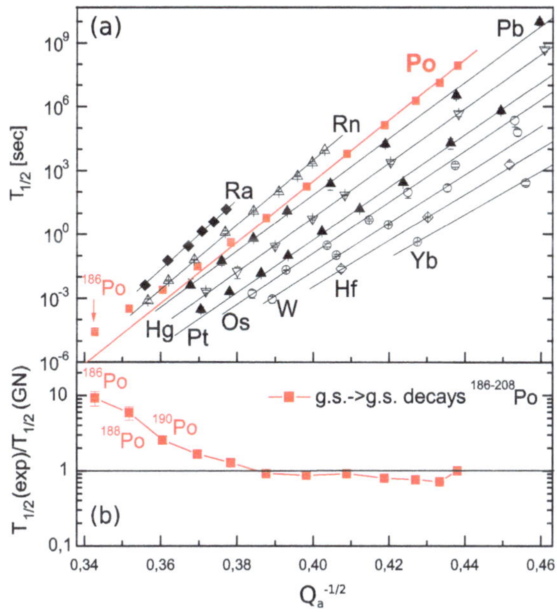

Figure 1.9. A plot of half-life, $T_{\frac{1}{2}}$ as a function of $Q_\alpha^{-\frac{1}{2}}$ for a range of isotopic chains illustrating the linearity associated with the Geiger–Nuttall relationship. The bottom section of the figure shows the case of the Po isotopes where the experimental data deviate slightly from the expected Geiger–Nuttall trend. (Reprinted from [7] with permission from Elsevier under Creative Commons License.)

where m_d is the mass of the daughter nucleus. Since α decay is a phenomenon characteristically observed in heavy nuclei for which $m_d \gg m_\alpha$, the bulk of the decay energy is transferred to the kinetic energy of the emitted alpha particle.

1.3.1 Fine structure in alpha decay

Alpha decay may also exhibit fine structure meaning that the decay proceeds in part to excited states (see section 1.6) in the daughter nucleus as well as to the ground state. An example of fine structure in the alpha decay of ^{242}Pu is presented in figure 1.10. Here, the decay proceeds around 76% of the time to the ground state of ^{238}U but some fraction of the decay also proceeds to excited states in the daughter. The reduction in decay intensity to the successive excited states represents to some extent the corresponding reduction in the Q-value of the decay but is largely governed by the angular momentum, l, carried away by the emitted alpha particle. To reach the $J^\pi = 6^+$ excited state in ^{238}U, the alpha particle must carry 6 $\hbar$ of angular momentum ($l = 6$) which is significantly less probable than the decay to the ground state ($l = 0$) which leads to the very small decay branch observed. Where the alpha decay proceeds to excited states in ^{238}U, as in the present example, the excited states will subsequently decay by γ ray emission or internal conversion towards the ground state (as will be described in more detail in section 1.7).

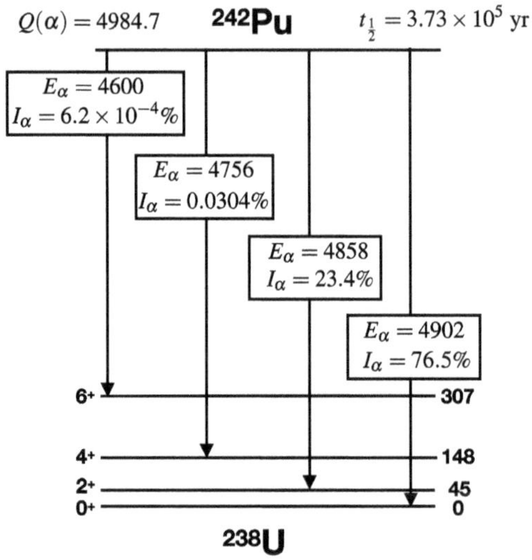

Figure 1.10. Depiction of the α decay of ^{242}Pu (a 100% alpha-decaying nucleus). The Q value for the decay is 4984.7 keV and the half-life of the isotope is 3.73×10^5 yr. The decay proceeds to the ground state of ^{238}U as well as to the first three excited states. The individual alpha transitions are labelled with their energy in keV and the corresponding decay intensity (summing to 100%). Data are taken from ENSDF.

1.3.2 Prompt proton emission

For very exotic proton-rich nuclei, prompt proton emission can take place from the ground state [8]. The physics of this process is very similar to alpha decay but lacks the pre-formation aspect since the proton is a fundamental particle. The probability for proton emission is therefore largely related again to the Q value of the decay as well as details of the nuclear structure i.e. the orbital from which the proton is emitted.

1.3.3 Conclusions on alpha decay and implications for measurements

1. Alpha decay concerns the spontaneous emission of alpha particles (^{4}He nuclei) from heavy nuclei.
2. The physics describing alpha decay is called 'quantum tunnelling' governed by the Q value of the decay.
3. The energetic conditions for quantum tunnelling mean that typical alpha particles emitted from the ground states of nuclei are in the approximate range, 3–10 MeV.
4. There is a strong correlation between alpha decay half-life and Q-value—the so-called Geiger–Nutall relationship.

1.4 Beta decay

Rutherford first identified beta particles in 1899 as a more penetrating form of radiation than alpha particles. By studying how they interacted with a magnetic field, he later showed that they behaved the same as electrons and therefore that they

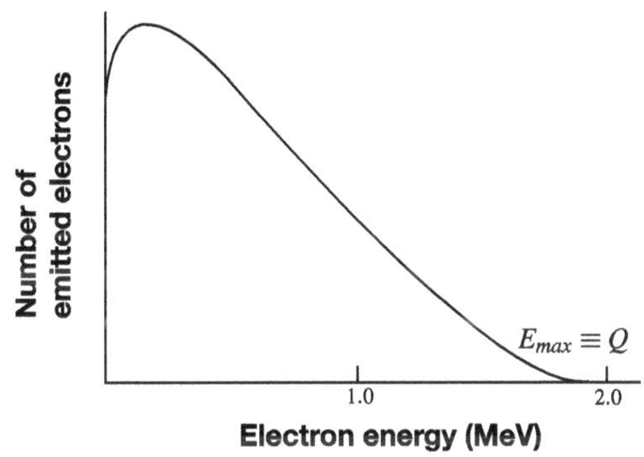

Figure 1.11. A schematic of a typical Fermi–Kurie distribution of β^- particle energies.

should be identified as high-energy electrons. The story is, in fact, a little more complex since what Rutherford studied were β^- particles which are, indeed, electrons, but there is also a parallel mode of β^+ decay where the emitted particles are positrons—the antimatter counterpart of electrons.

From the measurement point of view, the key difference between alpha particles and beta particles is that the former are emitted from the nucleus with discrete energies, while beta-particle energies are typically found to follow a broad characteristic distribution, known as the Fermi–Kurie[5] distribution [9], up to some maximum energy governed by the Q value of the decay (see figure 1.11). Majorana explained that the fact that the energies of beta particles were not discrete was related to the emission of a third particle the neutrino (or antineutrino). For all practical applications, neutrinos are undetectable due to the extremely low cross-section for their interaction with matter. However, the neutrino as a fermion does have an intrinsic spin[6] and the coupling of this spin with the spin of the emitted electron or positron turns out to imply important selection rules in beta decay, as will be described below.

Beta decay has three main types: β^-, β^+ (see figure 1.12) and electron capture (EC) (see figure 1.13). The latter two processes can compete with one another depending on the energetics of the decay process. Beta decay can only proceed if the decay is energetically favourable. Figure 1.14 shows an example of this with the energetically-favoured decays in the A = 131 isobaric chain. Let us now distinguish the main characteristics of the different β decay processes.

1.4.1 β^- decay

β^- decay is relevant to neutron-rich nuclei where a neutron within the nucleus transforms into a proton accompanied by the emission of an electron and an antineutrino:

[5] Note that Kurie has a 'K' and is not the same surname as Pierre/Marie Curie.

[6] Spin is an intrinsic form of angular momentum carried by elementary particles. It does not have a classical counterpart.

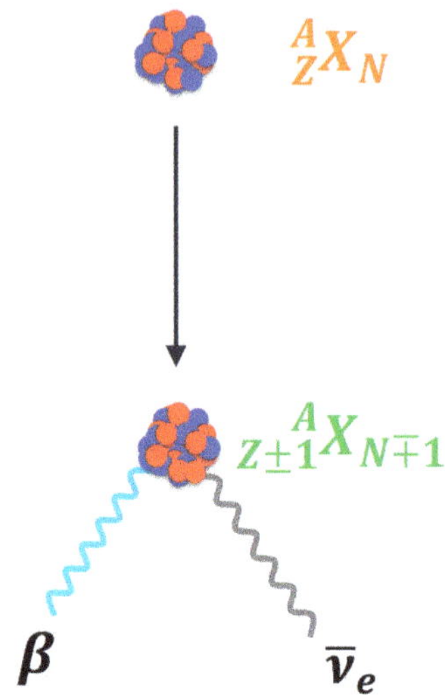

Figure 1.12. Cartoon of the beta decay process where an initial nucleus $^A_ZX_N$ transforms into $^A_{Z\pm1}X_{N\mp1}$ through emission of a beta particle (β^- or β^+) and an electron neutrino or antineutrino.

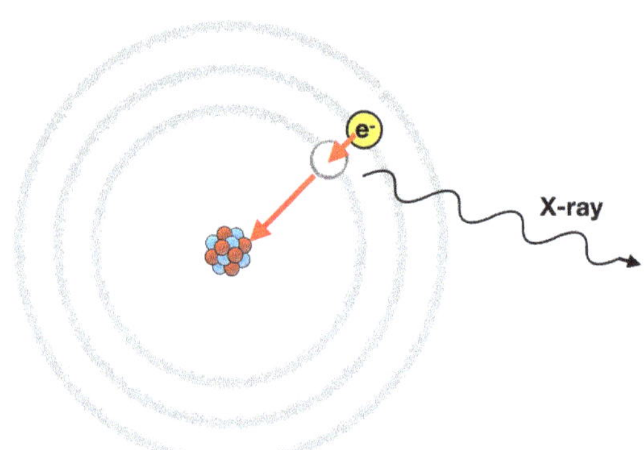

Figure 1.13. The process of electron capture. In the example shown, an electron is captured by the nucleus from the innermost (K) shell. This has the effect of transforming the nucleus $^A_ZX_N$ into $^A_{Z-1}X_{N+1}$. Characteristic x-rays are then emitted as electrons cascade down to the fill the vacancy created by the capture electron.

$$^A_ZX_N \rightarrow ^A_{Z+1}X_{N-1} + e^- + \bar{\nu}, \tag{1.11}$$

The Q value (in mass units: $1u = 931.49\ \text{MeV}/c^2$) for the process is given by the following:

$$M\left(^A_ZX_N\right) = M\left(^A_{Z+1}X_{N-1}\right) + Q. \tag{1.12}$$

The Q-value represents the energy shared between the kinetic energies of the emitted electron and antineutrino.

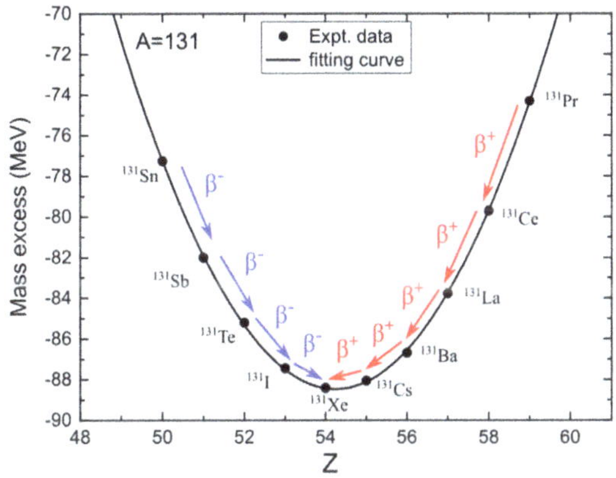

Figure 1.14. The mass parabola for the A = 131 isobaric chain: the mass excess is plotted as a function of Z. Observed beta decays are labelled with blue arrows (β^-) and red arrows (β^+/EC).

1.4.2 β^+ decay

β^+ decay is associated with proton-rich nuclei and represents the transformation of a proton into a neutron within the nucleus mediated by the emission of a positron and a neutrino:

$$^{A}_{Z}X_N \rightarrow ^{A}_{Z-1} X_{N+1} + e^+ + \nu, \tag{1.13}$$

The Q value is given by:

$$M\left(^{A}_{Z}X_N\right) = M\left(^{A}_{Z-1}X_{N+1}\right) + 2m_e + Q, \tag{1.14}$$

where m_e is the rest mass of the electron.

The positron (the antimatter counterpart of the electron) is a fundamentally unstable particle which may annihilate with an electron releasing twice the rest mass energy of the electron (1.022 MeV) in the form of photons. The principal mode of annihilation is the generation of two photons emitted near back-to-back each with an energy of 511 keV (or sometimes into three photons which share the decay energy of 1.022 MeV). The positron can annihilate in flight or it can come to rest and form positronium—an exotic atom comprised of an electron and a positron [10].

1.4.3 Electron capture

Electron capture is a process in proton-rich nuclei (see figure 1.13) where an atomic electron is captured transforming a proton in the nucleus into a neutron according to the following equation:

$$^{A}_{Z}X_N + e^- \rightarrow ^{A}_{Z-1}X_{N+1} + \nu. \tag{1.15}$$

The Q value for the process is:

$$M\left(^{A}_{Z}X_N\right) = M\left(^{A}_{Z-1}X_{N+1}\right) + Q. \tag{1.16}$$

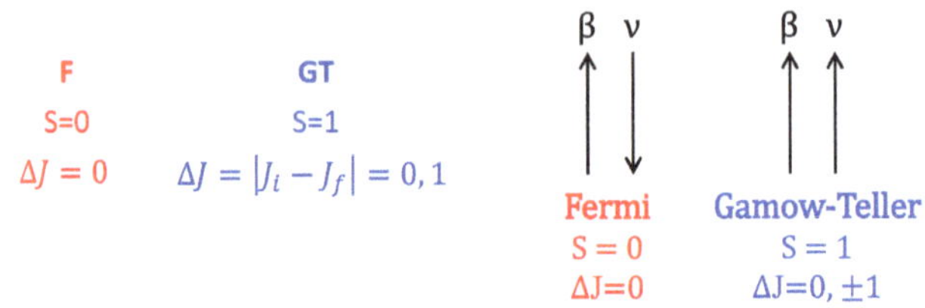

Figure 1.15. Illustration of Fermi and Gamow–Teller decay on the basis of whether the spins of the electron (positron) and (anti-) neutrino have their spins aligned or anti-aligned. The implied angular momentum selection rules are shown on the left-hand side.

For smaller Q values, the most probable interaction is with the electron K-shell. Comparing equations (1.14) and (1.16) shows that while all proton-rich unbound nuclei will have an electron capture decay branch, a β^+ decay branch will open up for Q values above 1.022 MeV and become increasingly dominant as a function of increasing Q value.

1.4.4 Fermi and Gamow–Teller decay

Two classes of beta decay are distinguished depending on whether the spins of the electron (positron) and (anti-) neutrino are aligned (Gamow–Teller) or anti-aligned (Fermi), as shown in figure 1.15. The aligned configuration corresponds to a total spin, $S = 1$, and the anti-aligned corresponds to $S = 0$. For Fermi decays ($S = 0$), the selection rule for the initial and final states is $\Delta J = (J_i - J_f) = 0$. For Gamow–Teller decays ($S = 1$), the selection rule is $\Delta J = 0$, 1 but $0 \rightarrow 0$ transitions are forbidden because of the total spin, $S = 1$. This means that $0^+ \rightarrow 0^+$ transitions are pure Fermi transitions. A subset of these are known as Fermi superallowed decays which are the fastest beta transitions known. Figure 1.16 provides two examples of Fermi superallowed transitions where a proton transforms into a neutron in the same shell model orbital.

Forbidden transitions

The discussion of beta decay transitions above focussed on the so-called *allowed transitions* where there is no associated orbital angular momentum being carried away by the emitted electron or positron, i.e. $l = 0$. There are a corresponding set of so-called *forbidden transitions* for $l > 0$. The forbiddenness of the decay is categorised by the angular momentum removed l, so that, for example, $l = 1$ decays are so-called first-forbidden and so on. Table 1.1 presents the various angular momentum selection rules depending on the order of forbidenness and whether the decay is Fermi or Gamow–Teller in character. The higher order forbidden transitions become increasingly slower leading to longer half-lives.

1.4.5 Conclusions on beta decay and implications for measurements

1. Beta decay is a collective term for several different decay modes including β^- decay in neutron-rich nuclei and the competing processes of β^+/EC decay for proton-rich nuclei.

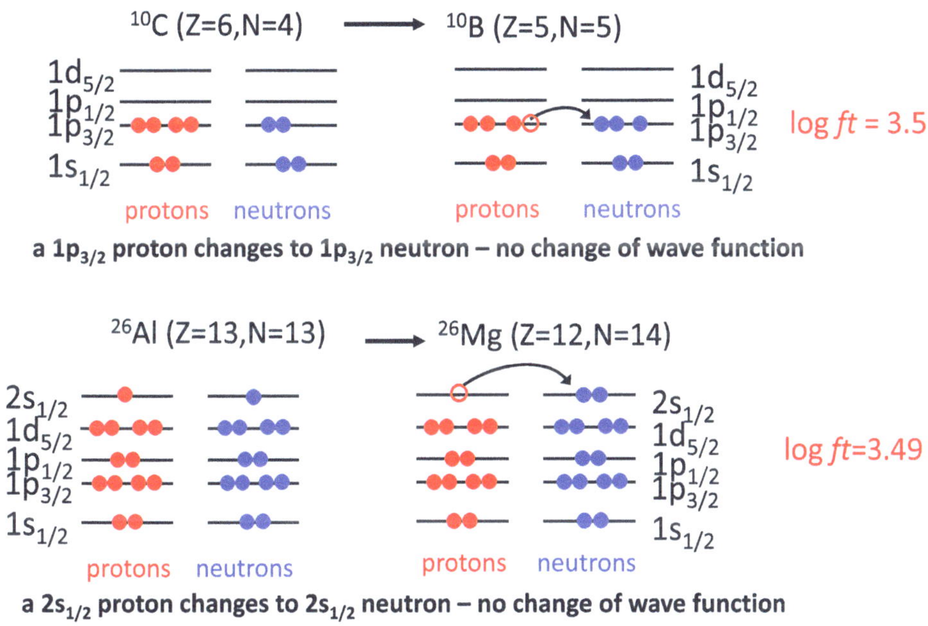

Figure 1.16. Two examples of Fermi superallowed decays: (top) the decay of ^{10}C ($Z = 6$, $N = 4$) to ^{10}B ($Z = 5$, $N = 5$) where a $1p_{3/2}$ proton transforms into a $1p_{3/2}$ neutron; (bottom) the decay of ^{26}Al ($Z = 13$, $N = 13$) to ^{26}Mg ($Z = 12$, $N = 12$) where an $2s_{1/2}$ proton transforms into an $2s_{1/2}$ neutron.

Table 1.1. Classification of beta transitions according to their order of forbiddenness. The relevant selection rules for ΔJ are given for the Fermi and Gamow–Teller case along with the parity selection rule ($\Delta\pi$). An indication of the typical range of half-lives is given.

	$\Delta\ell$	F ΔJ	GT ΔJ	Δπ?	$t_{1/2}$	log*ft*
Allowed	0	0	0, 1	No	Seconds	3.5 to 7.5
1st Forbidden	1	0, 1	0, 1, 2	Yes	Days	6.0 to 9.0
2nd forbidden	2	2	2, 3	No	Years	10 to 12
3rd forbidden	3	3	3, 4	Yes	1000 years	14 to 20

2. β^- and β^+ decays are three-body decays involving emission of an (anti) neutrino. The energies of the beta particles are therefore not discrete but rather comprise a continuous distribution (the Fermi–Kurie) distribution up to some maximum energy defined by the decay Q value.
3. Beta decay has two modes: Fermi and Gamow–Teller which depend on the relative spin direction of the emitted electron(positron) and (anti)neutrino.

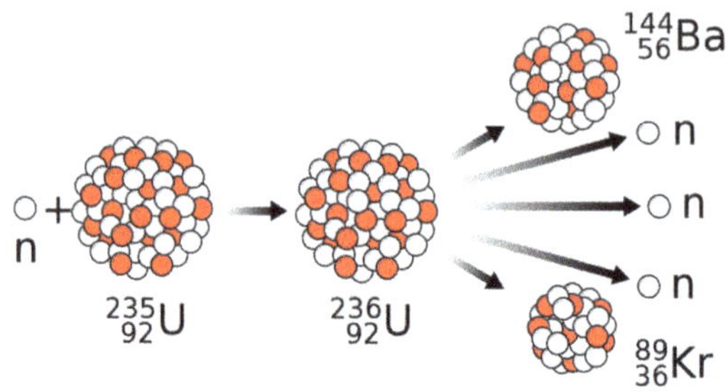

Figure 1.17. General schematic of fission showing one specific branch for the fission of ^{236}U into ^{144}Ba and ^{89}Kr accompanied by the emission of three neutrons. This image has been obtained by the author from the Wikimedia website where it was made available by MikeRun under a CC BY-SA 4.0 licence. It is included within this article on that basis. It is attributed to MikeRun.

4. Beta decay half-lives vary widely depending on the *Q* value of the decay, the decay mode i.e. Fermi or Gamow–Teller, and the degree of forbiddenness.

1.5 Fission

Fission is a process common to very heavy nuclei where the nucleus spontaneously breaks into two asymmetric fragments[7] plus a number of emitted neutrons. Some nuclei exhibit spontaneous fission from their ground state (and excited states) while other nuclei can be induced to fission by, e.g. capture of a low energy neutron. Figure 1.17 shows an example of this where one specific branch is shown for the fission of ^{236}U following capture of a low energy neutron onto ^{235}U. Here, the particular daughter fragments obtained are ^{144}Ba and ^{89}Kr but this is only schematic and, in fact, a wide distribution of possible daughter fragment nuclei are observed from this fission process. As an example, figure 1.18 shows the distribution in mass and kinetic energy of the daughter fission fragments from the β-delayed fission of ^{180}Tl [11]. In fact, this is a rather exotic example of the fission process where the *Q* value of the beta decay of ^{180}Tl is sufficiently large that the decay proceeds to some excited states in the daughter nucleus ^{180}Hg which spontaneously fission—hence, the name, β-delayed fission. This particular β-delayed fission process exhibits quite a distribution of fragment masses and fragment kinetic energies centred around a typical value of 140 MeV.

Returning to figure 1.17, it can be seen that three neutrons are also emitted as a by-product of the fission. The number of neutrons emitted can also vary depending on the fragment distribution (and indeed, the details of nuclear structure effects) and so an average number of neutrons per fission is commonly presented. Spontaneous fission can be understood as evolution through a deformed potential barrier. Figure 1.19 provides a cartoon of the process, while figure 1.20 shows some recent calculations of the fission of ^{240}Pu which illustrates the speed of the process.

[7] While the most commonly observed mode of fission is asymmetric, in certain special cases, symmetric or near-symmetric fission is observed for various exotic isotopes

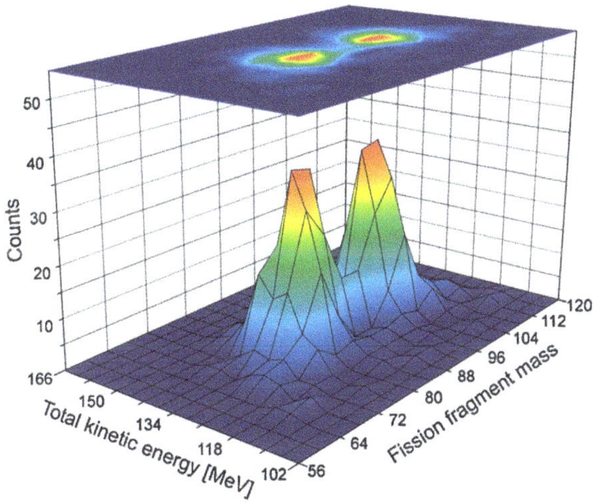

Figure 1.18. Fission fragment mass and kinetic energy distribution for the β-delayed fission of ^{180}Hg. (Reprinted with permission from [11]. Copyright (2010) by the American Physical Society.)

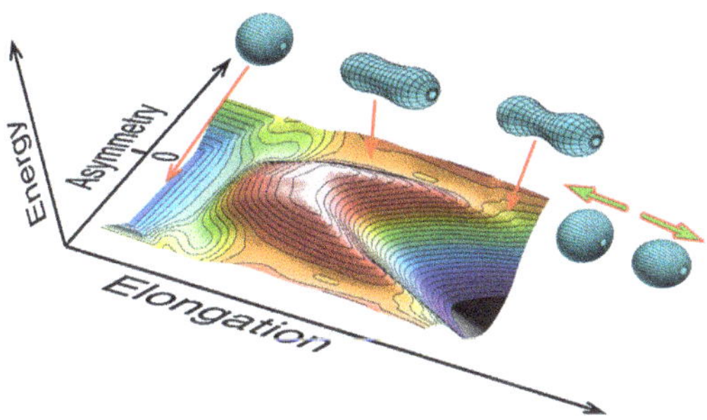

Figure 1.19. Cartoon of the fission process. (Reprinted with permission from [11]. Copyright (2010) by the American Physical Society.)

1.5.1 Conclusions on fission and implications for measurements

1. Fission is a process associated with heavy nuclei where the nucleus spontaneously splits into two fragments accompanied by the emission of a few neutrons.
2. Fission fragments have typical energies of the order of 100 MeV.
3. Some nuclei spontaneously fission but many more will fission following, e.g. capture of a thermal neutron.

1.6 Excited states

Nuclear excited states are quantised in a similar manner to electron orbitals in atoms, and are labelled by quantum numbers of total angular momentum and parity, usually written as J^{π}. For nuclei with one proton or one neutron outside a doubly-magic closed shell, the extreme independent particle model allows the wavefunction of the first few excited states to be described simply in terms of the shell

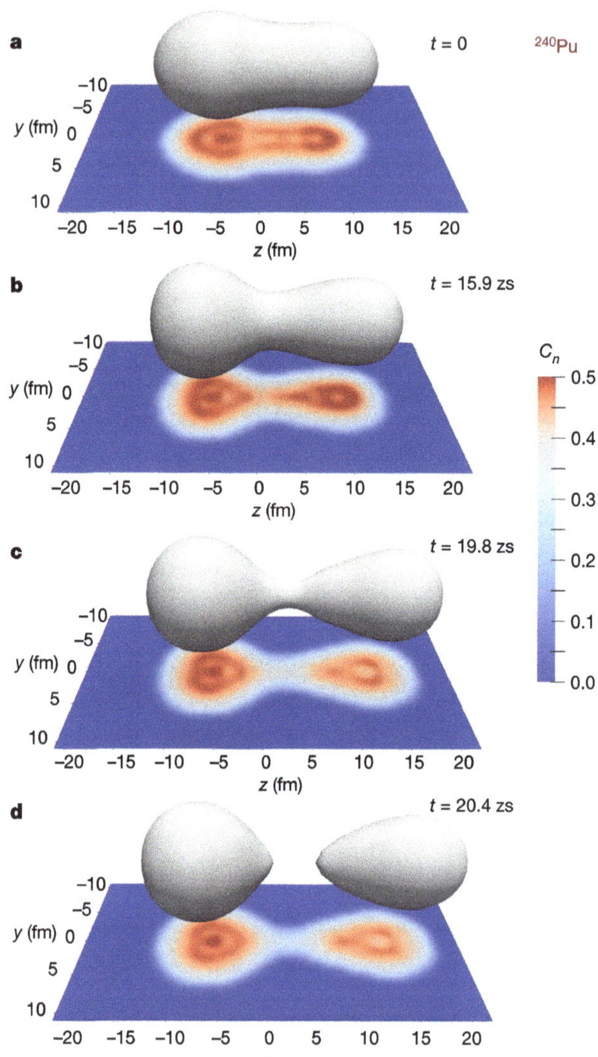

Figure 1.20. Calculated isodensity surfaces for fission of ^{240}Pu as a function of time. The timescale is in zeptoseconds (10^{-18} s) which demonstrates how fast the fission process takes place. (Reprinted by permission from Springer Nature [12].)

model orbitals presented in figure 1.5. In general, however, the wavefunctions of nuclear excited states are significantly complex with many individual components. The present e-book is not intended to provide a detailed overview of such aspects of nuclear structure[8]—this is covered in parallel e-books in this series. It is worthwhile, however, to discuss some basic aspects of nuclear structure so as to set the measurement problem in context. A first observation to make is on the energy

[8] A good online resource for evaluated nuclear structure data is found here: ENSDF.

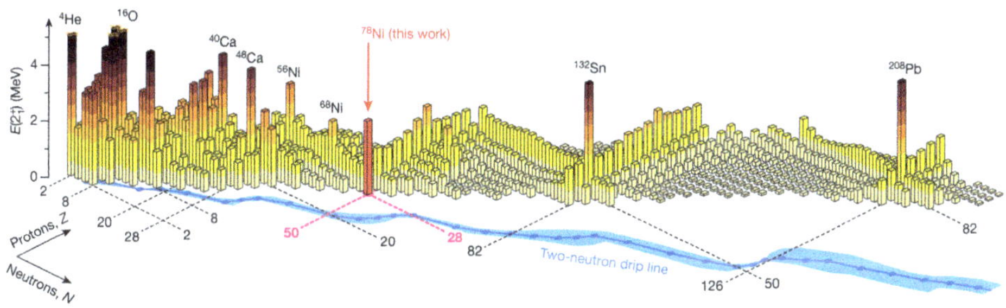

Figure 1.21. Energies of the first excited 2^+ state in even–even nuclei as a function of proton and neutron number. The positions of the magic numbers for protons and neutrons are indicated along with the two-neutron drip line. (Reprinted by permission from Springer Nature [13].)

scaling i.e. that the scaling and spacing of nuclear levels is of the order of 1 MeV, compared to eV in the case of an atomic system. Figure 1.21 illustrates this by plotting the energy of the first excited 2^+ states[9] in even–even nuclei as a function of proton and neutron number. We will return to the issue of energy scale when we discuss transitions between nuclear excited states typically mediated by gamma-ray emission (see section 1.7.1) or internal conversion (see section 1.7.3).

Let us now look in more detail at aspects of nuclear structure in terms of the first few excited states of three even–even (meaning even number of protons and neutrons) nuclei, ^{40}Ca, ^{110}Cd and ^{238}U. Figure 1.22 shows in an overview the diversity observed in these different nuclear systems. A general feature of the pattern of excited states is its compression in excitation energy as a function of atomic mass, A.

As discussed in section 1.1.4, a key feature of nuclear structure is the appearance of magic numbers for N and Z ($N = Z = 2, 8, 20, 28, 50\ldots$). These are particularly stable nuclear configurations representing closed shells and there is a high cost in energy to excite protons and neutrons across this shell closure. Such nuclei are near-spherical in their ground state. A good example of this would be ^{40}Ca (left-hand side of figure 1.22) where the first excited states are high in energy—around 3.5 MeV—with a large gap between them and the ground state. Figure 1.21 presents the energies of the first excited 2^+ state[10] in even–even nuclei across the chart of the nuclides. This clearly shows how strongly the energy of this first excited state varies and how it is large for chains of nuclei with a magic number of protons and neutrons, reaching maxima for cases where the nucleus has a magic number of both protons and neutrons.

Nuclei also exhibit collective modes such as vibration and rotation. Quadrupole vibration should lead to a distinctive pattern of states which are equally spaced in energy: the first excited state would have spin/parity 2^+, the next set of excited states (degenerate in energy in a pure vibrational model) would have spin/parity 0^+, 2^+, 4^+.

[9] The first excited state of an even–even nucleus is nearly always a 2^+ state.
[10] This is nearly always the first excited state although in some particular nuclei, as the example of ^{40}Ca given, a 0^+ state is the first excited state.

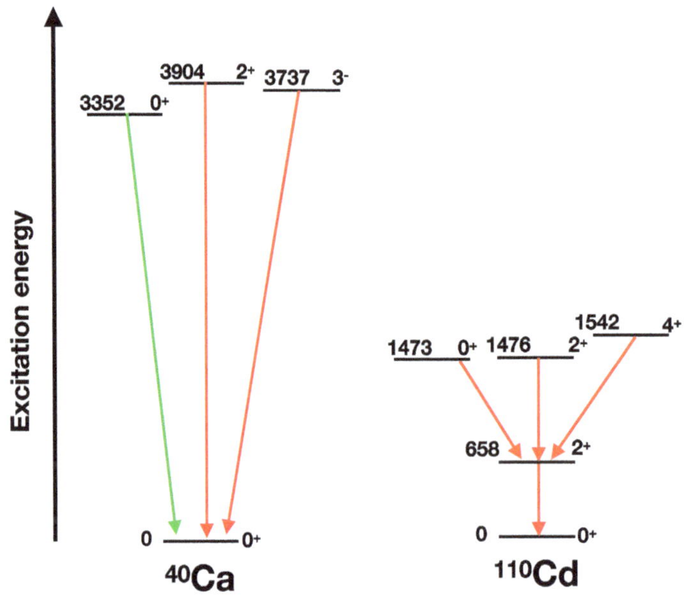

Figure 1.22. The first few excited states in some even–even nuclei: ^{40}Ca whose structure relates to being a doubly-magic ($Z = 20$, $N = 20$) nucleus, ^{110}Cd which has something close to the expected structure of a vibrational nucleus, and ^{238}U which is typical of a rotational nucleus. The states are labelled by their excitation energy in keV and their spin/parity (J^{π}). By convention, the excitation energy of the 'ground state' is shown as 0. Observed electromagnetic transitions are indicated with the coloured arrows: the red arrows are gamma-ray transitions while the green arrow is the special case of internal pair transition (see section 1.7.3) since $0^+ \rightarrow 0^+$ transitions are forbidden to occur by gamma-ray emission. Transitions between the low-lying states of ^{238}U are said to be highly converted meaning that there is a high probability that they occur through emission of conversion electrons (see section 1.7.3).

The centre panel of figure 1.22 shows the first excited states of ^{110}Cd which have traditionally been interpreted as being consistent with such a vibrational model. The second nuclear collective mode is rotation, which is associated with a deformed nuclear shape (predominantly a prolate—rugby-ball shape). Here, the pattern of excited states follows an excitation energy dependence proportional to $J(J + 1)$. This leads to the stretching out of the spectrum of rotational states as a function of spin. A textbook example of such rotational behaviour is the strongly-deformed nucleus, ^{238}U whose low-lying states are shown in the right-hand panel of figure 1.22.

In a sense, it is highly artificial to classify nuclei exclusively according to one of the three modes discussed above. In fact, more than one of these modes can usually be seen at work in the structure of a given nucleus. This is certainly true of the nuclei used as examples above. For example, while the low-lying states of ^{40}Ca appear to conform closely to that of a closed-shell nucleus, in reality the first excited 0^+ state is the 'band head' of a superdeformed rotational band [14]. Moreover, deformed, rotational configurations are observed in ^{110}Cd and it has been suggested that the entire low-lying structure may be better described as the coexistence of different deformed shapes and their respective rotational/vibrational excitation (see figure 1.23)

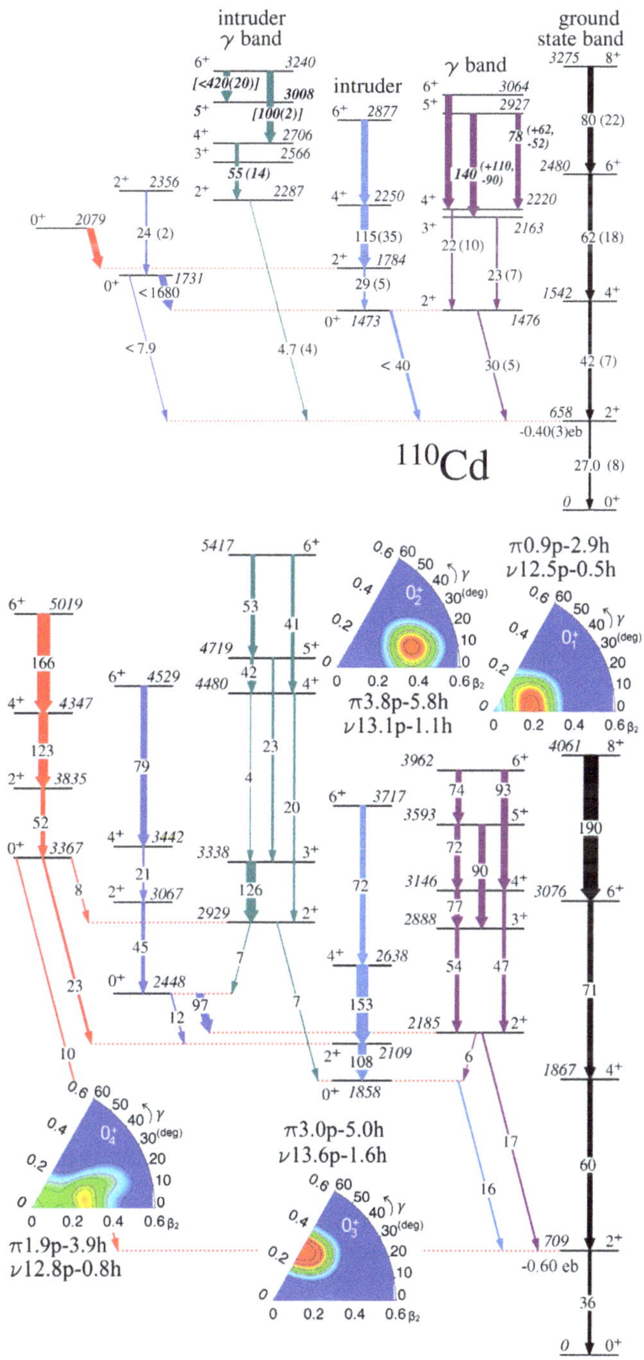

Figure 1.23. Partial experimental level scheme for ^{110}Cd (top) compared with beyond-mean-field (BMF) calculations (bottom). Levels are labelled with spin and parity, and excitation energy in keV. The transitions shown are labelled not with transition energy but $B(E2)$ transition strength (see section 1.7.2). (Reprinted with permission from [16]. Copyright (2019) by the American Physical Society.)

[15, 16]. This clearly illustrates the complexity of nuclear structure and the widespread occurrence of nuclear shape coexistence.

1.7 Transitions between nuclear excited states: electromagnetic decay modes

1.7.1 Gamma decay

Gamma rays are associated with a high energy (high frequency) part of the electromagnetic spectrum. Gamma-ray emission is commonly associated with an electromagnetic transition between quantised excited states of a given nucleus. The observations in the previous section on the structure of various even–even nuclei of differing mass and differing underlying structure show that the relevant transition energies will be in the range of keV to MeV. In principle, transition energies can range from even a few eV up to 10s of MeV. While this represents a substantial overlap (in the keV region) with the energy range of x-rays, the convention is to discriminate these two types of ionising radiation based on their origin viz the gamma ray is emitted by a nucleus and x-rays from rearrangement of electrons in the atomic shells. From the point of view of their detection—the focus of this book—the issues are the same.

In principle, the energy difference between the initial and final nuclear states is shared between the emitted gamma ray and the recoiling nucleus. The energy imparted to the recoiling nucleus is typically very small, however, due to the low momentum associated with the gamma ray. Conserving momentum in gamma ray emission shows that the recoil energy is:

$$E_R = \frac{E_\gamma^2}{2Mc^2}, \tag{1.17}$$

where M is the atomic mass in a.m.u. For a typical gamma ray with an energy of 1 MeV emitted from a nucleus with $A = 100$, the nuclear recoil is of order 5 eV. This means for most practical purposes, the gamma-ray energy as detected is almost equal to the transition energy between the two nuclear states and only in specialist applications would a correction be important. Although unbound states are also relevant for nuclear spectroscopy, in most gamma-ray spectroscopy studies in nuclear physics and in essentially all practical applications of gamma-ray spectroscopy, gamma rays of interest connect bound nuclear states. Bound nuclear states have lifetimes from the order of fs through to the timescale of years in the case of some nuclear isomers. The natural width of the gamma-ray transition is that of the emitting state. The width of the state in terms of energy can be extracted from the lifetime via the Heisenberg uncertainty relation:

$$\Delta E \cdot \Delta t \sim \hbar. \tag{1.18}$$

For a half-life as short as a femtosecond, the width of the state is around 1 eV. This is far smaller than the energy resolution of most practical detectors.

Multipolarity of transitions

Gamma ray photons may carry angular momentum away from the nuclear system. For a transition from a state with angular momentum vector $\vec{J}_i$ to a state with vector $\vec{J}_f$ the angular momentum vector of the photon $\vec{L}$ must complete the vector triangle:

$$\vec{J}_i = \vec{L} + \vec{J}_f. \tag{1.19}$$

This implies that the largest possible value of L is $J_i + J_f$ and the smallest value of L is $|J_i - J_f|$. Gamma rays also possess electric (E) or magnetic (M) character. This is related to the change in parity between the initial and final states mediated by the decay. The selection rule[11] is:

$$\pi(ML) = (-1)^{L+1} \tag{1.20}$$

$$\pi(EL) = (-1)^{L}. \tag{1.21}$$

For example, an $E1$ transition is an electric dipole transition where parity changes between the initial and final state, while an $M1$ transition is a magnetic dipole transition where parity is unchanged. For example, for a transition $J^\pi = 4^+ \rightarrow 2^+$, the allowed multipolarities are $E2$, $M3$, $E4$, $M5$, $E6$. However, the lowest-order multipolarity for a given transition will generally dominate. It is quite common for a transition to be of mixed multipolarity e.g. $M1/E2$, the extent of mixing is defined by a mixing ratio, δ.

Gamma ray angular distributions/correlations

A means of distinguishing the multipolarity of gamma-ray transitions is to make use of observed angular correlations. A single gamma-ray emitted, for example, in conjunction with the beta decay of a radioactive source will have no preferred direction of emission; the emission will be *isotropic*. However, this is not the case for gamma rays in a cascade where angular correlations between the emission direction of the gamma rays are observed and are related to the spin/parity of the excited states concerned[12].

The angular correlation between two gamma rays in a cascade has the general form

$$W(\theta) = \sum_{i=0,even}^{\infty} B_{ii} G_{ii}(t) A_{ii} P_i(\cos\theta), \tag{1.22}$$

where B_{ii} is a parameter specifying the initial nuclear orientation, $G_{ii}(t)$ represents time-dependent effects of extranuclear interactions, A_{ii} are coefficients which have a dependence on the spins of the states involved, the multipolarity of the

[11] Note that $E0$ transitions can occur but only by internal conversion not gamma-ray emission, while $M0$ transitions do not exist.

[12] A comprehensive review of the theory of gamma ray angular distributions is presented in reference [17].

individual transitions and mixing of each transition, and $P_i(\cos\theta)$ are Legendre polynomials [17].

For short-lived nuclear states, the time-dependent effect can be ignored and where the initial nuclear orientation is isotropic then $B_{ii} = 1$. In this case,

$$W(\theta) = \sum_{i=0,\text{even}}^{\infty} A_{ii} P_i(\cos\theta). \tag{1.23}$$

The angular distribution is therefore a sum of even Legendre polynomials weighted by appropriate factors reflecting the spins and partities of the excited states involved. Figure 1.24 shows the measured angular correlation of the two principal gamma rays from the decay of ^{60}Co which form a $4^+ \rightarrow 2^+ \rightarrow 0^+$ cascade, exhibiting the characteristic Legendre polynomial fit.

Gamma ray emission may also exhibit non-isotropic angular distributions where the emitting nucleus has its spin aligned either through the nuclear reaction mechanism, or by cooling a sample of nuclei to liquid helium temperatures and aligning their spins with a strong magnetic field. A further method is to measure angular distributions of gamma rays with respect to the angle of detection of an emitted charge particle e.g. an alpha particle—this is the eponymous 'Litherland and Ferguson Method II' [18].

Depending on the circumstances where they are produced, gamma rays (in common with photons in other parts of the electromagnetic spectrum) may also be linearly or circularly polarised [19].

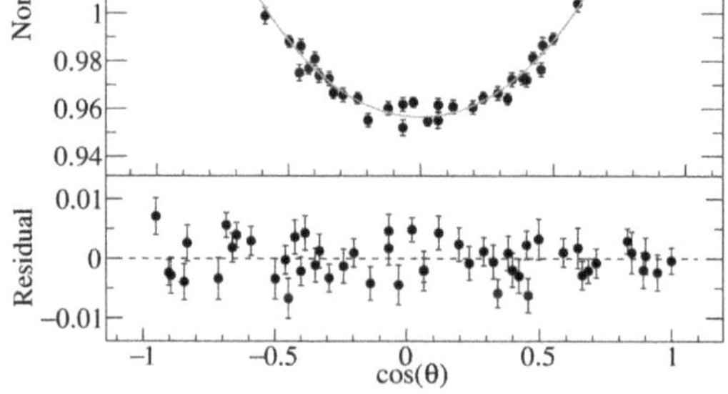

Figure 1.24. Angular correlation of gamma rays following the decay of ^{60}Co. The two gamma rays have energies of 1332 and 1173 keV corresponding to a $4^+ \rightarrow 2^+ \rightarrow 0^+$ cascade. (Reprinted from [20] with permission from Elsevier.)

1.7.2 Lifetime of nuclear excited states

The lifetime of excited nuclear levels is governed by energetic and nuclear structure considerations. This means that the lifetime of excited nuclear states can vary broadly from the order of fs to, in principle, millions of years. A useful classification for long-lived states is to describe them as *isomers* which are defined as a nuclear state whose decay is retarded. Formally, there is no lower limit for an isomeric half-life but practically speaking, minimal isomeric half-lives are of the order of 10 ns while some isomers have half-lives of years e.g. 31 yr in the case of ^{178m}Hf [21]. Figure 1.25 provides some examples of nuclear isomers which originate from different aspects of nuclear structure.

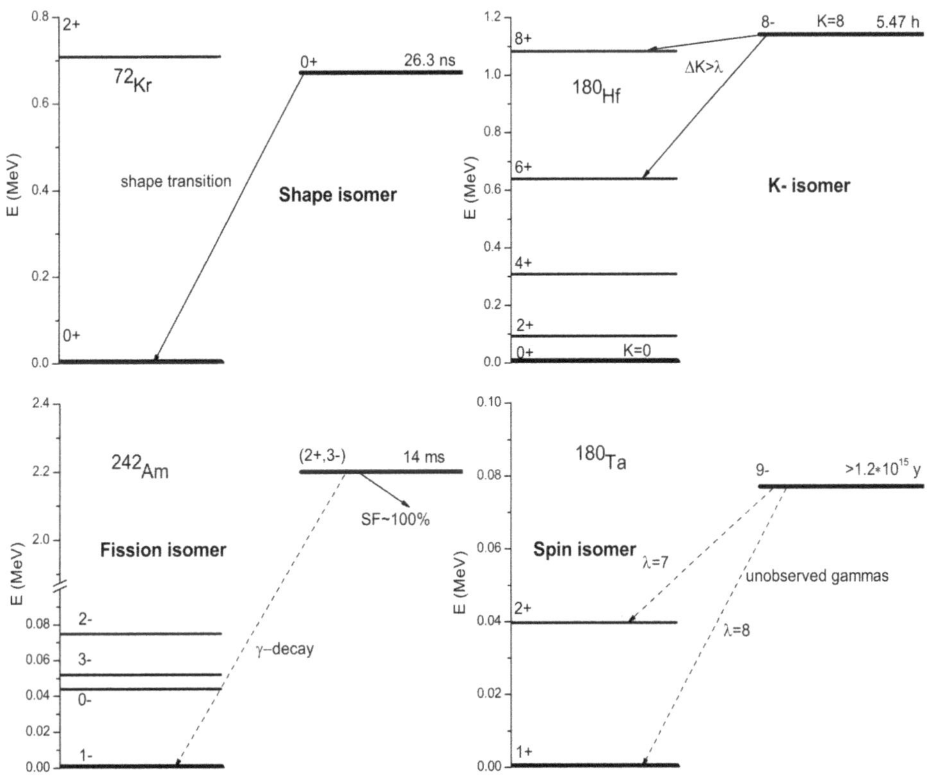

Figure 1.25. Four examples of isomeric states labelled with their respective half-life: (top left) the first excited state (0^+) in ^{72}Kr which appears, unusually, below the first 2^+ state due to shape coexistence in this nucleus. The $0^+ \rightarrow 0^+$ decay is slow because it can occur only through internal conversion (see section 1.7.3). (Top right) a so-called K isomer where the decay is retarded due to the nuclear structure of the isomer which has the total angular momentum vector oriented in a very different direction to that of the states which it can decay to. (Bottom left) a fission isomer in ^{242}Am which is a highly-deformed excited state that can fission more readily than undergo energetically allowed gamma decay. (Bottom right) a spin isomer in ^{180}Ta which cannot gamma decay because the associated angular momentum change implies a very high multipolarity (e.g. *E*7) for the gamma rays which implies a highly retarded decay (see section 1.7.2). (Reprinted from [22] with permission from Elsevier.)

The probability of a given nuclear transition is conventionally expressed as a transition strength, $B(EL)$ for electric transitions and $B(ML)$ for magnetic transitions, where L is the multipolarity of the transition. This factors out the large dependence of the decay rate/lifetime on gamma ray energy, E_γ. The transition strengths may be obtained from the following formulae:

$$\lambda(E1) = 1.587 \times 10^{15} E_\gamma^3 B(E1) \tag{1.24}$$

$$\lambda(E2) = 1.223 \times 10^{9} E_\gamma^5 B(E2) \tag{1.25}$$

$$\lambda(E3) = 5.698 \times 10^{2} E_\gamma^7 B(E3) \tag{1.26}$$

$$\lambda(M1) = 1.779 \times 10^{13} E_\gamma^3 B(M1) \tag{1.27}$$

$$\lambda(M2) = 1.371 \times 10^{7} E_\gamma^5 B(M2) \tag{1.28}$$

$$\lambda(M3) = 6.387 \times 10^{0} E_\gamma^7 B(M3), \tag{1.29}$$

where $B(EL)$ is in units of $e^2 fm^{2L}$, $B(ML)$ is in units of $\mu_N^2 fm^{2L-2}$, and E_γ is in units of MeV.

1.7.3 Internal conversion

Internal conversion is a process competitive to gamma-ray emission where the nuclear transition is mediated through an interaction with atomic electrons leading to the emission of an electron [23–25]. Since this leaves a vacancy in the electron shell, often the innermost, K-shell, electrons from higher orbitals may cascade down leading to emission of characteristic x-rays or *Auger electrons*, where some of the energy of the transition is transferred to another electron which is then ejected from the atom.

The extent of internal conversion for a given electromagnetic transition is commonly represented by a conversion coefficient, α_{ic} defined as the ratio of the electron emission rate (T_{ic}) to the gamma-ray emission rate (T_γ):

$$\alpha_{ic} = \frac{T_{ic}}{T_\gamma}. \tag{1.30}$$

Since internal conversion can take place through interaction with different atomic shells, the conversion coefficient is typically broken down into contributions from major shells and subshells (figure 1.26):

$$\alpha_{ic} = \alpha_K + \alpha_{L1} + \alpha_{L2} + \cdots. \tag{1.31}$$

The energy of the transition must exceed the binding energy of a given atomic shell/subshell for that decay mode to be possible. This leads to discontinuities in the magnitude of the total conversion coefficient as a function of transition energy as successive decay modes are turned off (see figure 1.27). The difference between the

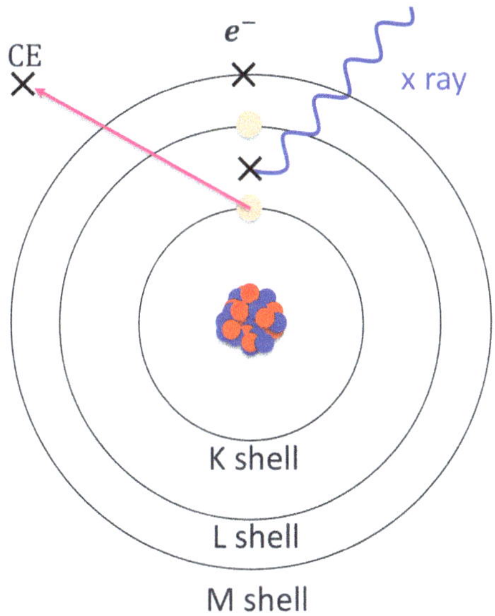

Figure 1.26. Cartoon of internal conversion process. An electron is ejected from the innermost K electron shell. X-rays are emitted as electrons cascade down to fill the vacancy created in the K-shell.

transition energy and the binding energy is imparted to the electron in the form of kinetic energy. An electron spectrum corresponding to a given transition generally therefore has multiple peaks corresponding to ejection of electrons from different shells/subshells. Internal conversion coefficients increase as Z^3, and decreasing transition energy and are therefore more prominently observed in the case of electromagnetic transitions in heavy nuclei (see figure 1.28). The total conversion coefficient also varies strongly as a function of multipolarity (see example in figure 1.27).

The magnitude of the various subshell contributions contributing to α_{ic} is dependent on the multipolarity of the electromagnetic transition. Internal-conversion spectroscopy yielding the total conversion coefficient and the different shell/subshell contributions can therefore, in some cases, serve to identify the dominant multipolarity and electric/magnetic character of the electromagnetic decay. This can be very useful spectroscopic information which is often more difficult to establish using gamma-ray spectroscopy alone.

The reader is directed to the web-based package Bricc developed by Tibor Kibédi at the Australian National University (ANU), which can calculate ICCs for a given transition in a specific nucleus as a function of multipolarity of the transition.

A special case to note relates to transitions connecting a state with $J^\pi = 0^+$ to a second state with $J^\pi = 0^+$. The photon has an intrinsic spin, $S = 1$ and cannot couple these states meaning that a gamma-ray transition is forbidden and only internal conversion is possible. Strictly speaking, two-photon emission is allowed (at around 10^{-4} level relative to internal conversion) and these are some of the few cases that spontaneous two-photon emission has been measured in a nuclear system [27, 28].

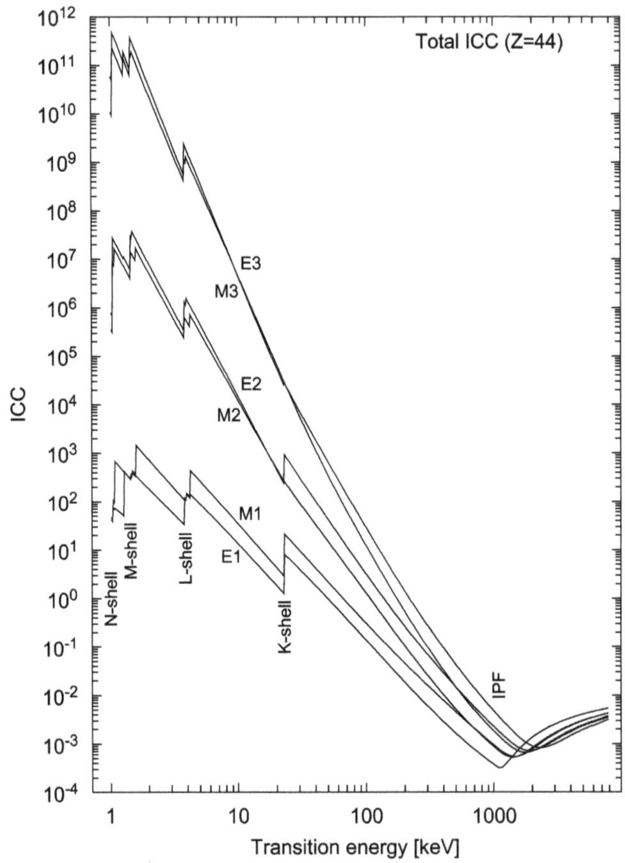

Figure 1.27. Total internal conversion coefficient (ICC) calculated for $Z = 44$ as a function of multipolarity. (Reprinted from [25] with permission from Elsevier.)

Internal pair transition

Where the transition energy is high enough (>1.022 MeV), it is possible for an emitted photon to convert into a positron–electron pair in the Coulomb field of the emitting nucleus [29]. This is known as an internal pair transition and can be considered as another form of 'internal conversion'.

1.7.4 Conclusions on gamma decay and implications for measurements

1. Gamma rays are a form of electromagnetic radiation. They correspond to transitions between excited states in nuclei.
2. Gamma rays can, in principle, have a wide range of energy but the typical energy range of interest is of the order 10 keV–few MeV.
3. In the region below 100 keV, gamma ray energies overlap with the energies of x-rays and they are physically indistinguishable. The only difference is their origin—x-rays relate to transitions between atomic states.

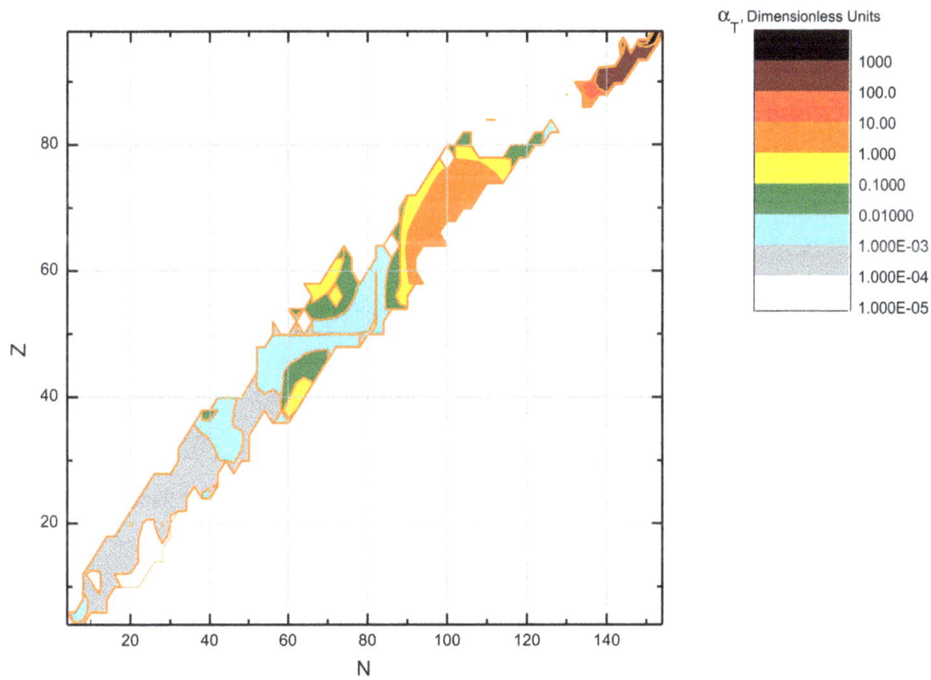

Figure 1.28. Total internal conversion coefficient for the $2^+ \rightarrow 0^+$ transition in even–even nuclei in the range $Z = 2 \rightarrow 104$. The variation in internal conversion coefficient as a function of Z is evident. Conversion is very high for high-Z elements and largely negligible in low-Z elements. (Reprinted from [26] with permission from Elsevier.)

4. For high-Z nuclei and low-energy transitions, internal conversion (ICC) is a competing process corresponding to ejection of atomic electrons.
5. Depending on the scenario, gamma-ray emission may be isotropic or may exhibit angular distributions/correlations. Gamma rays may also be linearly or circularly polarised. Polarisation and angular correlation measurements may provide insight to the underlying multipolarity of the transition.

References

[1] Rutherford E 1911 The scattering of alpha and beta particles by matter and the structure of the atom *Philos. Mag.* **21** 669–88

[2] Chadwick E 1932 The existence of a neutron *Proc. R. Soc.* A **136** 692–708

[3] Rutherford E and Royds T 1901 The nature of the alpha particle from radioactive substances *Philos. Mag.* **17** 281–6

[4] Rose H J and Jones G A 1984 A new kind of natural radioactivity *Nature* **307** 245–7

[5] Qi C 2016 Alpha decay as a probe for the structure of neutron-deficient nuclei *Rev. Phys.* **1** 77–89

[6] Geiger H and Nuttall J M 1911 The ranges of the alpha particles from various radioactive substances and a relation between range and period of transformation *Philos. Mag.* **23** 439–45

[7] Qi C, Andreyev A N, Huyse M, Liotta R J, Van Duppen P and Wyss R 2014 On the validity of the Geiger-Nuttall alpha-decay law and its microscopic basis *Phys. Lett.* B **734** 203–6
[8] Woods P J and Davids C N 1997 Nuclei beyond the proton drip-line *Annu. Rev. Nucl. Part. Sci.* **47** 541–90
[9] Kurie F N D, Richardson J R and Paxton H C 1936 The radiations emitted from artificially produced radioactive substances *Phys. Rev.* **49** 368–81
[10] Berko S and Pendleton H N 1980 Positronium *Annu. Rev. Nucl. Part. Sci.* **30** 543–81
[11] Andreyev A N *et al* 2010 New type of asymmetric fission in proton-rich nuclei *Phys. Rev. Lett.* **105** 252502
[12] Scamps G and Simenel C 2018 Impact of pear-shaped fission fragments on mass-asymmetric fission in actinides *Nature* **564** 382–6
[13] Taniuchi R *et al* 2019 ^{78}Ni revealed as a doubly-magic stronghold against nuclear deformation *Nature* **569** 53–8
[14] Ideguchi E *et al* 2001 Superdeformation in the doubly magic nucleus ^{40}Ca *Phys. Rev. Lett.* **87** 222501
[15] Garrett P E 2014 The evolving structure of the Cd isotopes *EPJ Web Conf.* **66** 02039
[16] Garrett P E *et al* 2019 Multiple shape coexistence in ^{110,112}Cd *Phys. Rev. Lett.* **123** 142502
[17] Rose H J and Brink D M 1967 Angular distributions of gamma rays in terms of phase-defined reduced matrix elements *Rev. Mod. Phys.* **39** 306–47
[18] Litherland A E and Ferguson A J 1961 Gamma-ray angular correlations from aligned nuclei produced by nuclear reactions *Can. J. Phys.* **39** 788–824
[19] Fagg L W and Hanna S S 1959 Polarization measurements on nuclear gamma rays *Rev. Mod. Phys.* **31** 711–58
[20] Smith J K, MacLean A D, Ashfield W, Chester A, Garnsworthy A B and Svensson C E 2019 Gamma-gamma angular correlation analysis techniques with the GRIFFIN spectrometer *Nucl. Instrum. Methods Phys. Res.* A **922** 47–63
[21] Helmer R G and Reich C W 1968 Decay of an isomer state in ^{178}Hf with $K \geqslant 16$ *Nucl. Phys.* A **114** 649–62
[22] Jain A K, Maheshwari B, Garg S, Patial M and Singh B 2015 Atlas of nuclear isomers *Nucl. Data Sheets* **128** 1–130
[23] Church E L and Weneser J 1960 Nuclear structure effects in internal conversion *Annu. Rev. Nucl. Part. Sci.* **10** 193–234
[24] Band I M, Trzhaskovskaya M B, Nestor C W, Tikkanen P O and Raman S 2002 Dirac–Fock internal conversion coefficients *At. Data Nucl. Data Tables* **81** 1–334
[25] Kibedi T, Burrows T W, Trzhaskovskaya M B, Davidson P M and Nestor C W 2008 Evaluation of theoretical conversion coefficients using BrIcc *Nucl. Instrum. Methods* A **589** 202–29
[26] Pritychenko B, Birch M, Singh B and Horoi M 2016 Table of E2 transition probabilities from the first 2^+ states in even-even nuclei *At. Data Nucl. Data Tables* **107** 1–139
[27] Kramp J, Habs D, Kroth R, Music M, Schirmer J, Schwalm D and Broude C 1987 Nuclear two-photon decay in $0^+ \rightarrow 0^+$ transitions *Nucl. Phys.* A **474** 412–50
[28] Schirmer J, Habs D, Kroth R, Kwong N, Schwalm D, Zirnbauer M and Broude C 1984 Double gamma decay in ^{40}Ca and ^{90}Zr *Phys. Rev. Lett.* **53** 1897–900
[29] Schlüter P, Soff G and Greiner W 1981 Pair creation by internal conversion *Phys. Rep.* **75** 327–92

IOP Publishing

Radiation Detection for Nuclear Physics
Methods and industrial applications
David Jenkins

Chapter 2

Interaction of ionising radiation with matter

This chapter provides an overview of the underlying physics involved in the stopping of different kinds of ionising radiation in materials. These can be broadly classified into directly ionising radiation such as alpha particles, heavy ions and electrons, and indirectly ionising radiation such as gamma rays and neutrons.

Concepts: stopping of protons, alpha particles and heavy ions; stopping of electrons; gamma-ray interactions: photoelectric absorption, Compton scattering and pair production; fast neutron interactions

2.1 General remarks

Understanding how different kinds of ionising radiation interact with matter, whether it be a solid, liquid or gas, are fundamental to an appreciation of how detectors are chosen and operate in the context of experimental nuclear physics and its applications. Many new to this field may still likely recall the high-school-level description of radiation where alpha particles are stopped by a piece of paper, beta particles by a thin sheet of metal e.g. aluminium and gamma rays by a significant thickness of lead (see figure 2.1). In this chapter, we attempt to go beyond this simplistic description to understand some of the essentials of the underlying physics which is fundamentally quantum mechanical in nature. We can discriminate between *directly ionising* radiation such as heavy ions, alpha particles and electrons which in their slowing down in matter, cause atoms to be ionised, and *indirectly ionising* radiation such as gamma rays and neutrons which scatter e.g. electrons and protons, respectively, in matter and these secondary particles lead to ionisation. Prior to considering the details of the stopping in matter, an initial question might be: are these different ionising radiations stopped in a single interaction or do they interact many times with the components making up the material—nuclei and electrons? The answer is that gamma rays, for example, may effectively lose all their energy in a single (or a few) interaction(s) but heavy ions like alpha particles lose energy semi-continuously through many smaller interactions with the components

doi:10.1088/978-0-7503-1428-2ch2

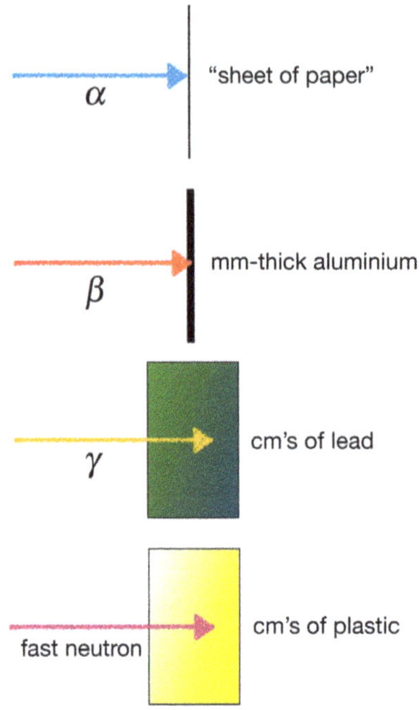

Figure 2.1. Illustration of the relative ranges of different types of ionising radiation in different materials (here, we assume incoming radiation with an energy of around 1 MeV). Alpha particles are stopped by even 10s of microns of material e.g. a sheet of paper, beta particles are stopped by around 1 mm of aluminium. Gamma rays are stopped by a few cm of lead. Neutrons thermalise in cms of plastic.

of the material. Monte Carlo simulation such as GEANT4 (see section 4.2.1) has reached a level of sophistication that when modelling a detector, all these interactions are taken account of as a matter of course. This allows the detector physicist to appreciate e.g. the efficiency of such a detector for a given class of ionising radiation. However, without a basic appreciation of the interaction processes, their dependence on Z and other physical properties of the material, it would be difficult to know where to start. We begin then with the interactions of protons, alpha particles and other heavy ions with matter.

2.2 Protons, alpha particles and heavy ions

2.2.1 Fundamentals of stopping power theory

The slowing down of ions in material is an issue which is theoretically complex and has occupied the attention of scientists from the earliest studies of the interactions of alpha particles with thin foils. The treatment of the underlying physics has evolved to the extent that the stopping power at a given energy, $S(E) = -\frac{\mathrm{d}E}{\mathrm{d}x}$, i.e. energy loss per unit length for a given energy, can be calculated to an accuracy of the order of a few % for light ions such as protons, deuterons and alpha particles in the energy range >1 MeV/u. The accuracy of the underlying theory can be confirmed through comparison with experimental data. An excellent review of this topic is given by

Ziegler [1]. The reader is directed there for a full derivation of the relevant theory and a historical overview of the topic in terms of the recognition of the importance of increasingly detailed elements of physics in the interaction process. Briefly, a key early realisation was that of Bohr that the stopping process could be divided into two contributions from nuclear stopping, i.e. interaction with nuclei in the stopping material and electronic stopping where energy is transmitted to the atomic electrons making up the stopping material. The latter turns out to be the more important process. Given that the interactions are electromagnetic, a strong relationship to the atomic number (Z_1) of the stopping ion and that of the stopping material (Z_2) should be expected. Higher Z ions should be more effectively stopped and therefore have shorter ranges compared to lower Z ions of an equivalent energy.

Charged particles travelling through matter lose energy principally by interacting with atomic electrons within the material. This leads to excitation of atomic electrons and ionisation. The energy loss of charged particles such as protons, alpha particles in matter is given by the Bethe–Bloch formula [1]. We present it in the relativistic form derived by Fano [2]:

$$-\frac{\mathrm{d}E}{\mathrm{d}x} = \frac{4\pi e^4 Z_2}{m_e v^2} Z_1^2 \left[\ln \frac{2mv^2}{\langle I \rangle} - \ln(1-\beta^2) - \beta^2 - \frac{C}{Z_2} - \frac{\delta}{2} \right], \tag{2.1}$$

where m_e is the rest mass of the electron, v is the velocity of the particle, β is v/c, $\langle I \rangle$ is the averaged excitation potential per electron. The two final terms in equation (2.1) are the *shell correction* term, $\frac{C}{Z_2}$, which corrects for the assumption that the ion velocity is much larger than the velocity of electrons in the target material, and the *density effect* correction term, $\frac{\delta}{2}$, which corrects for polarisation effects in the target i.e. the difference between electromagnetic fields in free space and in the dielectric medium of the target.

Latterly, corrections were suggested to the theory which account for higher order terms. Defining a prefactor, $\kappa = 4\pi r_0^2 m_e c^2$, the Bethe–Bloch formula can be expressed as:

$$-\frac{\mathrm{d}E}{\mathrm{d}x} = \frac{\kappa Z_2}{\beta^2} Z_1^2 [L_0(\beta) + Z_1 L_1(\beta) + Z_1^2 L_2(\beta) + \cdots]. \tag{2.2}$$

Here, the term $L_0(\beta)$ incorporates the terms in the right-hand side of equation (2.1) but two additional corrections are considered: L_1—the *Barkas correction* and L_2—the *Bloch correction*. The Barkas correction scales as Z_1^3 and so is sensitive to the sign of the charge of the stopping particle accounting for differences found experimentally between stopping of negative and positively charged ions. This small difference arises from the distortion to the electron cloud in the target medium induced by the charge of the stopping particle. The Bloch correction represents higher order terms of order Z_1^4 which are necessary to fit data and represent violation of the assumption that the interacting particles can be fully treated as free particles.

2.2.2 Low-velocity ions

While the basic Bethe–Bloch theory described above is valid for high velocities, at lower velocities account must be taken of increasing neutralisation of ions, i.e. as they slow down they acquire atomic electrons reducing their effective charge when considered as an ensemble. This must be considered in a statistical way because clearly ions cannot individually have fractional charges. Northcliffe [3] parametrized this in terms of a statistical net charge, Z_1^*, which can be thought of as the mean value of the charge-state distribution:

$$\frac{Z_1^*}{Z_1} = 1 - \exp\left\{\frac{-v_1}{v_0 Z_1^{\frac{2}{3}}}\right\}, \tag{2.3}$$

where v_1 is the velocity of the stopping particle and v_0 is the Bohr velocity $\left(\frac{e^2}{\hbar}\right)$. Clearly, at high velocities, $Z_1^* \simeq Z_1$. This effect is much more important for heavy ions than light ions; it can be thought of as replacing Z_1 with Z_1^* in equation (2.2) in order to reproduce the experimentally observed stopping power.

2.2.3 Range

The range of a particle stopping in a material, Δx can be evaluated by integration of the reciprocal of the stopping power, $S(E)$, from the incident energy down to zero:

$$\Delta x = \int_0^{E_0} \frac{1}{S(E)} \mathrm{d}E, \tag{2.4}$$

where E_0 is the initial particle kinetic energy.

2.2.4 Bragg peak

Inspection of equation (2.1) shows that the stopping power, $-\frac{\mathrm{d}E}{\mathrm{d}x}$ has a dependence of $\frac{1}{v^2}$ meaning that the stopping power strongly increases at the end of the stopping process when the ion has a low velocity. Accordingly, most of the energy is deposited close to the end of the range leading to the occurrence of the so-called *Bragg peak*. Figure 2.2 shows an example of a 290 MeV/u ^{12}C beam stopping in a plastic (PMMA); the Bragg peak is clearly visible[1]. Interestingly in this example, further energy dissipation mechanisms are visible in terms of fragmentation of some fraction of the carbon ions along their track into lighter ions with differing energies and correspondingly varying ranges.

[1] The characteristic Bragg peak is exploited in particle-beam therapy as a means of treating cancers avoiding surrounding tissue damage. This relies on the fact that the energy can be deposited at depth in the body in a controlled manner, rather than fully along the track in the case of conventional radiotherapy.

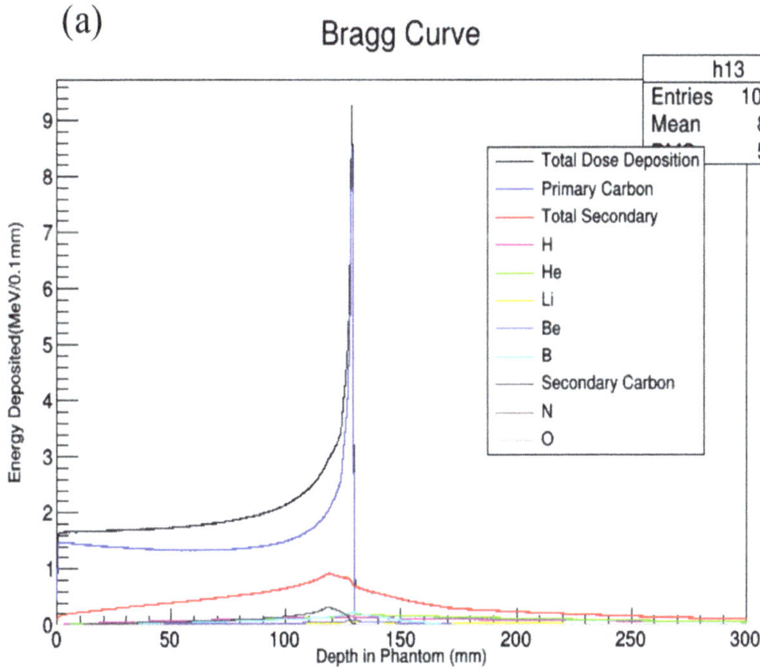

Figure 2.2. GEANT4 simulation of the Bragg peak for 290 MeV/u ^{12}C beam incident into a polymethyl methacrylate (PMMA) phantom. (Reprinted from [4] with permission of IOP Publishing under Creative Commons License.)

2.2.5 Straggling

The theory discussed above is statistical in nature, meaning that each stopping particle will have an individual trajectory based on the details of the collisions it undergoes. This leads to a *straggling* in energy and range, i.e. a statistical distribution around a central value predicted by the relevant stopping theory. Straggling can be defined as the square root of the variance in the particle range. Straggling will occur both in the direction of the incident particle beam and laterally. Measuring straggling precisely by experiment is challenging due to the likely non-uniformity and inhomogeneity of realistic target materials.

2.2.6 Channeling

A special case to note in terms of the interaction of heavy ions with materials is *channeling*[2]. Channeling occurs in oriented crystals such as the Si $\langle 110 \rangle$ surface which has a honeycomb structure. If ions from, e.g. an accelerator can be accurately aligned with the regular holes in the crystal structure then they can follow trajectories where they weakly interact with the electron gas in the channels and

[2] For a review of heavy-ion channeling, the reader is directed to chapter 5 of vol 6 of *Treatise on Heavy Ion Science*, Springer, 1985, ed D A Bromley (ISBN: 978-1-4615-8105-5).

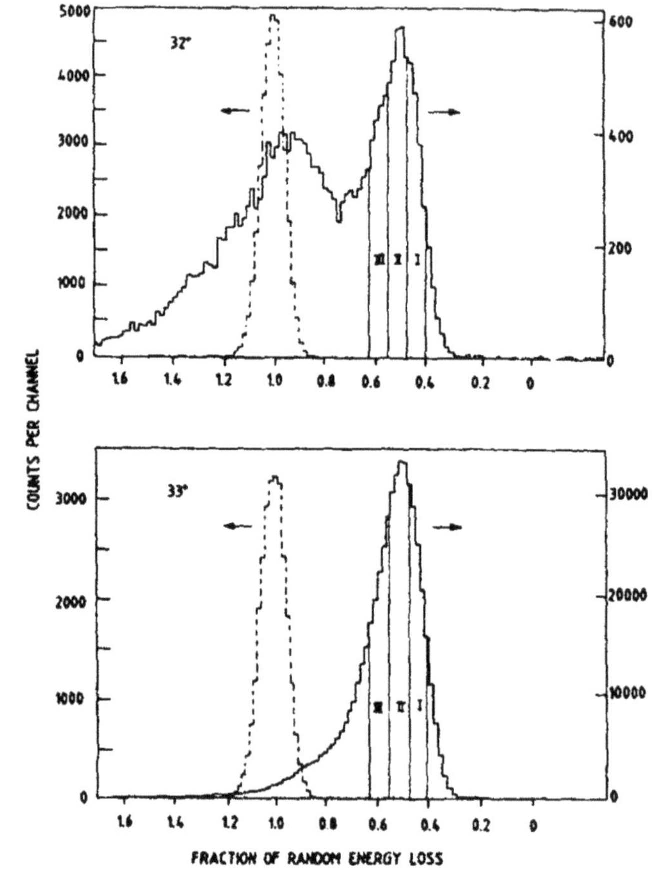

Figure 2.3. Energy loss spectra for 32+ (top) and 33+ (bottom) ions of ^{79}Br incident on thin silicon foils at an energy of 15.6 MeV/u. The dashed line is the data for a randomly-oriented Si crystal of 1 μm thickness while the solid line is the data for an aligned 8 μm thick Si ⟨110⟩ crystal. The channeling effect in the oriented crystal leads to a reduction in the energy loss to around 50% the value for the randomly-oriented foil (Reprinted from [5] with permission of Elsevier.)

not encounter atoms head-on. In this case, energy loss can be strongly reduced. Figure 2.3 shows an example of accelerated highly-charged ^{79}Br ions being channeled through an oriented Si ⟨110⟩ crystal [5]. Here, the energy loss experienced in transiting the crystal is around 50% of the nominal value expected for a randomly-oriented surface. Moreover, over 90% of the ions exit the oriented crystal in their initial 33+ charge state [5], whereas a much broader final charge distribution is obtained with the randomly-ordered crystal.

2.2.7 Delta electrons

Many nuclear physics experiments involve an accelerated beam of light or heavy ions striking a thin target foil. In these interaction, *delta electrons* will be generated which are secondary electrons knocked out from atoms in the target foil by interaction with the energetic beam. Such electrons are in the typical energy range 10–100 keV depending on the beam/target species and beam energy [6]. Since these electrons and their spectrum contain no nuclear physics information they are generally regarded as a nuisance to measurement. They can be particularly disturbing where the application of interest is detection of low energy conversion electrons; examples of how to mitigate their effect on measurement are discussed in section 8.4.2.

2.2.8 Simulation of energy loss, range and straggling

While it is helpful to understand the basics of the underlying theory of stopping powers, it is not necessary to resort to the basic equations to make calculations of stopping power and ranges. Indeed, there a number of helpful software packages that can calculate such parameters for different particle types in a wide range of materials. For example, stopping powers for protons may be obtained using the PSTAR online tool and for alpha particles using the ASTAR tool. A widely used software package in experimental nuclear physics is SRIM. This powerful software can calculate stopping powers and ranges for light and heavy ions over a wider energy range. Moreover, it can treat elements and compounds (pre-defined and user-defined) as well as isotopically-enriched materials. It can also handle layers of different materials which is a feature commonly encountered in experimental nuclear physics e.g. a thin target foil on a thick target backing, or a silicon detector with a thin aluminium/silicon oxide layer on the front.

SRIM carries out simulations over N particle trajectories and can calculate a mean projected range, R_p defined as:

$$R_p = \sum_i \frac{x_i}{N} = \langle x \rangle, \tag{2.5}$$

where the particle beam is incident along the x-axis and each individual trajectory has a length, x_i. Similarly a lateral projected range, R_y can be defined

$$R_y = \sum_i \frac{|y_i|}{N} = \langle |y| \rangle. \tag{2.6}$$

The straggling in range, σ, can also be extracted as

$$\sigma = \sqrt{\frac{\left(\sum_i x_i^2\right)}{N} - R_p^2} = \sqrt{\langle (\Delta x_i^2) \rangle}, \tag{2.7}$$

where $\Delta x_i = x_i - R_p$.

An alternative to obtaining tabulated data on stopping powers through e.g. SRIM, is to set up a GEANT4 simulation and allow GEANT4 to treat the physics of interaction between particles and matter. GEANT4 is described more extensively in section 4.2.1.

2.3 Electrons and positrons

The interaction of electrons and positrons with matter is similar in character in that the particles concerned are point-like and have equal (but opposite charges). The principal difference is that the positron is an unstable particle which can annihilate in flight or at the end of its range when it forms the exotic positronium system with an electron. These differences will be reviewed in detail below in section 2.3.2.

2.3.1 Electrons

The interaction of electrons with matter is qualitatively different to that of light and heavy ions discussed in section 2.2.1. A heavy ion has a relatively direct path through matter, losing energy in a large number of individual collisions/interactions. The interaction of electrons is very different in that atomic electrons are important in the stopping process and these collisions clearly involve two particles of the same mass, rather than one massive and one light particle as in the case of the stopping ions. This gives rise to a high probability for large-angle multiple scattering and, indeed, back scattering. An electron (or positron) entering a slab of matter therefore has a non-zero probability of scattering back out again. A second feature is that collisions between the stopping electron and atomic electrons may lead to a large transferred energy per collision. The de-excitation process may not therefore necessarily comprise a large number of smooth steps but may reflect a relatively small number of major deflection/energy loss events leading to significant straggling. Figure 2.4 illustrates the complexity of electron/positron interactions with materials for the case of 600 keV positrons stopping in a 5 mm thick plastic scintillator.

The energy loss of electrons in matter is comprised of two components: energy loss through excitation and ionisation due to inelastic Coulomb collisions with

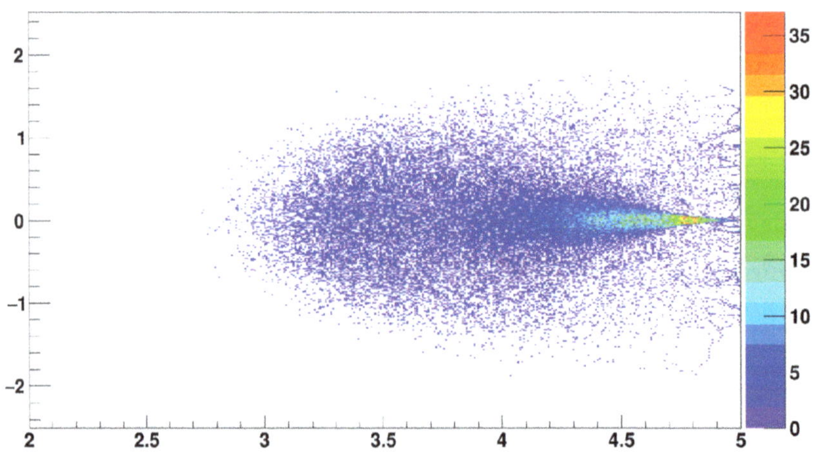

Figure 2.4. A GEANT4 simulation of the maximum range of 600 keV positrons fired as a pencil beam into a 5 mm thick plastic scintillator. The range in depth is on the x-axis (scale in mm's) while the lateral range is on the y-axis (scale in mm's). The maximum range of the positrons is about 2.2 mm while most positrons stop within the first 0.5 mm.

bound atomic electrons of the stopping medium, and energy loss due to the emission of *bremsstrahlung*[3] radiation in the field of nuclei and atomic electrons. These two components sum to give the total energy loss:

$$\left(\frac{\mathrm{d}E}{\mathrm{d}x}\right)_{tot} = \left(\frac{\mathrm{d}E}{\mathrm{d}x}\right)_{exc} + \left(\frac{\mathrm{d}E}{\mathrm{d}x}\right)_{rad}, \tag{2.8}$$

where $\left(\frac{\mathrm{d}E}{\mathrm{d}x}\right)_{exc}$ is a modified version of the Bethe–Bloch formula discussed in section 2.2.1:

$$\left(\frac{\mathrm{d}E}{\mathrm{d}x}\right)_{exc} = \frac{2\pi e^4 NZ}{m_0 v^2}$$
$$\left(\ln\left(\frac{m_0 v^2}{2I^2(1-\beta^2)}\right) - \ln\left(2(2\sqrt{1-\beta^2} - 1 + \beta^2)\right) + \frac{(1-\sqrt{1-\beta^2}^{\,2})^2}{8}\right), \tag{2.9}$$

where $\beta = \frac{v}{c}$, and I is the mean ionisation potential. The bremsstrahlung energy loss can be expressed as

$$\left(\frac{\mathrm{d}E}{\mathrm{d}x}\right)_{rad} = NE_0\Phi_{rad}, \tag{2.10}$$

where

$$\Phi_{rad} = \frac{1}{E_0}\int_0^{\nu_0} h\nu\frac{\mathrm{d}\sigma}{\mathrm{d}\nu}\mathrm{d}\nu, \tag{2.11}$$

and E_0 is the initial electron energy, ν is the frequency of the emitted bremsstrahlung photon.

In broad terms, the radiative (bremsstrahlung) energy loss is proportional to the initial electron energy and is proportional to Z^2 for the stopping material, while the energy loss from collision/ionisation is linearly related to Z, and increases logarithmically with electron energy.

It is useful to define a critical energy (E_c) where the stopping powers from the two processes are equal, i.e. where

$$\left(\frac{\mathrm{d}E}{\mathrm{d}x}\right)_{exc} = \left(\frac{\mathrm{d}E}{\mathrm{d}x}\right)_{rad}. \tag{2.12}$$

This critical energy is given approximately by:

$$E_c \approx \frac{1600 m_e c^2}{Z}. \tag{2.13}$$

[3] Bremsstrahlung from the German meaning 'braking radiation'.

2.3.2 Positrons

Positrons can annihilate in flight (mostly into two 511-keV gamma rays) [7] or come to rest and form *positronium*—effectively an electron–positron atom [8]. Positronium has two forms: *para-positronium* which has a lifetime of 125 ps and annihilates into two 511 keV gamma rays, and *ortho-positronium* which has a much longer lifetime of 142 ns and annihilates into three gamma-ray photons which share the 1.022 MeV positronium mass. The former process tends to dominate in most cases but the ratio between the two annihilation modes is sensitive to the environment where the positronium is formed [8].

2.3.3 Tools for evaluating electron/positron ranges

Stopping powers for electrons may be obtained using the ESTAR online tool. GEANT4 (see section 4.2.1) can also be used to model the range of electrons and positrons in different materials.

2.3.4 Cherenkov radiation

Cherenkov radiation corresponds to emission of photons of visible/UV light emitted when a high energy charged particle transits a medium with a velocity greater than the phase velocity (or 'speed of light') in that medium. The energy loss of the particle may not be particularly significant in most cases but Cherenkov radiation is sometimes exploited for particle detection. The threshold velocity for Cherenkov to occur can be defined in terms of the refractive index of the medium, n as

$$v_{\mathrm{th}} = \frac{c}{n}. \tag{2.14}$$

Cherenkov radiation is emitted into a cone, with half-angle with respect to the particle's direction, θ, given by

$$\cos\theta = \frac{1}{\beta n}, \tag{2.15}$$

where $\beta = \frac{v}{c}$.

The kinetic energy of a relativistic charged particle is given by:

$$KE_{\mathrm{rel.}} = (\gamma - 1)m_0c^2 = \left(\frac{1}{\sqrt{1 - \frac{v^2}{c^2}}} - 1\right)m_0c^2, \tag{2.16}$$

where m_0 is the particle rest mass. Let us evaluate the corresponding kinetic energies for electrons and protons at the threshold velocity for Cherenkov emission, v_{th} for particles transiting glass, whose refractive index, n, is around 1.5. Substituting values of 511 keV/c^2 and 931 MeV/c^2 for the rest mass of the electron and proton, respectively, into equation (2.16) we will see that the threshold velocity corresponds to a kinetic energy of around 170 keV for electrons and 310 MeV for protons. Based on this calculation, it is reasonable to assume that Cerenkov radiation is possible in a solid medium with a typical refractive index of 1.5 for a significant fraction of electrons/positrons emitted in beta decay, as well as photoelectrons and Compton-scattered electrons associated with gamma-ray interactions (see section 2.4). Due to the proton having a mass a factor of 1800 larger than the electron, we would not expect Cherenkov light to be produced unless protons (or alpha particles etc) have a significantly high kinetic energy which are energies not usually encountered in low energy nuclear physics but are more commonly encountered in high energy particle physics. Cherenkov detectors are therefore comparatively common in particle physics but a somewhat niche topic in nuclear physics and its applications.

2.4 Gamma rays

In contrast to alpha and beta particles, gamma rays interact *discontinuously* with matter. There are three main types of interaction of interest for gamma-ray interaction in the typical energy domain of gamma-ray spectroscopy i.e. gamma rays in the energy regime from around 10 keV to a few MeV: *photoelectric absorption*, *Compton scattering* and *pair production*. These are all processes that transform some or all of the energy of the incident gamma ray into kinetic energy of an electron or create positron–electron pairs and it is these electrons(positrons) that interact with a detector and produce a signal. There are other low-energy modes such as *Rayleigh scattering* i.e. elastic scattering of gamma-ray photons but these are not especially interesting from the point of view of gamma-ray detection. The probability of the different interaction processes is highly material (i.e. density and effective Z of the material) and energy dependent. Figure 2.5 illustrates the total cross sections for these different processes as a function of incident energy for carbon ($Z = 6$) and lead ($Z = 82$).

A different way of interpreting the cross section data of figure 2.5 and a more practical way to describe how readily a material can be penetrated by gamma rays is to define a photon mass attenuation length (or mean free path), λ, corresponding to the decrease in the intensity of the incident gamma-ray flux, I_0 given by

$$I = I_0 \exp\left(-\frac{t}{\lambda}\right), \tag{2.17}$$

where t is a thickness. Unusually, here, the thickness is in units of mass per unit area because λ is defined as

$$\lambda = \frac{1}{\frac{\mu}{\rho}}, \tag{2.18}$$

where ρ is the density of the material (in units of g cm^{-3}) and μ is a linear attenuation coefficient specific to the material (in units of cm^{-1}). The shorter the absorption length the more absorbing the material is. Figure 2.6 shows the absorption length, λ as a function of incident gamma ray energy; this figure indicates clearly how high-Z materials are significantly more absorbing of gamma radiation, confirming in a more

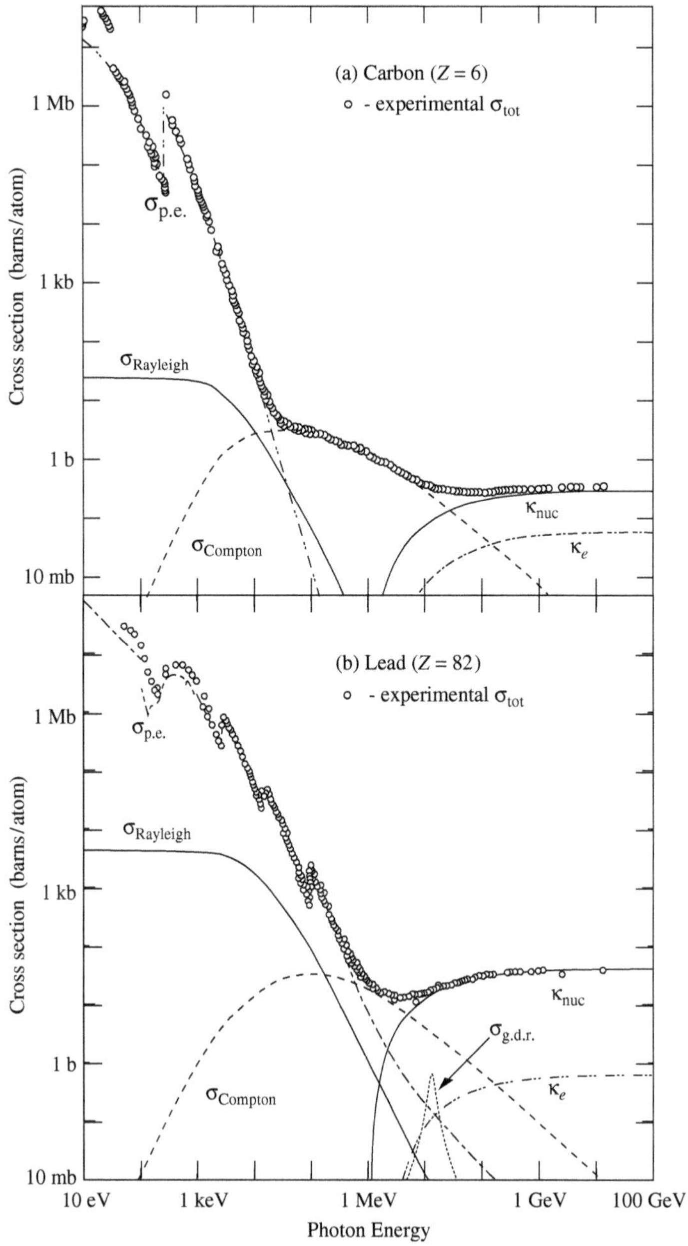

Figure 2.5. Total cross sections for various photon types of photon interaction as a function of energy in carbon—$Z = 6$ (top) and lead—$Z = 82$ (bottom). The respective cross sections are labelled 'p.e.' for photoelectric absorption, 'Compton' for Compton scattering (i.e. inelastic scattering), 'Rayleigh' for Rayleigh scattering (i.e. elastic scattering) and 'g.d.r.' for photonuclear reactions (Giant dipole resonance). κ_{nuc} is pair production in the field of a nucleus and κ_e is pair production in the field of an electron. (Reprinted with permission from [9]. Copyright (2018) by the American Physical Society.)

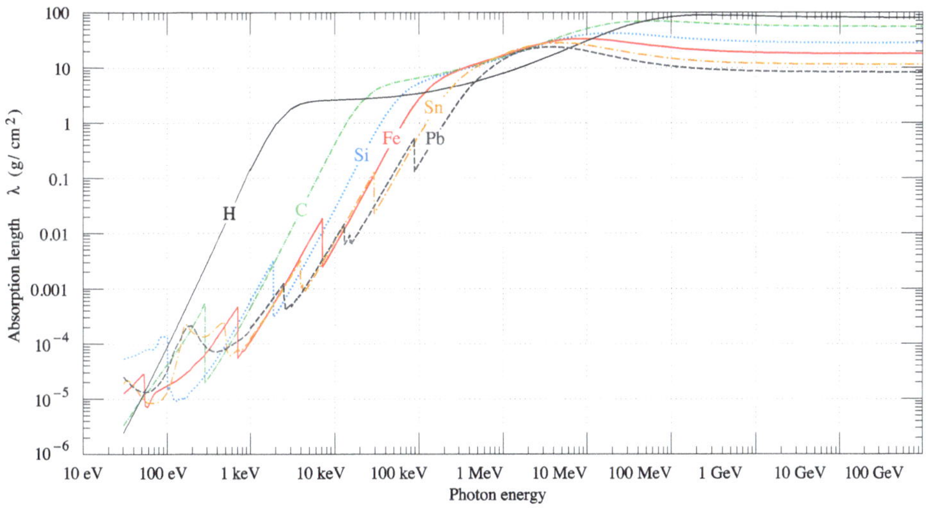

Figure 2.6. Photon mass attenuation length, λ, as a function of incident gamma-ray energy plotted for several different elements from hydrogen to lead (see text for definition of λ). (Reprinted with permission from [9]. Copyright (2018) by the American Physical Society.)

rigorous manner the high-school level description of lead as the best absorber for gamma radiation. Similarly, a detector made from a material of higher effective Z will have a higher efficiency than a detector of equivalent volume made from materials with a low effective Z.

2.4.1 Photoelectric absorption

Photolectric absorption is the most important interaction mechanism for low energy gamma rays (below 100 keV). In a photoelectric absorption interaction (see figure 2.7), an incoming gamma-ray transfers its energy to an atomic electron, usually the most tightly bound (K-shell) electron of an atom if the gamma-ray energy exceeds the binding energy of the K-shell electron[4]. If the gamma-ray energy is lower than the K-shell binding energy, then interactions are possible with higher shells (e.g. L shell). The ejected atomic electron, now termed a photoelectron, has an energy equal to that of the initial gamma-ray minus the binding energy for the atomic electron:

$$E_e = E_\gamma - E_b, \tag{2.19}$$

where E_b is the binding energy of the electron. The influence of the electron binding energies as representing thresholds for photoelectric absorption in conjunction with different electron shells, which vary as a function of Z, gives rise to the discontinuities in the cross sections in figure 2.5 and absorption lengths in figure 2.6.

[4] Photoelectric absorption can only occur with a bound electron not a free electron since a third body, the nucleus, is required to conserve momentum.

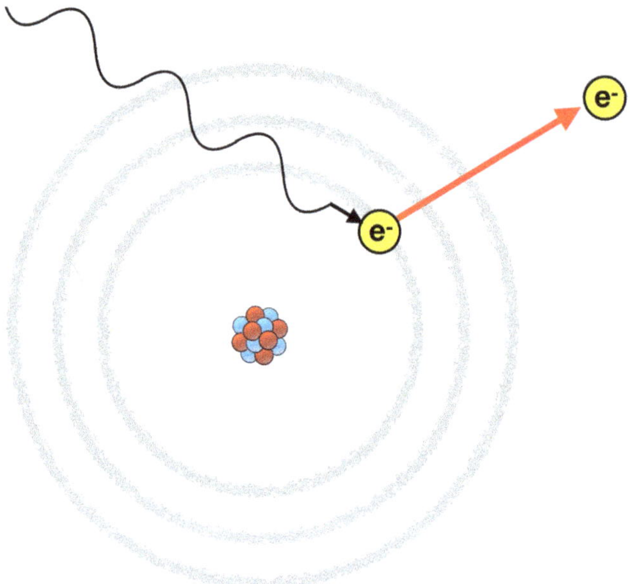

Figure 2.7. Schematic of photoelectric absorption.

Following photoelectric absorption, the atom is left with a vacancy in one of its atomic shells, e.g. the K-shell. Electrons will drop down through the atomic energy levels to fill this vacancy leading to the emission of characteristic x-rays and/or Auger electrons. Given that the range of the photoelectron, Auger electrons and x-rays is expected to be short compared to the range of the initial gamma-ray, the likelihood for any practical-sized gamma-ray detector is that the majority of these secondary particles/photons will be subsequently absorbed. The probability for photoelectric absorption is approximately proportional to Z^5 for a given stopping material.

2.4.2 Compton scattering

Compton scattering is where a gamma-ray photon scatters off an electron imparting some of its energy to the electron. As can be seen from figure 2.5, Compton scattering becomes the most probable interaction above about 50 keV in carbon and 800 keV in lead. For a typical material used as a gamma-ray detector such as germanium, this threshold is around 100 keV. Compton scattering then dominates through the range of energies most interesting to nuclear physics and its applications before giving way to pair production around 10 MeV or higher.

A schematic of the Compton scattering process is given in figure 2.8. The photon is scattered through an angle, θ. Since, the electron is initially effectively at rest, the initial energy and momentum of the photon must be conserved and partitioned between the energies and momenta of the scattered photon and electron. A photon with initial energy $h\nu_0$ is scattered off an atomic electron emerging with the reduced energy, $h\nu_1$, according to:

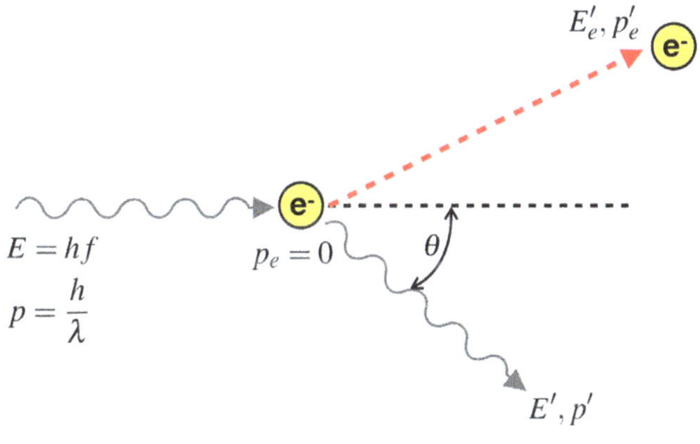

Figure 2.8. Schematic of Compton scattering.

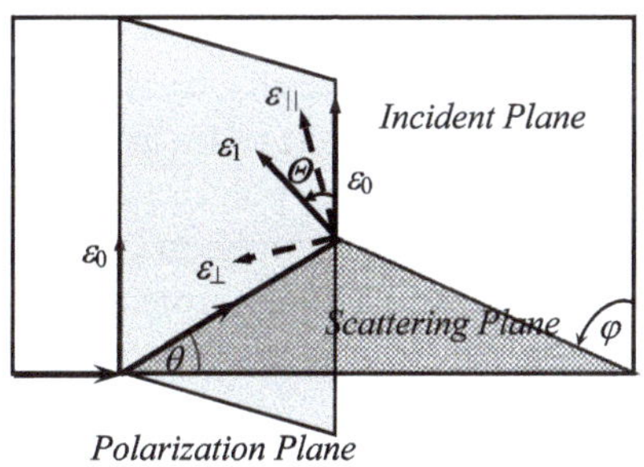

Figure 2.9. Definition of planes, photon electric vectors and angles for Compton scattering. (Reprinted from [11] with the permission of Elsevier.)

$$\frac{h\nu_1}{h\nu_0} = \frac{1}{1 + \frac{h\nu_0}{m_0c^2}(1 - \cos\theta)}, \tag{2.20}$$

where m_0c^2 is the rest mass energy of the electron and h is the Planck constant.

Clearly, from consideration of the scattering process, maximum energy will be imparted to the electron through back-scattering where the photon is scattered at 180° i.e. when $\cos\theta = -1$ (see equation (2.20)). In order to determine the probability for Compton scattering we need to consider the differential cross-section given by the Klein–Nishina formula [10]. This formula appears in many forms, here, we follow the presentation of reference [11]:

$$\frac{d\sigma}{d\Omega} = \frac{r_0^2}{4}\left(\frac{\nu_1}{\nu_0}\right)^2\left[\frac{\nu_1}{\nu_0} + \frac{\nu_0}{\nu_1} - 2 + 4\cos^2\Theta\right], \tag{2.21}$$

where Θ is the angle between the incident photon electric vector $\boldsymbol{\varepsilon}_0$ and the scattered photon electric vector $\boldsymbol{\varepsilon}_1$ (see figure 2.9). Figure 2.10 presents the Klein–Nishina cross section for three different gamma-ray energies as a function of scattering angle, θ.

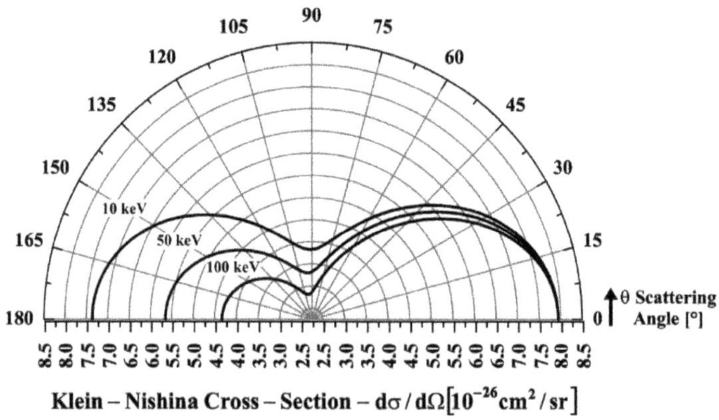

Figure 2.10. Cross-section as function of scattering angle for various gamma-ray energies. (Reprinted from [12] with the permission of Elsevier.)

Since the scattering cross section depends on the angle between the electric vectors, it depends on the linear polarisation of the initial photon. In fact, this is one of the main ways to determine the linear polarisation of gamma rays and hence determine multipolarity of gamma ray transitions. This is seen by resolving the intensity of the incoming photon into two components:

$$I_0 = I_{\|} + I_{\perp}, \tag{2.22}$$

where $I_{\|}$ is the component parallel to the polarisation plane and $I_{\perp}$ is the component perpendicular to that plane. The differential cross-section for scattering parallel to the incident component is given by:

$$\frac{d\sigma_{\|}}{d\Omega} = \frac{r_0^2}{4}\left(\frac{\nu_1}{\nu_0}\right)^2\left[\frac{\nu_1}{\nu_0} + \frac{\nu_0}{\nu_1} - 2 - 4\sin^2\theta\cos^2\phi\right], \tag{2.23}$$

where the relevant angles are indicated on figure 2.10. The differential cross-section for scattering perpendicular to the incident component is given by:

$$\frac{d\sigma_{\perp}}{d\Omega} = \frac{r_0^2}{4}\left(\frac{\nu_1}{\nu_0}\right)^2\left[\frac{\nu_1}{\nu_0} + \frac{\nu_0}{\nu_1} - 2\right]. \tag{2.24}$$

Example: 662 keV transition following beta decay of ^{137}Cs

Let us explore the implications of some of the theory outlined above for Compton scattering of a typical gamma ray of 662 keV which is that emitted from a ^{137}Cs radioactive source. Figure 2.11 presents the energies of the scattered photon for a 662 keV γ ray as a function of scattering angle up to 120°. The most energy that can be imparted to the electron is when back-scattering (180°) occurs. Substituting the incident energy (662 keV) into equation (2.20) shows that the minimum energy the scattered photon can have is 184 keV and hence, the maximum electron energy is

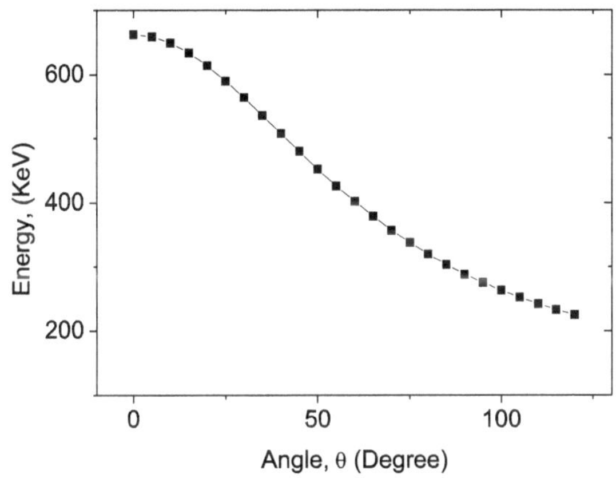

Figure 2.11. Energies of the scattered photon for a 662 keV γ ray as a function of scattering angle, θ. (Reprinted from [13] with permission from Academic Journals under Creative Commons License.)

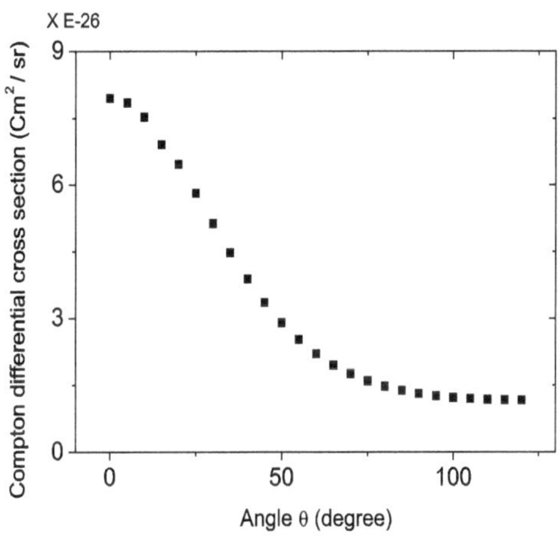

Figure 2.12. Differential cross-section for Compton scattering of a 662 keV γ ray as a function of scattering angle, θ. (Reprinted from [13] with permission from Academic Journals under Creative Commons License.)

478 keV. In general, for higher gamma ray energies where $E_\gamma \gg m_o c^2$, the minimum scattered photon energy will tend to $\frac{m_0 c^2}{2}$ i.e. 255 keV. This concept of a minimum energy for the scattered photon leads to the appearance of a Compton edge in gamma-ray spectra—see e.g. figure 5.1.

Figure 2.12 shows the differential cross section for Compton scattering of a 662 keV gamma ray as a function of scattering angle. This shows that the most probable scattering is for small angles.

2.4.3 Pair production

External pair production is where a positron–electron pair is generated in the field of an atomic nucleus [14] (see figure 2.13). The energy threshold for such a process is

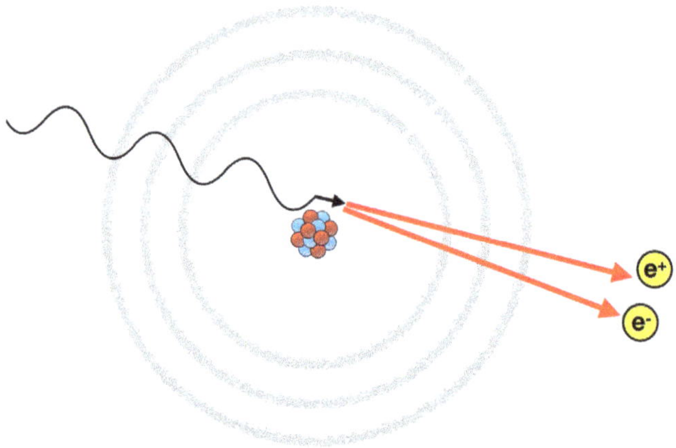

Figure 2.13. Schematic of external pair production.

$2m_ec^2$ where m_e is the mass of the electron. This corresponds to 1.022 MeV. Above $4m_ec^2$, the process can also take place in the field of an atomic electron. The former process is the more important for nuclear physics applications.

The underlying theory for pair production was given by Bethe and Heitler [15]; the total cross-section for pair production scales as Z^2 so a large material dependence is expected. The predictions of theory are compared with experimental data in figure 2.14.

2.5 Neutrons

As uncharged particles, neutrons in common with gamma rays, interact discontinuously in matter and may be highly penetrating. The manner in which they interact depends strongly on their energy (see figure 2.15) and the material with which they are interacting. *Fast neutrons*, e.g. MeV energies, will lose energy through elastic and inelastic collisions. In addition, if sufficiently energetic, they may induce nuclear reactions in materials leading to activation. Data on neutron cross-sections for interactions such as elastic/inelastic scattering and nuclear reactions is conveniently tabulated in the JENDL-4.0 library. The endpoint for fast neutrons, i.e. the fraction not removed through nuclear reactions, is as thermal neutrons. A thermal neutron is a free neutron with a kinetic energy equivalent to that of the surrounding medium—approximately 0.025 eV. Thermal neutrons will diffuse through a material in a manner similar to a gas.

2.5.1 Fast neutrons

Elastic and inelastic scattering

Fast neutrons will typically interact by elastic or inelastic scattering with nuclei in a material. Inelastic scattering means that the struck nucleus is left in an excited state which will subsequently relax by gamma-ray emission. Due to kinematics, most energy is imparted in a collision with a proton, i.e. the hydrogen nucleus which means that hydrogen-rich materials are the most effective in stopping fast neutrons. This is the reverse of the behaviour of gamma rays which are most readily stopped by high-Z materials like lead (cf section 2.4).

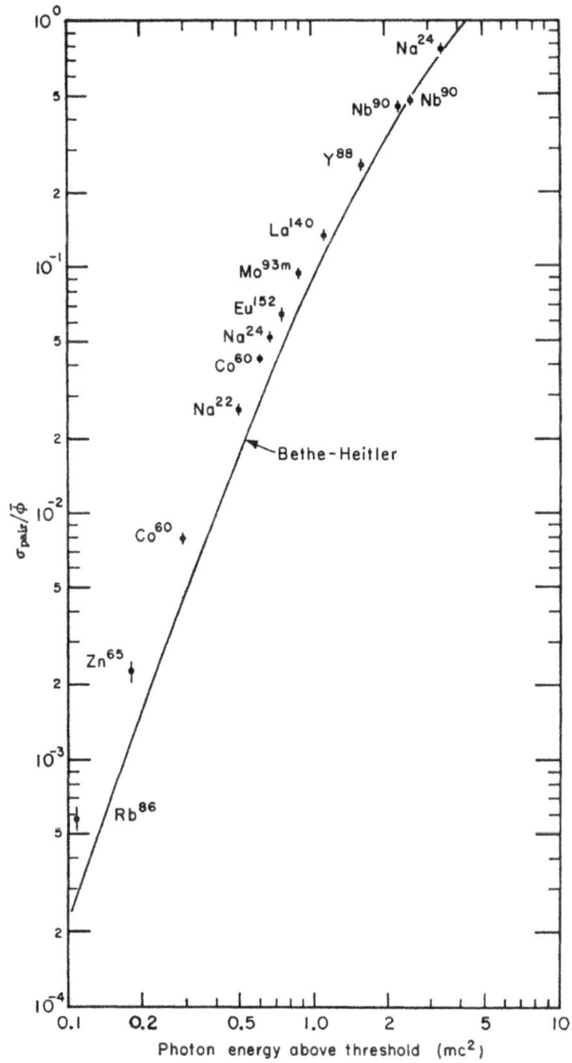

Figure 2.14. Cross-section for pair production as a function of photon energy above the threshold (in units of $m_e c^2$). Measured data for gamma rays from different radioisotopes (labelled on the figure) is compared with the Bethe–Heitler theory. (Reprinted with permission from [16]. Copyright (1965) by the American Physical Society.)

Neutron-induced reactions

If the energy of the neutrons is sufficiently high then some fraction of them can undergo nuclear reactions such as (n,2n) or (n,p) with nuclei making up the material they are passing through. In this case, a nuclear transmutation has taken place which can lead to activation of the material if the nucleus produced in the reaction is unstable. In the case of an (n,2n) reaction, the number of neutrons is effectively multiplied.

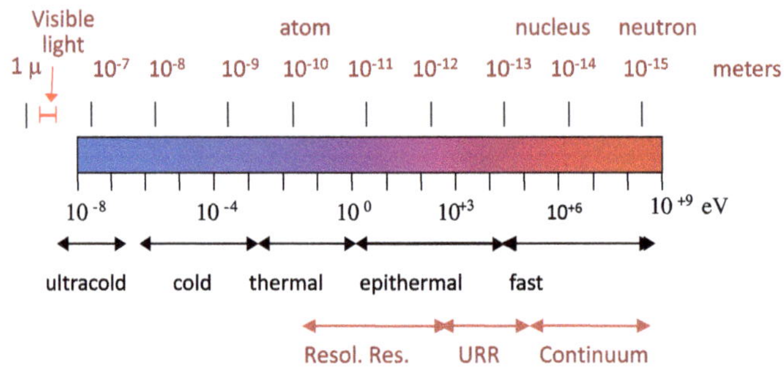

Figure 2.15. Spectrum of neutron energies from ultracold neutrons to fast neutrons. The upper section of the figure shows a parallel axis of the de Broglie wavelength of the neutrons and how this compares to the region of visible light, and the dimensions of the atom, nucleus and neutron. (Reprinted from [17] with permission from IOP Publishing under Creative Commons License.)

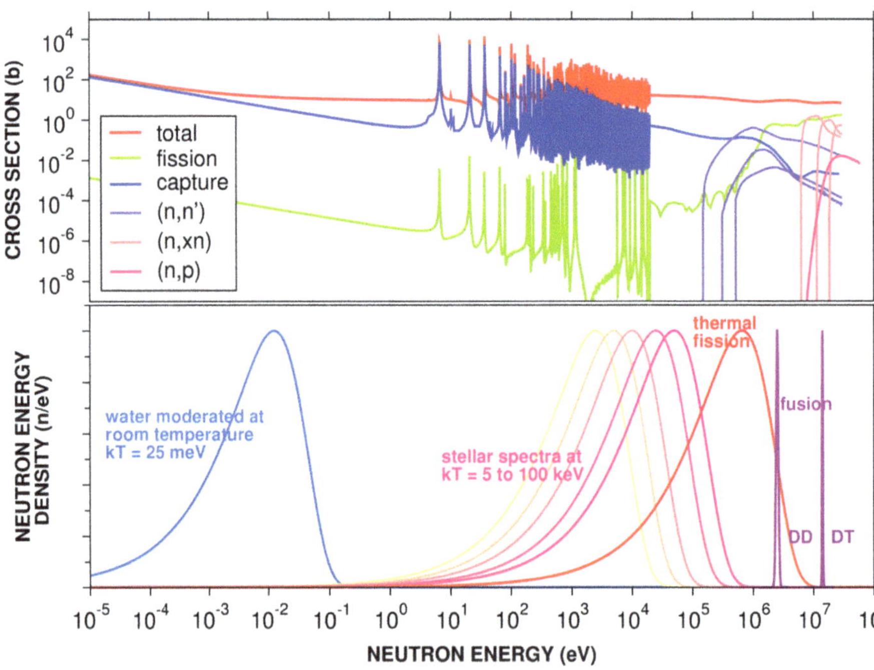

Figure 2.16. Neutron-induced reaction cross sections for ^{238}U, as an example of a typical heavy nucleus, as a function of neutron kinetic energy (upper panel), together with characteristic neutron energy distributions present in stellar environments and in technological applications like fission and fusion (lower panel). All distributions have been normalised to their maximum value. (Reprinted from [18] with permission from Elsevier.)

2.5.2 Thermal neutrons

The behaviour of thermal neutrons in materials is very different to fast neutrons and highly material specific in terms of its absorption (see figure 2.17). There are a few different mechanisms for the absorption of thermal neutrons. The

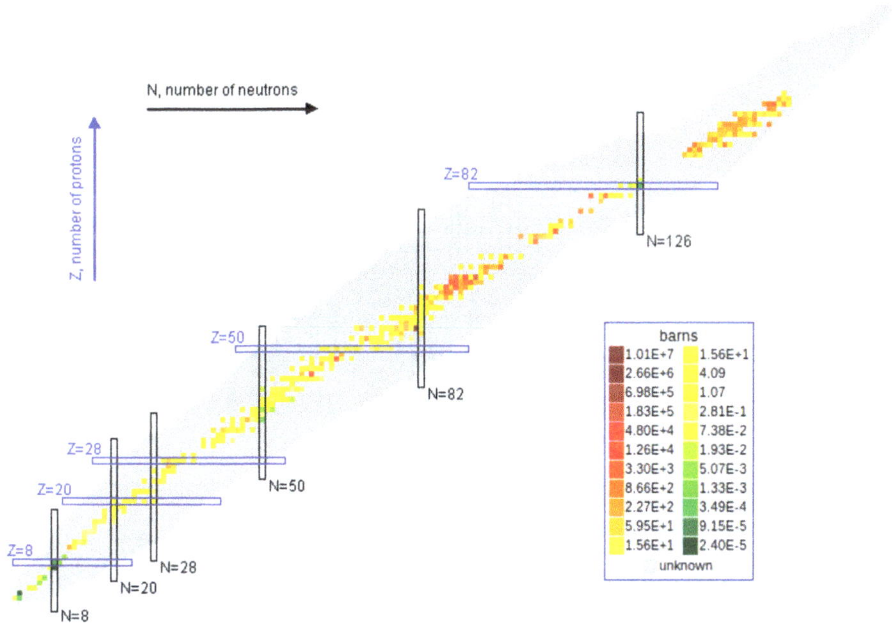

Figure 2.17. Thermal neutron capture cross-sections for stable isotopes. The colours indicate (see key) the cross-section. This image has been obtained by the author from the Wikimedia website https://commons.wikimedia.org/wiki/File:Chart_of_Nuclides_-_Thermal_neutron_capture_cross_sections.png, where it is stated to have been released into the public domain. It is included within this article on that basis.

principal mechanism is *thermal neutron capture*, i.e. capture of a neutron onto a nucleus leaving it in an excited state which can decay by prompt gamma-ray emission, and subsequently by other decay modes if the resulting nucleus is unstable. A second mechanism relevant to very heavy nuclei, e.g. the region around uranium, is neutron-induced fission where a neutron is captured by a nucleus producing an excited state which fissions[5]. Figure 2.16 presents the neutron-induced cross sections for ^{238}U over the whole neutron energy range from fast to thermal.

Cross-sections for thermal neutron capture vary strongly and can be as large as of the order of 100 000 barns, in the case of some isotopes of gadolinium, for example, compared to typical nuclear reaction cross sections which are of order millibarns–barn. This motivates the inclusion of gadolinium in the matrix of some types of thermal neutron detector. Figure 2.18 provides an example of the neutron capture cross-section for ^{73}Ge. In this example, the capture process is highly resonant. Above energies of around 10^4 eV, the pattern of resonances becomes so dense that they cannot be resolved experimentally—the so-called *unresolved resonance region*.

[5] Neutron-induced fission governs the underlying physics of nuclear reactors.

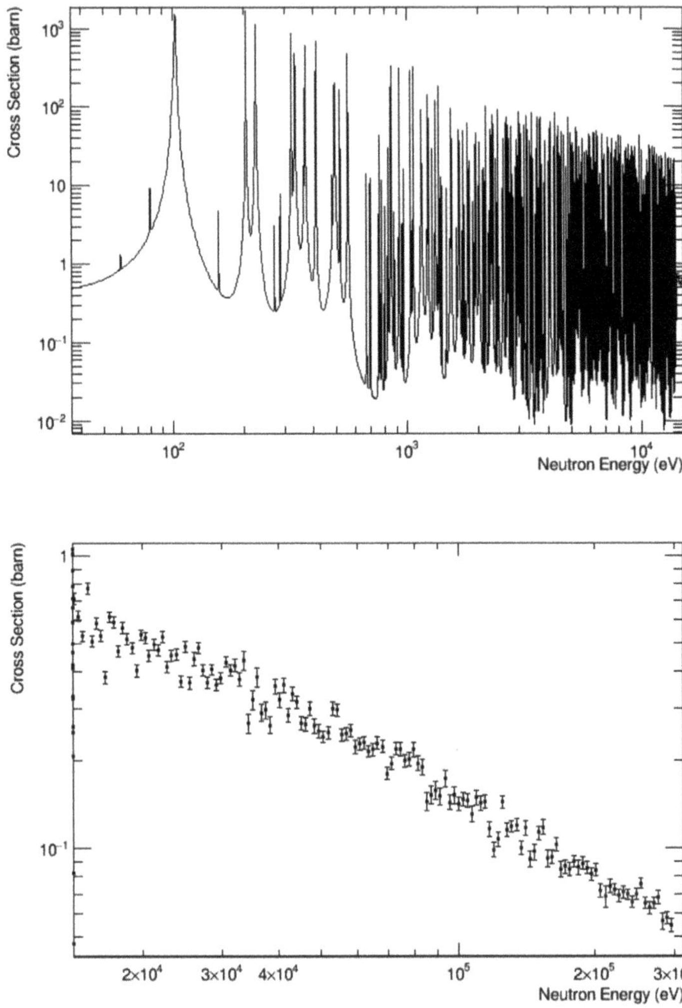

Figure 2.18. Neutron capture cross-section for ^{73}Ge for neutrons in the energy range 10^2–10^5 eV. The resonances shown are fits using the R-matrix code, SAMMY, to data taken with the nToF experiment at CERN. (Reprinted from [19] with the permission of Elsevier.)

References

[1] Ziegler J F 1999 The stopping of energetic light ions in elemental matter *J. Appl. Phys./Rev. Appl. Phys.* **85** 1249–72

[2] Fano U 1963 Penetration of protons, alpha particles and mesons *Annu. Rev. Nucl. Sci.* **13** 1–66

[3] Northcliffe L C 1960 Energy loss and effective charge of heavy ions in aluminium *Phys. Rev.* **120** 1744–57

[4] Ying C K, Bolst D, Tran L T, Guatelli S, Rosenfeld A B and Kamill W A 2017 Contributions of secondary fragmentation by carbon ion beams in water phantom: Monte Carlo simulation *J. Phys. Conf. Ser.* **851** 012033

[5] Andersen J U, Ball G C, Chevallier J, Davies J A, Davies W G, Forster J S, Geiger J S and Geissel H 1996 Heavy ion channeling *Nucl. Instrum. Methods Phys. Res.* B **119** 292–307

[6] Bell F, Trollmann G, Böckl H and Betz H-D 1982 Double-differential cross sections for delta-electrons from heavy ion-atom collisions *Nucl. Instrum. Methods* **194** 423–7

[7] Kendall H W and Deutsch M 1956 Annihilation of positrons in flight *Phys. Rev.* **101** 20–6

[8] Berko S and Pendleton H N 1980 Positronium *Annu. Rev. Nucl. Part. Sci.* **30** 543–81

[9] Tanabashi M *et al* (Particle Data Group) 2018 Review of particle physics *Phys. Rev.* D **98** 030001

[10] Klein O and Nishina Y 1929 Über die Streuung von Strahlung durch freie Elektronen nach der neuen relativistischen Quantendynamik von Dirac *Z. Phys.* A **52** 853–68

[11] Hamzawy A 2016 Compton scattering I: Angular distribution and polarization degree *Radiat. Phys. Chem.* **119** 103–8

[12] Scannavino F A and Cruvienl P E 2012 A graphical tool for an analytical approach of scattering photons by the Compton effect *Nucl. Instrum. Methods Phys. Res.* A **674** 28–38

[13] Hossain I, Hamzi N A B, Hoque E and Viswanathan K K 2011 Compton scattering of 662 keV gamma rays proposed by Klein-Nishina formula *Sci. Res. Essays* **6** 6312–6

[14] Motz J W, Olsen H A and Koch H W 1969 Pair production by photons *Rev. Mod. Phys.* **41** 581–639

[15] Bethe H A and Heitler W 1934 On the stopping of fast particles and on the creation of positive electrons *Proc. R. Soc.* A **146** 83

[16] Yamazaki T and Hollander J M 1965 Measurement of pair-production cross section near threshold *Phys. Rev.* B **140** 630–7

[17] Haight R C 2012 Fast-neutron detectors for nuclear physics experiments *J. Instrum.* **7** C05002

[18] Colonna N, Gunsing F and Käppeler F 2018 Neutron physics with accelerators *Prog. Part. Nucl. Phys.* **101** 177–203

[19] Lederer-Woods C *et al* 2019 Measurement of ^{73}Ge(n,γ) cross section and implications for stellar nucleosynthesis *Phys. Lett.* B **790** 458–65

IOP Publishing

Radiation Detection for Nuclear Physics
Methods and industrial applications
David Jenkins

Chapter 3

Radioactive sources in the laboratory

This chapter provides an overview of some of the means by which ionising radiation is produced in the laboratory. The chapter is broken into two parts. The first reviews the common classes of radioactive sources which are widely employed in the calibration of detectors and determination of their relative and absolute efficiency. The second part of the chapter reviews the manner in which exotic nuclei are produced and their properties probed in the laboratory. In nearly all cases, the detection of various types of ionising radiation is essential to probe the underlying physics. In contrast to well-characterised calibration sources, the charged particles, gamma rays and neutrons detected in experimental nuclear physics measurements may be emitted from nuclei travelling at relativistic speeds. There may be complicating experimental backgrounds and charged-particle/gamma-ray spectra may be dense and complex. This chapter provides some insight into these challenges to set the measurement problem in context.

Concepts: Radioactive sources; sealed and open sources; alpha sources; beta sources; gamma sources; fast neutron sources; laboratory methods for producing exotic nuclei: stable beams, radioactive beams; fusion-evaporation reactions; ISOL and in-flight techniques; secondary accelerators; Coulomb excitation and single-particle transfer; neutron-induced reactions

3.1 Radioactive sources

Radioactive sources are an essential tool for working with radiation detectors as a means of calibrating them in terms of their efficiency and, if they are spectroscopic detectors, to calibrate them in terms of energy and evaluate their energy resolution. Radioactive sources come in two main forms: open and sealed sources. Sealed sources are the most appropriate to detector calibration since the source is intrinsically safe[1]

[1] We give no specific guidance here on the use and safe employment of radioactive sources. We expect that the reader should seek information from the appropriate authority in their home institution and follow all relevant regulations applying in their home country.

doi:10.1088/978-0-7503-1428-2ch3

meaning that the activity is somehow encapsulated so it cannot leak out under normal use. Moreover, the radioactive material in the source is usually well-localised, approximating to a point source. Such sources typically have half lives measured in years, meaning that they remain useful over a significant life-span. The manufacturer will have determined the activity of the source using apparatus calibrated to primary standards. Typical calibration sources may have activities of the order of MBq or higher.

We note, that for most of the radioactive sources discussed below, there is a near one-to-one mapping with challenges in homeland security (see section 7.1) and nuclear decommissioning (see section 7.2) since these are the isotopes produced as a by-product of nuclear energy generation. Where such isotopes are found in these applications, they may not necessarily be in the form of a sealed source and may correspond to contamination of plant and infrastructure or be spread in the environment. Nevertheless, the relevant detection challenges are very similar apart from the nature of the source which can be spatially dispersed. These issues will be discussed in more detail in chapter 7.

3.1.1 Alpha sources

Alpha sources are of high value in testing charged-particle detectors in terms of calibration and optimisation of their energy resolution. Given that alpha particles lose energy transiting any material including air, in which they only have a range of a few centimetres, all such sources are designed to be vacuum compatible so that they can be operated in an evacuated chamber. Moreover, the source is generally deposited in a thin layer very close to the surface of a metal disk to minimise self-absorption.

Triple-line alpha source

There are a range of different isotopes that alpha sources can be fabricated from. However, one of the most useful is a so-called 'triple line' source containing a mixture of the isotopes: ^{239}Pu, ^{241}Am and ^{244}Cm. This results in a near-equal spaced triple line source with the principal energies: 5.155, 5.486 and 5.805 MeV. These three points of reference provide an excellent energy calibration for a charged-particle detector such as a silicon detector. Moreover, each of the alpha decays exhibits fine structure (see section 1.3.1) meaning that a satellite peak appears on the left-hand side of each principal peak. Being able to resolve these satellite peaks in a spectrum is an immediate visual confirmation that a silicon detector (see section 8.2) is operating with good energy resolution even before efforts are made to optimise the resolution. Figure 3.1 illustrates the alpha-particle spectrum obtained with a typical silicon detector.

3.1.2 Beta sources

Distinguishing beta and gamma sources is slightly artificial as most classes of gamma-ray source encountered in the laboratory are where gamma rays are emitted from excited states in the daughter nucleus following beta decay. The closest to a pure β^- source commonly used for calibration is ^{90}Sr. ^{90}Sr itself has a half-life of

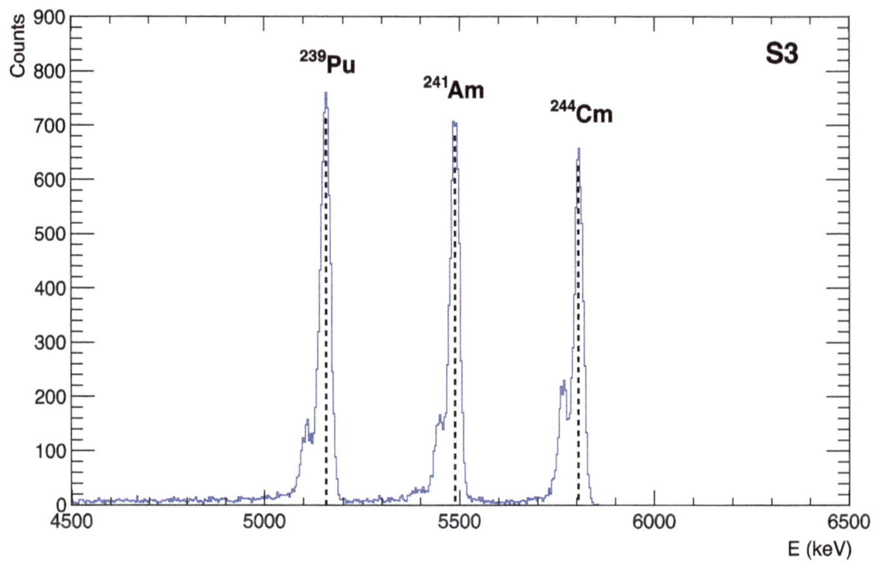

Figure 3.1. Spectrum of a triple alpha source taken with a Micron Semiconductor S3 annular silicon detector (see section 8.2). The energy resolution (FWHM) is of order 0.5%. Satellite peaks are visible on the left-hand side of the main peaks corresponding to alpha decay to excited states in the daughter nucleus.

around 29 years. It decays to the daughter ^{90}Y, which in turn decays to stable ^{90}Zr through β^- decay with an end-point energy of 2.28 MeV (see figure 3.2). The beta decay to excited states of ^{90}Zr is negligible meaning that the source essentially emits only beta particles with two different end-point energies corresponding to the two successive decays.

In terms of β^+, i.e. positron-emitting sources, the most commonly used sealed source is ^{22}Na whose half-life is 2.6 years. Such sources can readily be used to test the sensitivity of detectors to positrons. In addition, the positron will come to rest in a short range, often within the source itself, form positronium (see section 1.4) and annihilate, principally into two photons of 511 keV emitted near back-to-back. This can be employed to determine the timing resolution of a detector with respect to a fast reference detector through coincident detection of the two annihilation gamma rays (see section 10.2.5).

3.1.3 Gamma-ray sources

Gamma-ray sources are available in a wide range of strengths and source configurations for different applications [2]. As discussed above, these are all essentially gamma rays from excited states in a daughter nucleus populated via beta decay.

^{137}Cs

The fission-derived radioisotope ^{137}Cs is particularly suitable for characterising detectors as following the beta-decay of this isotope, a single 662 keV γ ray is emitted. The full-width at half-maximum (FWHM) of the 662 keV photopeak in a

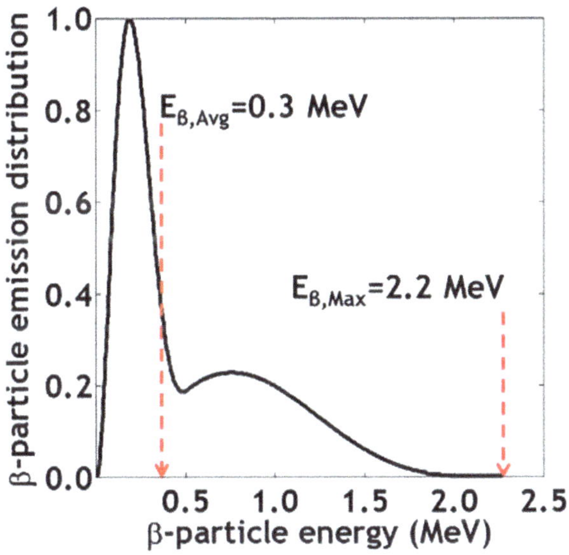

Figure 3.2. Beta-particle spectrum of a ^{90}Sr source. The endpoints of the successive beta decays are marked (see text). (Reprinted by permission from Springer Nature [1] under Creative Commons License.)

gamma-ray spectrum is conventionally used to describe the energy resolution of scintillator detectors (see chapter 5).

^{60}Co

The reactor-produced isotope ^{60}Co is also an important standard for characterising detectors. The gamma-ray spectrum is dominated by two transitions of 1173 and 1332 keV in a near one-to-one intensity ratio. The 1332 keV gamma ray is conventionally used for defining the efficiency of germanium detectors (see section 6.3.5).

^{152}Eu

The beta-decay of ^{152}Eu feeds a range of excited states in the daughter nucleus via a complex decay path. This leads to a broad range of intense gamma-ray lines from 121 keV up to 1.5 MeV (see figure 3.3). This makes it a highly suitable source for calibrating a detector over a wide energy range.

^{56}Co

The gamma-ray emission from a ^{56}Co source continues up to nearly 4 MeV, making it a good source for high energy calibration and efficiency studies [4]. Unfortunately, the source has to be produced using a proton beam via the ^{56}Fe(p,n)^{56}Co reaction and has a relatively short half-life of 77.3 days.

3.1.4 Conversion-electron sources

As discussed in section 1.7.3, internal conversion can compete with gamma-ray emission and the relative probability of the conversion electron branch depends on

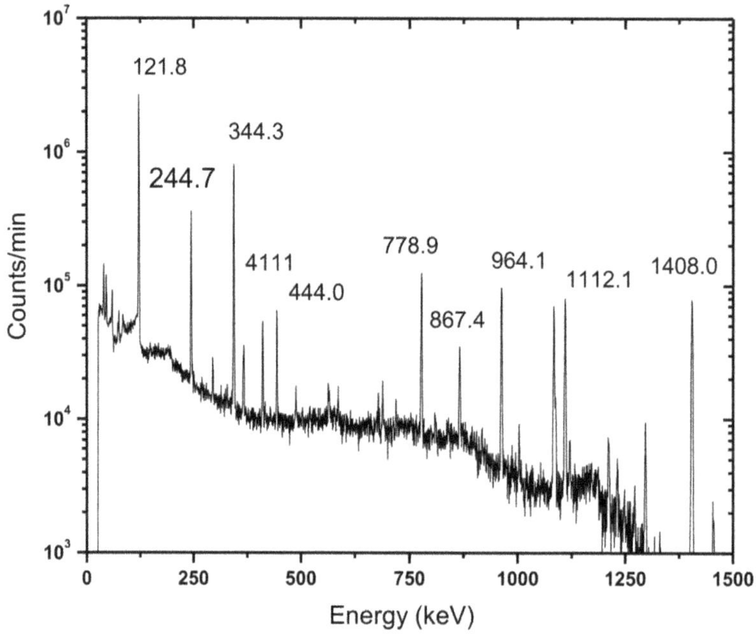

Figure 3.3. Gamma-ray spectrum from a ^{152}Eu source recorded with a HPGe detector. (Reprinted from [3] with the permission of Elsevier.)

details such as the atomic number, Z, the transition energy and the transition multipolarity. In this sense, all gamma-ray sources may emit some conversion electrons but there are two particular examples of such sources which are very useful for calibration of, e.g. silicon detectors, and which have several prominent conversion electron peaks: these are ^{133}Ba and ^{207}Bi. Figure 3.4 provides example spectra for these two sources: ^{133}Ba provides useful calibration peaks up to 400 keV while ^{207}Bi extends the energy range to just above 1 MeV.

3.1.5 Neutron sources

There are a range of typical neutron sources commercially available to assist in characterising detectors: comprising (α, n) sources, photonuclear sources and fission sources [6]. Most of these sources produce fast neutrons but they can readily be thermalized by surrounding the source with a moderating material of low Z such as plastic or heavy water.

(α, n) sources

A common class of neutron source is a so-called (α, n) source where an alpha-emitting nucleus is packaged in a homogeneous mixture with a material which has a low reaction threshold for an (α, n) reaction. The alpha particles which can initiate such a reaction need to exceed a threshold energy, E_{th} where

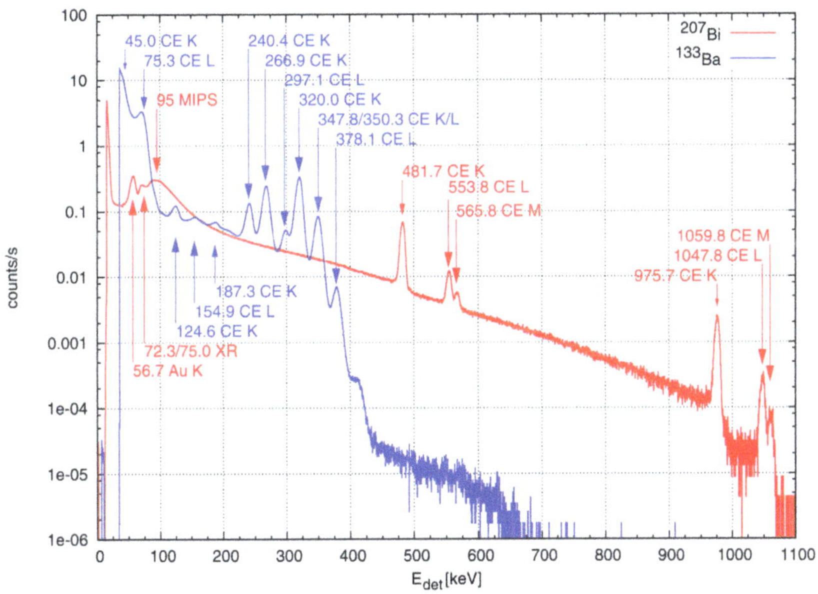

Figure 3.4. Conversion electron source spectra from ^{133}Ba (blue) and ^{207}Bi (red) obtained using a 300 μm thick PIPs silicon detector. Note that since these are conversion electron sources, there are several peaks corresponding to each transition for emission from the K, L and M shells. (Reprinted by permission from Springer Nature [5].)

$$E_{th} = -\frac{M + m_\alpha}{M} Q \quad (Q < 0); \quad E_{th} = 0 \quad (Q > 0), \tag{3.1}$$

where M is the mass of the target nucleus and m_α is the mass of the alpha particle; Q is the Q-value for the (α, n) reaction. For quite a number of isotopes, the Q-value is positive and hence there is no threshold energy leading to the best efficiency. A very commonly found combination is a so-called AmBe source where ^{241}Am and ^{9}Be are packaged together. The choice of ^{241}Am reflects its long half-life of 432 years which also makes it the radioisotope of choice in other applications such as smoke alarms. However, it should be noted that 70% of the time, the ^{9}Be$(\alpha, \mathrm{n})^{12}$C reaction proceeds to the excited state of ^{12}C and this results in emission of a 4.44 MeV gamma ray. Such a gamma ray is hard to shield and so the detector under test will need to be gamma-ray insensitive or have a mechanism to resolve gamma ray and neutron interactions. Note that a typical AmBe source is quoted in terms of the activity of the alpha emitter but the emission of neutrons is about four orders of magnitude less than this, i.e. around 5×10^4 neutrons s^{-1} per GBq (in alpha emission) of the source.

Photonuclear sources

A photonuclear source induces fast neutron emission through a (γ, n) reaction. The bombarding gamma energies need to be above a threshold corresponding to the Q value of the reaction in order to induce neutron emission. This is typically achieved by wrapping material such as ^{9}Be around a highly-active gamma source. Typically, the neutrons emitted are near monoenergetic which can be an advantage over (α, n) sources for specific applications where controlling the neutron energy is important. However, the yield of neutrons is low in relation to the high gamma activity which will need significant shielding.

Fission sources

As noted in section 1.5, there are some nuclei which exhibit spontaneous fission from their ground state. A common choice for such a fission neutron source is ^{252}Cf, which has a high neutron yield per unit mass. It decays 3% of the time by spontaneous fission and the remaining 97% of the time by alpha decay, with a half-life of around 2.6 years. Each spontaneous fission produces 2–4 neutrons and the neutron spectrum produced is similar to the fast neutron flux from a fission reactor with an average neutron energy of around 2 MeV.

Neutron generator tubes

Neutron-generator tubes are an interesting class of neutron source. These make use of d–d or d–t fusion—the types of fusion reaction explored as potential future sources of energy in tokamaks. Deuterons are accelerated through a potential difference of the order 100 kV onto targets containing deuterium or tritium to produce neutrons via one of the following nuclear reactions:

$$\mathrm{d} + \mathrm{d} \rightarrow \mathrm{t} + \mathrm{n} \tag{3.2}$$

$$\mathrm{d} + \mathrm{t} \rightarrow^{4} \mathrm{He} + \mathrm{n}. \tag{3.3}$$

Such neutron-generator tubes have the advantage over conventional radioactive sources of being able to be turned on and off on demand. In addition, due to the nature of the reactions employed, the neutrons produced are quasi-monoenergetic with energies of around 2.5 MeV for d–d fusion and 14 MeV for d–t fusion.

3.2 Laboratory methods for studying exotic nuclei, nuclear reactions and nuclear excited states

We now turn our attention to some of the techniques of experimental nuclear physics as a means of understanding what challenges radiation detection will face in this domain[2]. Experimental nuclear physics is a mature field and facilities across the globe can routinely produce highly exotic nuclei on demand using accelerator technology. Moreover, it is becoming increasingly common to have the ability to re-accelerate such exotic nuclei and use them to initiate nuclear reactions in turn. This e-book is not intended to review the history of nuclear reactions or accelerator science. There are many references to this elsewhere and the reader is directed to some useful further reading in appendix A. Instead, the focus of this section is to describe a limited subset of the types of reaction commonly encountered in

[2] This section may be of less interest to those working in nuclear applications and such readers are invited to skip ahead to chapter 4.

contemporary nuclear physics experiments and elucidate the issues around the detection of the charged particles, gamma rays and neutrons that result from such reactions. The challenges are often rather different to those encountered when detecting ionising radiation from fixed sources in the laboratory. Understanding these challenges in terms of the demands for detection will help us to appreciate the diversity of detectors encountered in the laboratory and will latterly focus our attention on their application outside the laboratory in societally-relevant applications. In this section, therefore, we will focus on some of the key contemporary techniques used for producing exotic nuclei in their ground and excited states. The detection of the resulting ionising radiation is often the last link in the chain following demanding advances in accelerator technology which make the creation of short-lived radioactive nuclei possible. Such nuclei may be so exotic that they are essentially never created on earth by natural processes and exist elsewhere in our Universe, in esoteric environments such as exploding stars and merging neutron stars. Examining the structure of such exotic nuclei and the manner in which they undergo nuclear reactions can therefore give direct access to the physics of dramatic astrophysical objects—this is the domain of *nuclear astrophysics*—the reader is directed to further reading on this topic in appendix A.

3.3 Stable beam methods

The first artificial nuclear reaction was initiated in 1932 by Cockcroft and Walton who bombarded lithium with an accelerated beam of protons, creating ^{8}Be which immediately breaks up into two alpha particles. By the 1950s, accelerator technology had moved on and it was possible to accelerate light heavy nuclei such as ^{12}C. By the 1970s, accelerator facilities were able to produce intense beams of very heavy nuclei such as ^{238}U, meaning that the whole nuclear chart, at least as far as stable nuclei were concerned, was able to be accelerated. These light and heavy-ion beams could be used to initiate various kinds of nuclear reactions. Here, we focus on an important exemplar of the available reactions—*fusion-evaporation*—which has played an important role in producing, for example, the superheavy elements.

3.3.1 Fusion-evaporation reaction

In a *fusion-evaporation* reaction (see figure 3.5), a heavy-ion beam is accelerated to energies around the Coulomb barrier and is incident on a fixed target, e.g. a thin metallic foil. The nuclei in the beam and target can fuse leading to a compound nucleus in a highly-excited state. This compound nucleus will cool by evaporation of light charged particles, e.g. protons and alpha particle, and fast neutrons. Such a reaction may lead to ten or more different final nuclei, associated with evaporation of varying numbers of charged particles and neutrons. The probability, i.e. cross-section, for the different reaction channels can generally be estimated in advance for a given bombarding energy using a statistical model. Following particle evaporation, the resulting nucleus will be left in a highly-excited (but particle-bound state). This state may initially cool by emission of high-energy gamma rays (of order 10 MeV) associated with nuclear giant resonances, e.g. the giant dipole resonance. This

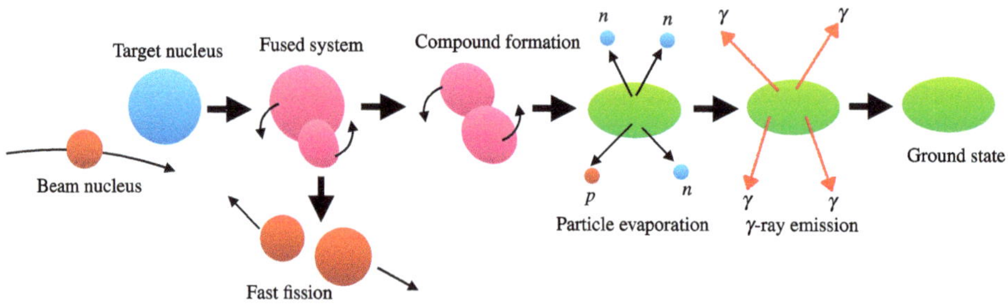

Figure 3.5. Cartoon of a fusion-evaporation reaction. An accelerated beam of nuclei is incident on a thin foil of target nuclei. The beam and target nucleus fuse forming a system with high angular momentum. For heavy systems, the process can continue nearly instantaneously to fission. If fission does not occur, the compound nucleus formed is left in a highly-excited state. Cooling takes place first by light charged-particle and neutron emission to reach a final nuclear state in an excited state which cools subsequently by gamma-ray emission towards the ground state (or long-lived isomeric state).

gamma-ray emission is statistical in nature with a broad energy distribution. Such emissions are most readily studied with large, high-efficiency scintillator detectors (see chapter 5). The gamma decay then continues as transitions between bound excited states (see section 1.6) where gamma-ray transitions have discrete energies and negligible (from the point of view of realistic detector technologies) widths. It is hard to generalise on fusion-evaporation reactions since the topic is so diverse, but nuclei are frequently produced in states of high angular momentum and excitation energy meaning that a high multiplicity cascade of gamma rays will cool the nucleus to its ground state. This field of study is traditionally known as *high spin spectroscopy*.

Issues to consider with fusion-evaporation

1. The number of emitted gamma rays (i.e. the multiplicity) may be high, even up to 30+ depending on the system under study. This suggests that there is value in having a large array of detectors (see section 6.4.2) in order to measure coincidences between gamma rays which will allow the construction of a *level scheme* based on the coincidence information and intensity of the gamma rays detected. Moreover, angular distributions and angular correlations of gamma-ray emission (see section 1.7.1) may be used to deduce the multipolarity of the gamma rays and hence, infer spins and parities for the excited states which they depopulate.
2. The recoiling nucleus from a fusion-evaporation reaction will typically have a velocity which is a non-negligible fraction of the speed of light. This velocity, $\beta = v/c$ can be of order 5%. This means gamma rays are Doppler shifted depending on the angle, θ, that they are emitted with respect to the recoil direction. The gamma energy detected in the laboratory frame, E_γ, can be expressed in terms of the emitted gamma-ray energy, E_0 as:

$$E_\gamma = \frac{E_0}{\gamma(1 - \beta\cos\theta)}, \tag{3.4}$$

where $\gamma = \frac{1}{\sqrt{1-\beta^2}}$ is the Lorentz factor. This angle-dependent factor needs to be corrected for in order to reconstruct a gamma-ray spectrum representing the combined detection of gamma rays in detectors at different angles or indeed, a large detector array. Since any realistic gamma-ray detector will present a finite solid angle (see section 4.1.4 for definition of solid angle) with respect to the recoiling nucleus, there will be a distribution of energies detected in a detector based on the angular range which it spans. This is known as Doppler broadening and can be obtained from equation (3.4) by straightforward differentiation:

$$\Delta E_\gamma = E_0\beta \sin\theta\Delta\theta, \tag{3.5}$$

where $\Delta\theta$ is the opening angle of the detector face. Inspection of equation (3.5) shows that the Doppler broadening effect will be maximised at 90° with respect to the recoil direction. It also demonstrates that segmentation of detectors will be the most effective in improving likely spectral quality for detectors close to 90°. These issues are generic for gamma rays emitted from nuclei moving with high velocity which includes some of the other classes of reaction described below.

3. A large number of different reaction channels result from a fusion-evaporation reaction with differing numbers of neutrons and light charged particles (e.g. protons, deuterons, alpha particles) emitted. An obvious means of channel selection would be to count the evaporated particles using a barrel of charged particle detectors around the target and/or fast neutron detectors. These classes of detectors are described in sections 8.1.1 and 9.1.1, respectively. In addition, if the energies and angles of the emitted particles are recorded then these can, in principle, be used to more accurately reconstruct the angle of the recoiling nucleus with respect to the beam direction, which would improve the Doppler-shift correction.
4. An alternative and powerful way to separate reaction channels is using recoil-decay-tagging (RDT) (see figure 3.6). Here, a recoil separator (a large system containing electric and magnetic dipole magnets) is used to separate the fusion products from the unreacted (scattered) beam. The flight time through the separator is of the order of 1 μs. The fusion products are implanted in a double-sided silicon strip detector (DSSD) (see section 8.2). The fusion products which are generally radioactive nuclei, may subsequently decay by emission of alpha particles, beta particles, gamma rays etc. In principle, all of these may (and have been) used to correlate with the implantation and identify the isotope which was implanted of the many different ones produced in the reaction. This is perhaps simplest to understand for an alpha-decaying isotope where, as noted in section 1.3, alpha particle energies are discrete and characteristic of the emitting isotope. Once an isotope has been correlated with its decay, a time correlation can be applied to discriminate the gamma rays which it emitted at the target position. The net effect is to allow spectroscopy of the excited states of individual isotopes even though many different isotopes are produced in the experiment.

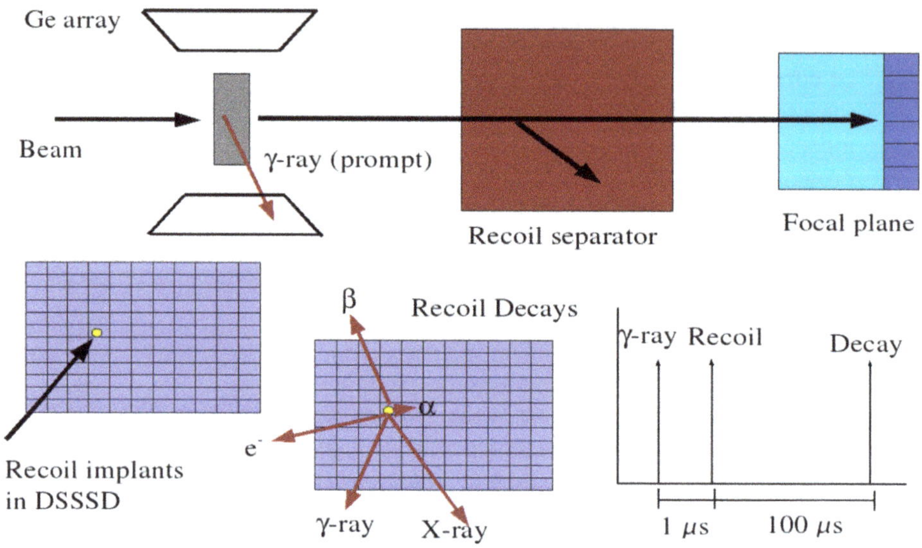

Figure 3.6. The recoil-decay-tagging technique for correlating gamma-rays emitted at the target position with the decay of fusion residues implanted at the focal plane. An accelerated heavy-ion beam incident from the top left strikes a thin target foil at the centre of an array of high-purity germanium detectors. The fusion residues are separated from the scattered beam in a recoil separator and implanted in a double-sided silicon strip (DSSD) detector at the focal plane. The implantation generates a signal in one pixel of the DSSD. Subsequently, the residue decays by e.g. emission of an alpha particle recorded in the same pixel. The alpha particle energy is characteristic of the implanted isotope and correlating the decay with the implantation event, allows a track back in time to identify the gamma rays emitted by this isotope at the target position.

3.4 Radioactive beams

Since the 1960s, there has been increasing focus on *radioactive beams* or *RI (rare isotope) beams*. There are two principal routes to production of radioactive beams: the *in-flight* technique and the isotope separation on-line (ISOL) technique (see figure 3.7):

3.4.1 In-flight technique

In simple terms, the in-flight technique starts with a heavy nucleus such as ^{238}U which is accelerated to very high energies then fragmented to produce many different radioactive isotopes. The presently world-leading facility using the in-flight technique is RIKEN-RIBF in Japan. New facilities coming online in the next few years include FRIB at Michigan State University and FAIR in Darmstadt, Germany.

3.4.2 ISOL technique

The ISOL (isotope-separation on-line) technique uses high-energy (~GeV), high-intensity beams of light ions such as protons to, literally, smash up heavy nuclei. For example, at the ISOLDE facility in CERN, 1.4 GeV protons from the PS-Booster

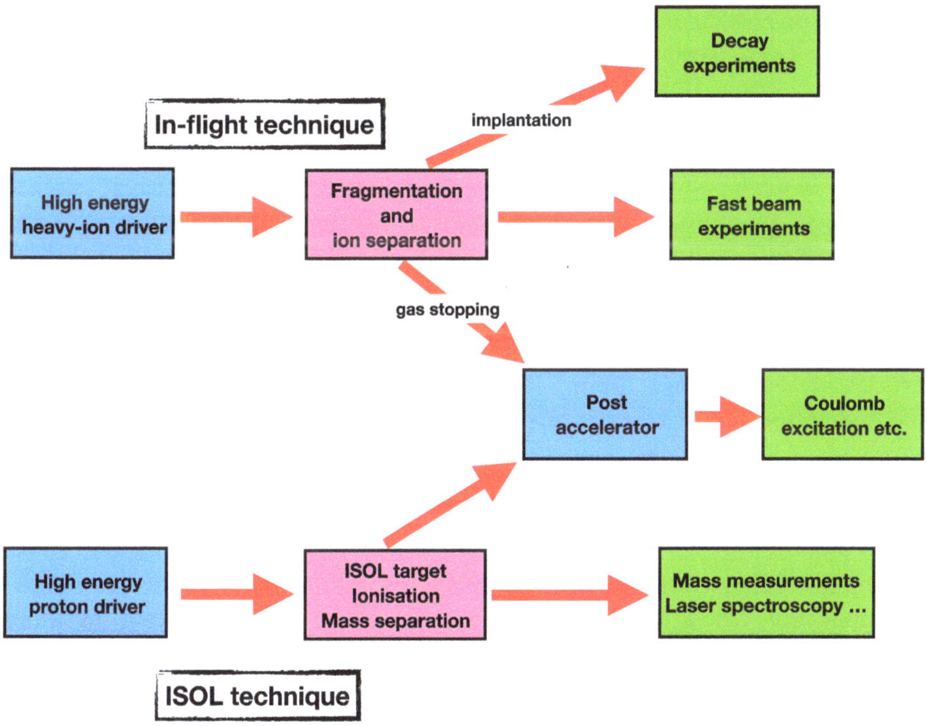

Figure 3.7. A comparison of the two main techniques for producing radioactive beams. The in-flight technique (top) takes a high energy, heavy-ion beam and fragments it on a solid target such as beryllium producing a wide range of nuclei with differing probability. These nuclei can then be separated and identified by mass and charge (if the ions are fully stripped at these energies then $Q = Z$). Experiments can be carried out using the resulting beams travelling at relativistic speed or the ions can be implanted into detectors. The ISOL technique (bottom) uses a high energy proton driver to produce intense beams that can break up nuclei in a solid, extended target. These nuclei are subsequently ionised and mass-separated. The resulting low energy beams can be probed via laser spectroscopy or captured in an atomic trap. Low energy ISOL beams and slowed, cooled in-flight beams (using gas stopping) can be re-accelerated (centre) to energies close to the Coulomb where they can initiate nuclear reactions in turn.

accelerator are incident on thick, extended targets of solid material e.g. uranium carbide. Depending on the nature, i.e. atomic number of the target material, different products, i.e. radioactive isotopes will be produced through processes such as *fragmentation*, *spallation* and *fission*. In order to study an individual isotope, it has, ideally, to be separated from the other isotopes produced or at least to dominate the resulting sample of radioactive nuclei. The manner in which this is achieved is highly dependent on the chemical element since techniques often rely on selective ionisation of a given element using, e.g. laser pumping.

3.4.3 Experimental techniques with low energy or stopped radioactive beams

Stopped radioactive beams are exploited in a wide range of experimental studies from mass measurements to atomic hyperfine measurements. These techniques, which do not particularly emphasise radiation detection, are beyond the scope of this e-book. Decay spectroscopy measurements are, however, highly relevant to the topic of this e-book. Here, the radioactive nuclei are transported and implanted in a measurement station. Depending on the radioactive species, a wide range of different studies can be conducted including: *beta-gamma*, *beta-delayed protons*, *beta-delayed fission* and *alpha decay*. The requirements of these different classes of measurement are quite diverse. Some require spectroscopic performance of charged-particle detectors such as alpha decay studies where energies are discrete (see section 1.3); this is best achieved with high-performing semiconductor detectors such as silicon detectors (see section 8.2). Such an approach would also be relevant to studies of beta-delayed protons. Beta decay itself is not spectroscopically interesting as beta particles are emitted in a Fermi–Kurie distribution of energies (see section 1.4). Silicon detectors could be used but there is also scope to use scintillator detectors (see chapter 5) which have good stopping power but low energy resolution. Often, beta decay populates excited states in the daughter nucleus and here, the addition of spectroscopic gamma-ray detectors to detect the gamma-rays emitted by these excited states is important. Germanium detectors (see chapter 6) are the best class of detector for such spectroscopic measurements. Scintillators have also been used in this application where the interest is completely surrounding the emitting sample and catching all of the gamma rays emitted—a technique known as total absorption spectrometry (see section 5.4.1).

Issues to consider in decay spectroscopy

1. Decay spectroscopy is carried out on stopped radioactive nuclei so conditions are perfect to achieve high energy resolution since there are no corrections to be applied for a moving source. This is not the case for studies using re-accelerated beams described below.
2. The radioactive nuclei may not be stopped in a tight geometry such as found with a conventional sealed point source. This spatial distribution may need to be taken into account in determining detector efficiencies. In some studies, high precision would be desirable in terms of decay branchings and here, particular attention has to be paid to detector efficiency and to effects such as true summing where, e.g. two or more gamma rays emitted simultaneously from the source end up in the same detector.
3. Stopping radioactive nuclei at one point will rapidly lead to a build-up of decay activity comparable to the incoming rate of the radioactive beam. This may impose challenges in handling the rates in the associated detectors. Many solutions exist to handle this such as implanting the ions in a moving tape or even a rotating detector (see example in section 8.4.1). The other option is to increase the granularity of the associated detector array.

3.4.4 Experimental techniques with re-accelerated radioactive beams

In the last twenty years, facilities have been developed which are able to re-accelerate ISOL beams. There are several such facilities worldwide such as SPIRAL [7],

REX-ISOLDE [8], HIE-ISOLDE and TRIUMF-ISAC [9]. The intensity of such beams can approach that of stable beams in some cases but typically beam intensities are much lower. Accelerated radioactive beams impose new challenges for experiment—any detectors we place around the beam will likely be recording ionising radiation from the radioactive decay of the beam in flight. This may well dominate the signal from the nuclear reaction under study and lead to a high background counting rate. Measurements of outgoing radiation *in coincidence* may be used to alleviate this. Let us look at a couple of examples of techniques employed with accelerated radioactive beams as they will set out better some of the challenges related to detection and measurement.

3.4.5 Coulomb excitation

Coulomb excitation is a technique commonly used to probe collectivity in nuclei, i.e. the manner in which the nucleus behaves as a coherent ensemble of nucleons rather than reflecting the excitation of individual nucleons. Coulomb excitation involves inelastic excitation of nuclei through the electromagnetic interaction. So long as the relative energy of the nuclei is constrained so that they do not approach closely enough to feel the effects of the nuclear force, the problem is fully solvable from the perspective of an electromagnetic formalism. Indeed, computer codes exist such as GOSIA which can be used to interpret and model Coulomb excitation experiments. Coulomb excitation can be used to extract information on the extent to which nuclei are deformed and the nature of that deformation e.g. prolate (rugby-ball shaped) or octupole (pear-shaped). Coulomb excitation is now commonly carried out using accelerated beams of exotic (i.e. short-lived radioactive) nuclei. A commonly encountered approach to reconstructing the Coulomb excitation process is to detect the scattered heavy ions, i.e. the beam and target nuclei with annular silicon detectors (see section 8.2) and to detect the emitted gamma rays with high-purity germanium detectors (see chapter 6) or in some cases, inorganic scintillator detectors (see chapter 5). Figure 3.8 shows a cartoon of a typical experimental set-up for Coulomb excitation studies using an accelerated radioactive beam.

Issues to consider with Coulomb excitation

1. The angle of detection of scattered heavy ions with respect to the beam axis is a critical parameter in the analysis of Coulomb excitation. The heavy-ion detectors used need high angular resolution, i.e. larger detectors, should be segmented and an annular detector shape is favoured to achieve rings spanning the same angular range (see section 8.2.4).
2. Scattered heavy ions may have energies up to the order of 1 GeV, therefore, heavy-ion detectors need a broad dynamic range. We note that silicon detectors are widely used in this application but their high intrinsic energy resolution is not particularly relevant here as the kinematics of the scattering couple the energy and angle of the scattered heavy ions. Angular resolution of the detector is therefore paramount and energy resolution is semi-redundant. The necessary performance can equally be achieved with

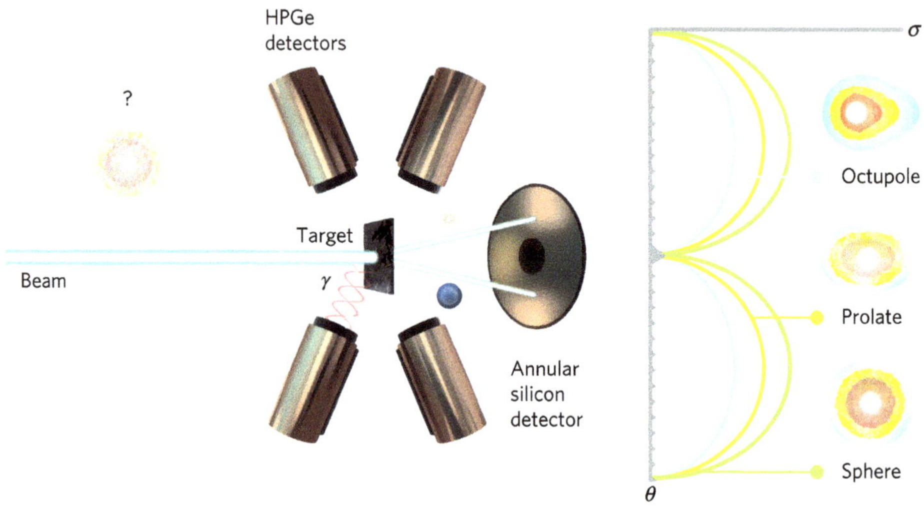

Figure 3.8. Schematic of Coulomb excitation: a heavy-ion beam is incident on a thin self-supporting target. The beam and target nuclei may scatter inelastically leading to production of excited states in these nuclei which subsequently decay by gamma-ray emission. In the example provided, the emitted gamma rays are detected in an array of high-purity germanium detectors and the scattered nuclei are detected in an annular silicon detector. Such an experimental study can be used, for example, to investigate nuclear shape. The right-hand side of the diagram shows, highly schematically, how Coulomb excitation cross-sections as a function of angle can be related to nuclear shape. (Reprinted by permission from Springer Nature [10].)

pixellated or segmented scintillator detectors, or gas-ionization detectors, with much poorer energy resolution.

3. The rate of scattered heavy ions can be high depending on the experiment concerned. Heavy ions damage silicon detectors relatively rapidly and so when used in this application, they may be seen, ultimately, as a consumable item although the initial effects of damage in degrading energy resolution will not materially affect their use in this application.
4. Gamma-ray detection is used to characterise the excitation of states in the nuclei undergoing scattering. Gamma-ray energy resolution may become important where the gamma-ray spectrum is particularly dense but in some cases, the poorer energy resolution of scintillator detectors will be sufficient.
5. Gamma-ray multiplicity can vary from one to tens of gamma rays in a cascade. The points made in section 3.3.1 about gamma-ray detector array design in the context of fusion-evaporation reactions also apply here.

3.4.6 Single particle transfer

Single-particle transfer reactions are an important class of reactions used in nuclear structure measurements. Historically, such studies were carried out using light-ion beams and a large spectrometer to momentum analyse the outgoing particles. For example, in a so-called (d,p) reaction, a deuteron beam is incident on a heavy target nucleus and a neutron is transferred to that nucleus, leaving a proton to be detected

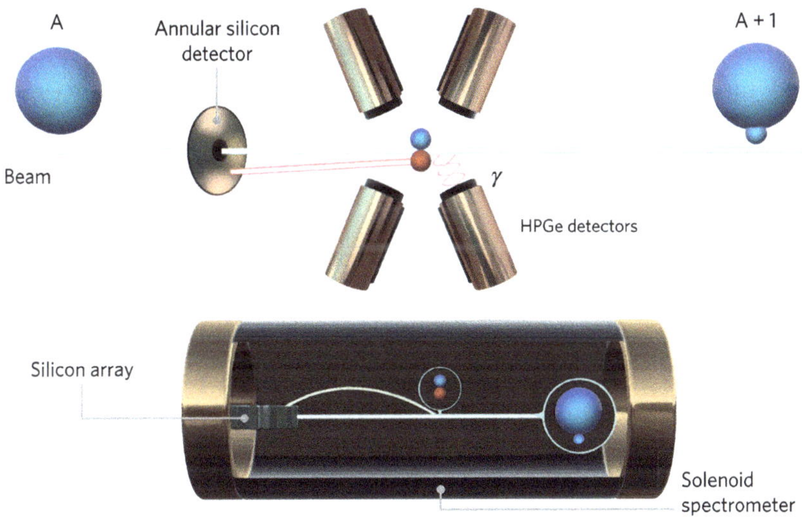

Figure 3.9. Schematic of single-particle transfer reactions with an accelerated radioactive beam. The top part of the figure illustrates a heavy-ion beam (blue circle) with atomic mass, *A*, incident on a deuteron (red circle) in a deuterated-plastic target. The beam nucleus picks up a neutron from the deuteron and the proton is ejected at backwards angles in the laboratory frame where it is detected in an annular silicon detector. The bottom part of the figure shows an alternative scheme where the reaction takes place inside a large solenoidal magnet with a homogeneous magnetic field in the centre. Here, the emitted proton follows a helical trajectory to be detected in a long narrow silicon array along the beam axis. (Reprinted by permission from Springer Nature [10].)

which can inform about the kinematics and probability (i.e. cross section as a function of angle) of the reaction. The final nucleus may be left in its ground state or excited states with varying probability based on the details of the reaction mechanism and underlying nuclear structure. This class of measurements was a major focus of study in the 1960s and 1970s at small university laboratories but the possible cases were largely exhausted. This class of reaction has come back into fashion with the advent of accelerated radioactive beams. In this case, the reactions are now done in *inverse kinematics*, i.e. a heavy beam incident on, e.g. a proton- or deuteron-rich plastic target. Figure 3.9 shows schematics for two alternative methodologies for studies of (d,p) reactions. The top of figure 3.9 illustrates an accelerated radioactive beam incident on a deuterated plastic target, where a neutron is picked up from the target and the proton is ejected at backwards angles in the laboratory, to be detected in an annular silicon detector (see section 8.2). If the recoiling heavy nucleus is left in an excited state then it will decay by gamma-ray emission which may be detected by the array of germanium detectors surrounding the target position. The lower panel of figure 3.9 shows an alternative scheme where the reaction takes place at the centre of a large, high-field (e.g. 3 T) solenoidal magnet[3]. Here, the ejected proton executes a helical trajectory in the magnetic field

[3] A suitable homogeneous field region is typical for an MRI magnet and examples of this type of spectrometer have made use of redundant MRI magnets for such purpose.

before detection on an extended linear silicon array close to the beam axis [11]. Gamma-ray detection could be added within the bore of the solenoidal magnet but is non-trivial with conventional detector technology as photomultiplier tubes do not operate in such high magnetic fields. This would be an excellent application for inorganic scintillators coupled to silicon photomultipliers (see section 5.3.3) since such detectors are insensitive to magnetic field.

Issues to consider with single-particle transfer reactions

1. The key parameters to measure are the angle and energy of emitted light charged particles. The detector employed needs high energy and angular resolution. In the example involving the solenoidal spectrometer, the motion through the magnetic field changes the key parameters to be linear distance, energy and time of flight.
2. Various reaction channels may be possible, leading to the probability of various light charged particles such as protons, deuterons, tritons, alpha particles being emitted. These may sometimes be separated based on the different kinematics of the reaction, e.g. some species of charged particles emitted in forward angles, others in backward angles. Other strategies to distinguish charged particles using detector telescopes or pulse-shape analysis are discussed in section 8.2.2.

3.5 Neutron-induced reaction studies

We reviewed bench-top sources of neutrons in section 3.1.5 and now turn our attention to reactions induced by neutrons generated by accelerators of varying scale. The challenge with neutrons is that they are neutral particles so we cannot build an accelerator which can increase their energy in a controlled manner as is possible for light and heavy ions. Similarly, they cannot be steered with electromagnetic elements such as magnets in the way that ion beams can, so a focussed beam is not achievable. A 'beam' of neutrons can only be generated using some kind of collimation, i.e. elements that block the neutrons in specific directions. It is therefore very challenging to produce a neutron beam with properties as ideal as can be achieved with an ion beam; reaction studies with neutrons therefore bring their own set of challenges and limitations on how radiation detectors can be used around the neutron beam.

3.5.1 Neutron time-of-flight measurements

There are a number of facilities worldwide engaged in neutron time-of-flight measurements; the premier such facility is nToF at CERN. Here, an intense pulse of 20 GeV protons from the CERN accelerator chain hit a 1.3 tonne cylindrical lead spallation target leading to the production of pulses of neutrons every 1.2 s [12]. These neutrons will have energies everywhere from GeV to meV (milli electron volt) and fly down a 185 m tunnel to reach a measurement station containing a secondary reaction target where studies on neutron-capture reactions, neutron-induced fission and other studies are carried out. The relevant detectors are gamma-ray detectors

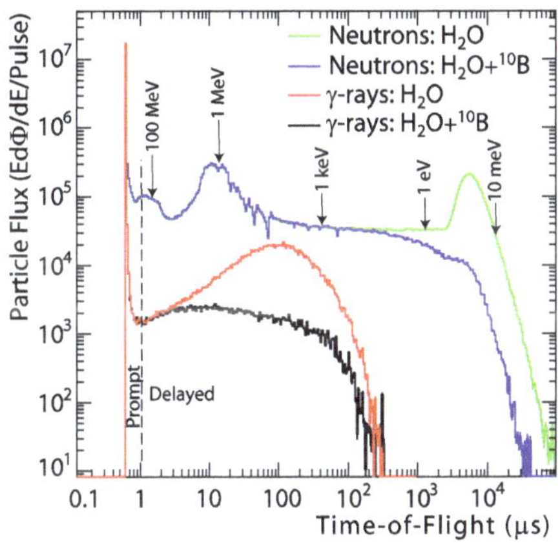

Figure 3.10. Flux of gamma rays and neutrons as a function of their time of flight at the nToF facility in CERN. The flux differs depending on whether the moderator used at the primary target is water, or borated water, hence, why four different flux spectra are shown (see legend on the figure). (Reprinted from [12] with the permission of Elsevier under Creative Commons License.)

such as inorganic scintillator detectors (see chapter 5) for capture gamma rays, and fission detectors based on the principles of gas ionization (see section 4.3.1).

Each neutron will have a different time-of-flight down the tunnel depending on its energy and so successive measurements can be made on neutrons of varying energy as they arrive at the secondary target station. This is very different to conventional ion-beam experiments where the beam energy is fixed. Neutron time-of-flight experiments produce a kind of 'history in time' sampling a wide range of neutron energies. The flux of neutrons and gamma rays produced from the nToF primary target is shown in figure 3.10. The challenge is immediately obvious in that the intense gamma-ray flux which arrives first, travelling at the speed of light, will overwhelm any gamma-ray detector placed around the secondary target and effectively blind them for some time while the first and most energetic neutrons are arriving[4]. All neutrons will arrive within around 10 ms with a time-varying flux. Neutrons may scatter in the secondary target and hit the detectors surrounding it. Moreover, depending on the cross-section, i.e. probability for a reaction to occur in the secondary target for a specific neutron energy, the flux of charged particles and gamma rays coming from the secondary target may also vary widely. In some studies, the secondary target is a sample of an actinide element which itself has a high level of activity, e.g. ^{241}Am. These effects taken together may lead to individual detectors counting very fast, e.g. MHz detector rates. This could lead to a pile-up of detector pulses (see section 4.1.3) and challenges in resolving individual events. The solution is to capture digital traces from each detector channel with the start of the

[4] It is interesting that similar challenges occur in high energy, high intensity laser/plasma interactions where nearby detectors are instantaneously blinded by intense fluxes of charged particles and gamma rays/x-rays, although electromagnetic pulse (EMP) is important here also. Such studies often employ passive radiation detection, e.g. activation of metallic foils which are then counted after the laser pulse. Laser pulse rates can be very low, e.g. a few shots per day.

data-taking triggered by the arrival of each proton pulse from the primary accelerator. The data is then analysed off-line, fitting the pulse shapes observed with appropriate algorithms to try to disentangle overlapping pulses (see section 10.3). Capturing digital traces has the consequence of producing very large datasets which need significant time to process.

3.5.2 Fast neutron beams

An interesting example of a controllable, fast neutron beam is that generated by the Licorne apparatus at IPN Orsay [13]. An *inverse kinematics* reaction is used to produce the neutrons where an accelerated ^{7}Li beam hits a hydrogen gas target (see figure 3.11). The inverse nature of the reaction, i.e. a heavy beam on a light-ion (hydrogen) target means that the fast neutrons are emitted into a narrow, forward-focussed cone with energies around 2 MeV. They can then interact with a secondary reaction target for studies of, e.g. fast neutron-induced fission on ^{238}U. Gamma ray detectors can be mounted around the secondary target because the neutrons are in a narrow cone and so will not hit the detectors directly. This is important because it is challenging to operate high-performing gamma ray detectors such as high-purity germanium detectors in a fast neutron field. Such detectors will be radiation damaged and their performance may rapidly degrade depending on the neutron flux (see section 6.3.6).

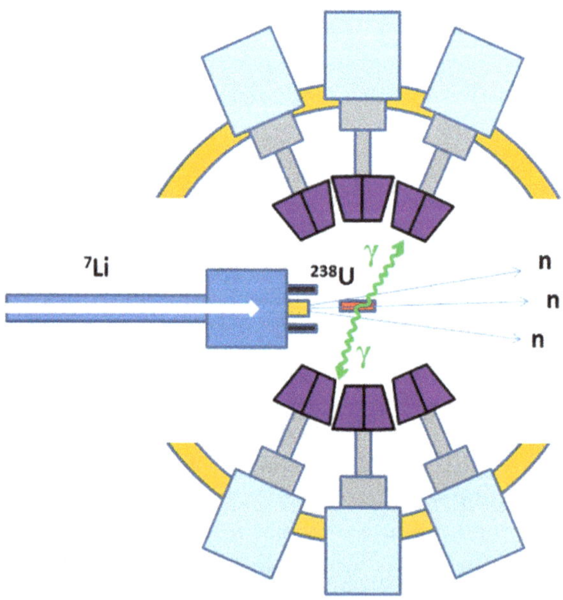

Figure 3.11. NU-BALL setup of gamma-ray detectors around the fast neutron beam produced by the Licorne apparatus. (Reprinted with permission from [13]. Copyright (2017) by the American Physical Society.)

Issues to consider with neutron-induced reactions

1. It is not possible to accelerate neutrons and tune their energy as can be done with ion beams. Where neutrons are produced in a nuclear reaction, they will often be accompanied by gamma rays. In neutron time-of-flight measurements this intense gamma flash can blind detectors for some time.
2. Neutrons scattered into gamma-ray detectors may cause radiation damage. Moreover, they may make it difficult to distinguish the gamma rays of interest.

References

[1] Dixon J *et al* 2016 Evaluation of a silicon ^{90}Sr betavoltaic power source *Sci. Rep.* **6** 38182

[2] Trzaska W H 1990 Recommended data on selected gamma-ray and conversion-electron calibration sources *Nucl. Instrum. Methods Phys. Res.* A **297** 223–9

[3] Naik Y, Tapase A S, Mhatre A, Datrik C, Tawade N, Kumar U and Naik H 2017 Determination of europium content in Li_2 SiO_3(Eu) by neutron activation analysis using am-be neutron source *Appl. Radiat. Isot.* **118** 7–11

[4] Barker P H and Connor R D 1967 ^{56}Co as a calibration source up to 3.5 MeV for gamma ray detectors *Nucl. Instrum. Methods* **57** 147–51

[5] Müller-Mellin R 2008 The solar electron and proton telescope for the STEREO mission *Space Sci. Rev.* **136** 363–89

[6] Roberts N J, Moiseev N N and Kralik M 2011 Radionuclide neutron source characterization techniques *Metrologia* **48** S239–53

[7] Navin A, de Oliveira Santos F, Roussel-Chomaz F and Sorlin O 2011 Nuclear structure and reaction studies at SPIRAL *J. Phys.* G **38** 024004

[8] Van Duppen P and Riisager K 2011 Physics with REX-ISOLDE: From experiment to facility *J. Phys.* G **38** 024005

[9] Ball G C *et al* 2011 Physics with reaccelerated radioactive beams at TRIUMF-ISAC *J. Phys.* G **38** 024003

[10] Jenkins D G 2014 Recent advances in nuclear physics through on-line isotope separation *Nat. Phys.* **10** 909–13

[11] Wuosmaa A H *et al* 2007 A solenoidal spectrometer for reactions in inverse kinematics *Nucl. Instrum. Methods Phys. Res.* A **580** 1290–300

[12] Guerrero C *et al* 2013 Performance of the neutron time-of-flight facility nTOF at CERN *Eur. Phys. J.* A **49** 27

[13] Wilson J N *et al* 2017 Anomalies in the charge yields of fission fragments from the ^{238}U(n,f) reaction *Phys. Rev. Lett.* **118** 222501

Part II

Detectors

Chapter 4

The right detector for the job

This chapter sets out to answer the question: what type of detector should I use for my measurement problem in experimental nuclear physics or in related applications? The answer to this question is quite complex and will not necessarily converge on a unique solution. The answer may also, more often than not, correspond not to a single detector but a combination of detectors. At some level, human factors such as cost must come into the equation. A detector system needs to provide insight into the underlying physics but at a level which is 'good enough'. The design of a particular detector system is obviously informed by the physics, by precedent and by the experience that comes in making such measurements. Here, we review some of the key criteria for choosing the components of a detector system and discuss Monte Carlo simulation which is rapidly becoming essential to validate the design of specific apparatus. We will briefly discuss the main types of radiation detector based on the principles behind their operation. We then go on to map the different detector technologies to specific classes of ionising radiation: light and heavy ions, e.g. α particles, β particles and electrons more generally, γ rays and neutrons.

Concepts: energy resolution; timing resolution; angular coverage; detector efficiency; counting rates; detector design and modelling: GEANT4, MCNP and FLUKA; gas ionisation detector; scintillator detectors; semiconductor detectors; detection strategy for different classes of ionising radiation.

4.1 Considerations in designing a detector setup

When planning a nuclear physics experiment or designing an apparatus for related applications, some of the key considerations at the outset are whether the detector system is required to simply record a level of activity, or whether it also needs to discriminate between different types of radioactivity and/or whether *spectroscopic*

doi:10.1088/978-0-7503-1428-2ch4

performance is required[1]. For example, a Geiger–Müller tube provides individual clicks when events are recorded or a rate on a meter of instantaneous activity. Such a device only provides a gross count of the level of radioactivity. Depending on the design of the Geiger–Müller tube, it may be sensitive to several different types of ionising radiation, e.g. alpha, beta and gamma radiation with varying efficiency. In a nuclear physics experiment, detectors are frequently optimised to be sensitive to a specific type of radiation, be it gamma rays, neutrons etc. Often this selectivity comes by default—for example, in most cases, situating detectors outside the vacuum chamber where the nuclear reaction under study takes place will ensure that it is insensitive to, e.g. alpha particles as such particles will be effectively stopped by the walls of the chamber.

A detector with *spectroscopic* performance is one where the amplitude of signals generated in the detector is, in some way, proportional to the energy deposited in the detector. We will examine the mechanisms which lead to this approximate proportionality for some general classes of detector including scintillators and semiconductors in the discussion below (see sections 4.3.2 and 4.3.3, respectively). If we have proportionality (not necessarily linear proportionality) in the signal with respect to energy deposited then, assuming we have access to an appropriate source for calibration purposes (see section 3.1), we can convert the detector signal into a spectrum where, depending on the physics and the subject of the study, it may be possible to resolve a series of peaks or other structure in the spectrum.

Turning to the issue of design of the detector setup, figure 4.1 introduces some of the relevant considerations. For example, the complexity of the spectrum anticipated

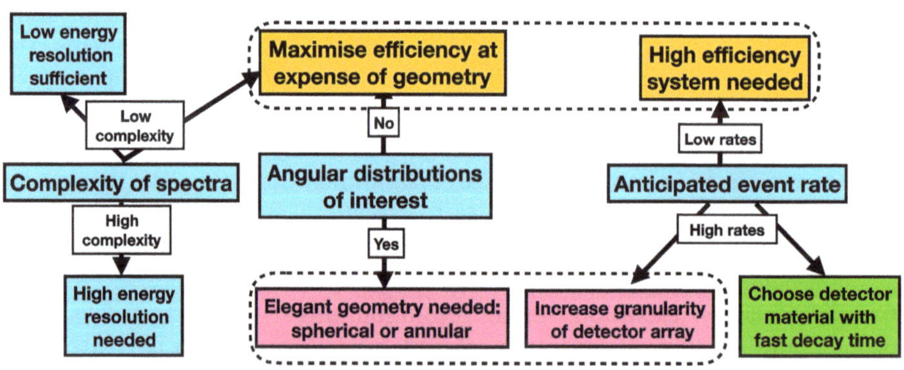

Figure 4.1. Flow-chart regarding some of the overarching issues on choice of detector(s) and design of an experimental setup designed for spectroscopy. These are subdivided into three key points: the complexity of the spectra which dictates the required energy resolution (see section 4.1.1), whether angular distributions are of interest which dictates the geometry of the setup (see section 4.1.5), and the expected event rate which has implications for the granularity of the setup and the choice of detector (see sections 4.1.2 and 4.1.3) and read-out electronics (see section 10.1).

[1] A further question might be: where is the source of the radiation located? This is more relevant to nuclear applications (and some examples are given in section 7.1.2) since in a nuclear physics experiment the source of the radiation is ordinarily well-defined.

defines the necessary *energy resolution*, the anticipated counting rates have implications for the desired *granularity* and *timing resolution* of the system. Whether angular distributions and/or angular correlations of the particles/gamma rays are important has implications for the design and geometry of the setup.

4.1.1 Energy resolution

Having determined that the measurement to be made is spectroscopic, the first key question is what energy resolution is required for the detector(s) to be used? In answering that, we should first define what we mean by energy resolution. It is common to convert the pulse height output from a detector into a histogram (or *spectrum*) which has the number of counts measured on the y-axis and the pulse height on the x-axis. (If we want to be precise, this is a *differential pulse height spectrum*, i.e. a plot of $\frac{\mathrm{d}N}{\mathrm{d}H}$ against H, where H is the pulse height and $\mathrm{d}N$ is the number of counts in the interval $\mathrm{d}H$.) Through energy calibration, we can relate the pulse height to energy converting the pulse height spectrum to an energy spectrum. If the detector was ideal and could measure the energy recorded in it with infinite precision and the radiation emitted from the source was fixed and was emitted with a single discrete energy defined with infinite precision then we might expect to see an infinitely thin spike in our spectrum, i.e. a mathematical delta function. Sadly, no realisable detector operates in such a way and, moreover, nuclear energy levels (even ground states) have a finite width in energy as they have a finite lifetime—it is not possible to avoid the Heisenberg uncertainty principle that $\Delta E \Delta t > \hbar/2$. As we will see later in this chapter, a realistic detector, depending on its working principle, converts the energy deposited in it into a finite number of electrons following ionisation, electron–hole pairs, excited atoms or molecules which emit scintillation light, etc. The number of, e.g. charge carriers produced, N, will not be fixed but will vary event-by-event in a random way following Poisson statistics. As we process this detector signal through an electronics chain, a certain amount of noise will be added to the signal event-by-event which also varies in a random way[2]. The *central limit theorem* in probability theory implies that when independent random variables are added in this way, their normalised sum tends to the normal (or Gaussian) distribution:

$$f(x) = \frac{1}{\sigma\sqrt{2\pi}} \exp\left[-\frac{(x - x_0)^2}{2\sigma^2}\right], \tag{4.1}$$

where x_0 is the mean and σ is the standard deviation. This distribution appears all across physics and in more familiar real-life situations such as plotting the heights of a large number of men or women, and is shown in figure 4.2.

[2] A more concrete example will be given in terms of the contributions to the energy resolution of an inorganic scintillator detector in section 5.1.1.

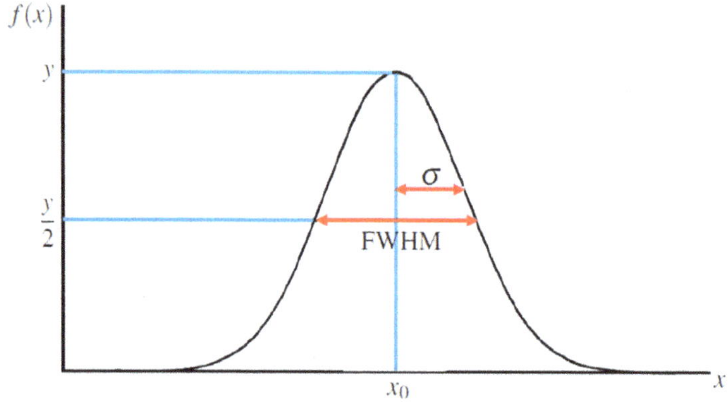

Figure 4.2. Gaussian distribution and the definition of the full width at half-maximum (FWHM).

To establish energy resolution, we define the width of the Gaussian distribution (shown in figure 4.2) in terms of the full width at half-maximum (FWHM), i.e. the width of the peak at half its maximum height:

$$\mathrm{FWHM} = 2\sqrt{2\ln 2}\,\sigma \approx 2.355\sigma. \tag{4.2}$$

Having calibrated a spectrum, i.e. related pulse height to energy, the FWHM can be expressed in units of energy e.g. keV or as a % of the centroid value[3]. For example, a sodium iodide, NaI(Tl), detector typically has a FWHM of around 7% at 662 keV.

Let us now examine the criteria of sufficient energy resolution in relation to gamma-ray detection[4]. Figure 4.3 compares the gamma-ray spectra obtained from the same mixed radionuclide source using a set of different classes of detectors: semiconductor detectors (hyperpure germanium and CZT) and scintillator detectors (traditional materials: CsI(Tl) and next-generation materials: $CeBr_3$ and $LaBr_3$). The detectors employed vary strongly in terms of their energy resolution and efficiency but also they vary significantly in terms of cost. In the example given, if the objective were to resolve the peaks in the spectrum and clearly identify their origin then it is obvious that this would most easily be achieved with a germanium detector which has the best energy resolution, i.e. the peaks in the spectrum, corresponding to gamma-ray transitions associated with the decay of several different isotopes, are the narrowest. However, for the example shown, the same objective could be achieved with the other classes of detector considered. The exercise only becomes problematic in the case of CsI(Tl) where some of the peaks are seen to overlap. This could still be

[3] Note that without specifying the energy centroid for which the resolution is quoted, then the energy resolution is meaningless. Note also that conventions differ in terms of the energy specified, e.g. resolutions are usually stated for scintillator detectors in terms of the 662-keV gamma-ray peak from a ^{137}Cs source, while resolutions for germanium detectors are conventionally presented for the 1.332-MeV gamma ray from a ^{60}Co source.

[4] It is important to note that the achievable energy resolution in a nuclear physics experiment may often not be as good as the intrinsic energy resolution of the detector such as the case of gamma rays emitted from a moving source.

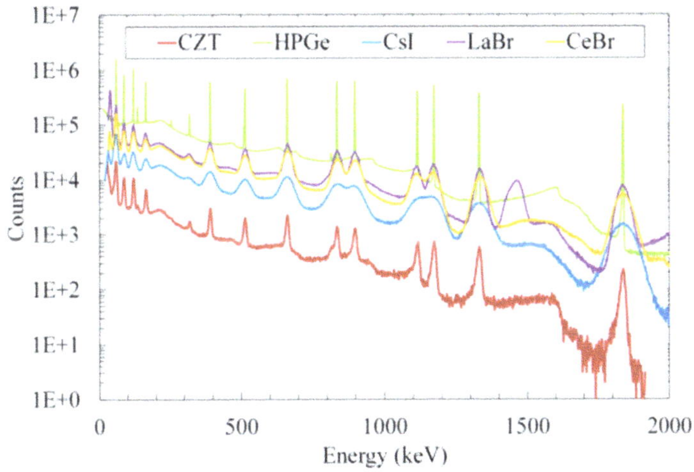

Figure 4.3. Comparison in performance of different gamma ray detectors in measuring the same radioactive source. The spectra were obtained over 24 h using a 29 kBq mixed radionuclide filter source. The measurements are labelled by the detector employed: CZT, high-purity germanium (labelled HPGe), CsI (Tl) scintillator (labelled CsI), $LaBr_3$(Ce) (labelled LaBr) and $CeBr_3$ (labelled CeBr). The spectra are essentially the same but with differing energy resolution. Note, however, the extraneous peak around 1400 keV in the $LaBr_3$(Ce) spectrum which arises from internal radioactivity within the scintillator crystal itself (see discussion of this issue in section 5.2.1). (Reprinted from [1] with permission from IOP Publishing under Creative Commons License.)

satisfactory for the application under consideration if the details of the mixed source used are well known, i.e. that the gamma rays to be detected have well established energies and relative intensities. In this case, the spectrum could be fitted, even in the case where the peaks overlap, based on Gaussian (or more complex functional form) peak shapes with a FWHM established for a given detector as a function of energy.

In the above example, we know *a priori* that we can fully reproduce the spectral form as all the information on the peaks which should appear is under control. Contrast this situation with a nuclear physics experiment. Here, the spectrum of alpha particles, heavy-ion energies, gamma rays, etc is usually not known *a priori*. This is discovery science where the answer is not known in advance. However, the situation in terms of designing the experiment is not hopeless since there is, in all likelihood, a good expectation as to the character of the spectrum likely to be observed. This is because an experimental study is not done in a vacuum[5], rather it is informed by background knowledge in the field of nuclear structure physics. Spectra may, however, prove to be dense and complex. There may be *degeneracies* in the sense that different gamma-ray transitions or charged-particle groups may happen to be very close in energy simply because the nuclear structure dictates that they do.

Figure 4.4 shows the state-of-the-art in terms of nuclear structure studies with first spectroscopic information on excited states in the exotic doubly-magic nucleus, ^{78}Ni [2]. The challenge here was that the exotic nucleus of interest was extremely difficult

[5] Although it may well be carried out in a vacuum chamber!

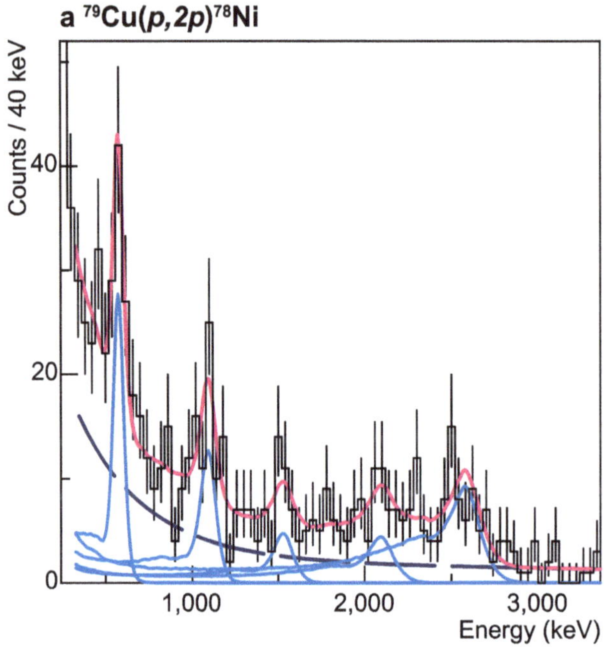

Figure 4.4. Spectrum of gamma rays associated with the decay of excited states in ^{78}Ni produced in the ^{79}Cu (p,2p)^{78}Ni reaction in an experiment carried out at the RIKEN-RIBF facility. The gamma rays were detected with the DALI-2 array of sodium iodide detectors. The fitted peak shapes for individual gamma ray transitions are shown in blue with the total fitted spectrum shown in red. (Reprinted by permission from Springer Nature [2].)

to produce, hence, the yield of gamma rays was low. However, only a handful of excited states are expected to be produced, meaning that the resulting gamma ray spectrum should not be particularly dense. To detect gamma rays, a high-efficiency NaI(Tl) (see section 5.1.2) detector array, DALI-2, was therefore used. This material has a low intrinsic energy resolution, made worse by Doppler broadening due to the high velocity of the emitting nucleus. Locating gamma ray peaks in this spectrum is challenging and the expected peak shapes are so positioned and so wide that they overlap (see figure 4.4). There is no little ambiguity in disentangling the peaks 'discovered' in this spectrum. Figure 4.5 shows a very different kind of gamma-ray spectrum to that of figure 4.4. Here, the nucleus of interest is ^{152}Dy that exhibits so-called 'superdeformation' and can be produced much more easily than ^{78}Ni so high statistics can be obtained. This highly collective nucleus exhibits many rotational bands with cascades of 10s of individual gamma rays. The spectra presented in figure 4.5 were obtained with Gammasphere, a large array of high-purity germanium detectors [3]. It is clear from the complexity of the spectra that using NaI(Tl) detectors as in the example of ^{78}Ni above, would simply lead to an unresolved 'bump' with all the detailed spectroscopic information lost. Improvements in resolving power with large arrays of germanium detectors will be considered in more detail in section 6.4.2.

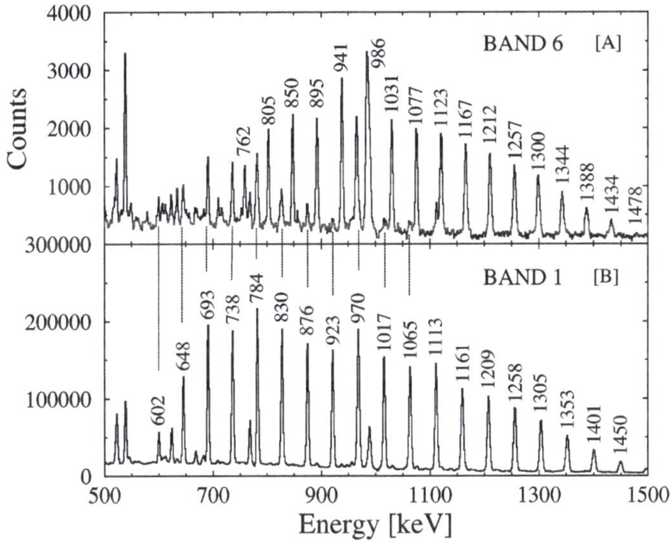

Figure 4.5. Gamma-ray spectra corresponding to superdeformed bands in ^{152}Dy measured with the germanium detector array, Gammasphere. (Reprinted with permission from [3]. Copyright (2002) by the American Physical Society.)

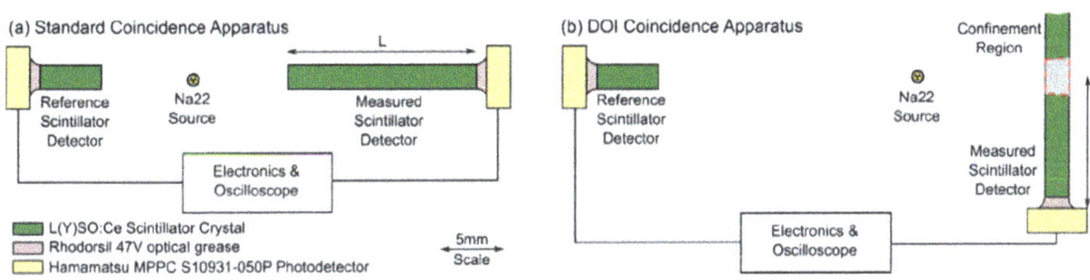

Figure 4.6. Two apparatus for making timing coincidence measurements: (a) a scintillator detector of varying length, *L*, measured with respect to a reference detector (b) a scintillator detector mounted side-on so that the effect of depth of interaction on timing resolution can be explored. (Reprinted from [4] with permission of PLOS under Creative Commons License.)

4.1.2 Timing resolution

The concept of timing resolution follows more naturally from thinking about coincidence events. Consider two identical scintillator detectors positioned opposite to one another, oriented face-to-face with a ^{22}Na source positioned at the centre (see figure 4.6). A ^{22}Na source is chosen because it decays by positron emission (see section 1.4). The positron most commonly annihilates into two 511 keV photons in a near back-to-back geometry. These two photons are naturally produced at the same *time*; if the two detectors are equally spaced then it might naively be supposed that the two gamma rays would be detected 'at the same time'. However, they will not be in practice because of the finite intrinsic timing resolution of the detectors and the uncertainties in fixing the timing of each pulse using electronics. To understand the effect of the timing resolution, an electronic (or digital) window can be set on the photopeak recorded in the two detectors, to remove Compton-scattered events. The time difference between the events recorded in the two detectors, which might be detected by the signal rising above a certain threshold[6], can be histogrammed. This time difference would be expected to have a Gaussian shape. The full-width-at-half-maximum (FWHM) of the time difference peak is called the coincidence time resolution (CTR) of the system,

$$CTR = 4\sqrt{\ln 2}\sigma, \tag{4.3}$$

where σ is the timing resolution of the individual detectors [4]. This relationship reflects the fact that the time-difference peak is a convolution of the two Gaussians associated with the individual detector timing resolutions. Such coincidence measurements are typically used to derive the (unknown) timing resolution of a specific detector since they can be measured with respect to a reference detector of well-established timing resolution, σ_{ref}. The unknown timing resolution can be extracted assuming that the CTR depends on summing the timing resolution of the two detectors in quadrature.

Timing resolution can depend on many factors. The first factor relates to the physical principles by which the detector operates. For an extended detector, its linear dimensions are important, i.e. a large detector will, in general, have a poorer timing resolution than a smaller one of the same type (see figure 4.7). In the example shown on the right-hand side of figure 4.6, the timing resolution may depend on the depth of interaction within an extended scintillator detector. These issues will be explored in more detail and with reference to specific examples in later chapters.

Timing resolution of detectors is important from an experimental perspective for a number of reasons:

- Good timing resolution can be used to reduce experimental background. In many nuclear physics experiments, the accelerated ion beam used to initiate

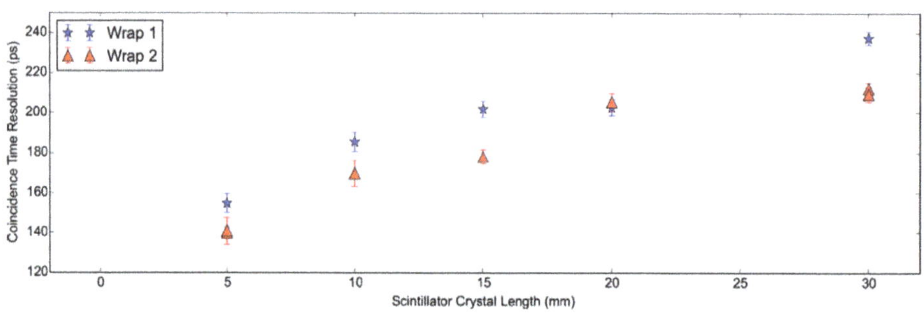

Figure 4.7. Coincidence time resolution (CTR) as a function of crystal length for two LYSO:Ce scintillator crystals with cross-section of 2×2 mm^2 under the geometry shown on the LHS of figure 4.6. (Reprinted from [4] with permission from PLOS under Creative Commons License.)

[6] Methods by which this can be achieved using relevant electronics modules are discussed in section 10.2.5.

the reaction is pulsed, e.g. with a width of the order of a nanosecond, with pulses separated of order 100 ns. *Prompt* radiation, e.g. gamma-ray emission can be selected by establishing a coincidence with the arrival of the beam either through an analogue *trigger* (see section 10.2.6) or more commonly nowadays, where data is time-stamped, through reconstruction of coincidences off-line in software.

- Different species of ionising radiation are often emitted in coincidence, e.g. gamma rays in coincidence with each other or gamma rays following alpha or beta decay to excited states in the daughter nucleus. Random coincidences can be avoided by setting a strong *timing coincidence gate* between the two detected events.
- Time differences between detected events can be established where events are not in *prompt coincidence* and this can be used in certain instances, e.g. to establish the lifetime of excited states.

4.1.3 Counting rates

A key question in the design of any experimental setup is *can my detector(s) count fast enough?* In other words, will individual pulses in the detector signal restore to the baseline before the next pulse comes along? Where signals come on top of each other, this is described as *pile-up* (see figure 4.8). This should be distinguished from *true summing* where two particles or gamma rays associated with the same decay event end up in the same detector. An example of this would be the *sum peak* commonly seen in a gamma ray spectrum from a ^{60}Co source where the two gamma rays of 1.173 and 1.332 MeV are emitted essentially simultaneously and may enter the same detector. This relates to the issue of detector geometry described below in section 4.1.5 since the probability of summing depends on the solid angle coverage of the gamma-ray detector.

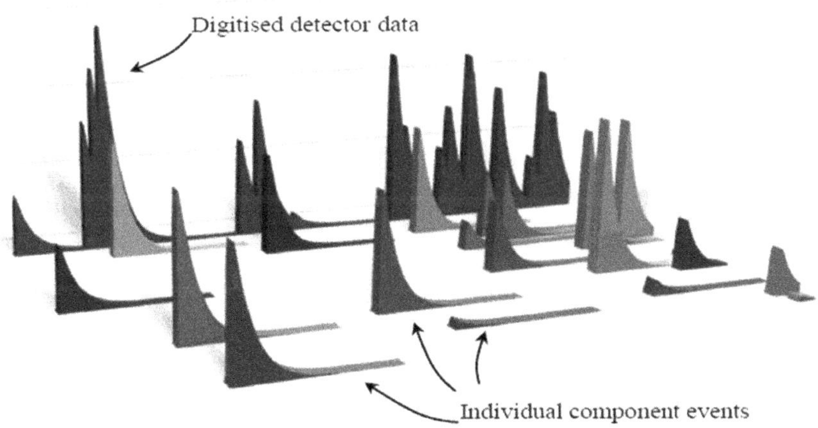

Figure 4.8. Schematic of pile-up: individual decay events arrive within the resolving time of the detector and sit on top of each other. The final signal is shown in the rear of the cartoon. (Reprinted from [5], with the permission of AIP Publishing.)

If the decay time of a pulse is of the order of μs then when such a detector is counting at a rate of e.g. 1 kHz, the probability of pile-up will be very low. As the counting rate increases, the probability of pile-up will increase until it begins to impair the measurement. There is a related issue of *dead time* in analogue data acquisition systems which will be discussed in more detail in section 10.2. There are several strategies which can be employed to allow a detector system to count faster. The first would be to choose a detector material which has an intrinsically fast decay time. A second approach to mitigating pile-up would be to increase the *granularity* of the detection system, i.e. the number of individual detectors employed. Segmenting a single detector may not necessarily achieve the same thing if the readout is common. A third approach is to employ a digital data acquisition system which can mitigate the dead time issue (see section 10.3).

4.1.4 Detector efficiency

The statistics needed in an experiment or application within a given measurement time will determine the necessary *efficiency* of the experimental setup and its component detectors. Let us first consider radiation of some type, e.g. alpha, beta, gamma rays emitted isotropically from a point source. We can define an *intrinsic efficiency* for a detector, $\varepsilon_{\text{intr.}}$, as the ratio of the number of particles/gamma rays detected to the number of particles/gamma rays incident on the detector. Some classes of detector in specific applications can approach 100% intrinsic efficiency. For example, silicon detectors with a thickness much wider than the stopping range of a given energy of alpha particle may essentially register the full energy of every alpha particle that falls on them. For indirectly ionising radiation such as gamma rays, it is much harder to approach such high intrinsic efficiencies due to the long mean free path for typical gamma-ray energies (see figure 2.6). Making a gamma-ray detector thicker will generally increase the photopeak efficiency as there will be more probability for the gamma ray to fully interact, e.g. see figure 5.3.

Intrinsic efficiency is only one element of the equation, since in most practical applications, the detector(s) will not fully surround the source. This leads to the concept of *absolute efficiency*, $\varepsilon_{\text{abs.}}$, as the ratio of the number of particles/gamma rays detected to the number of particles/gamma rays emitted from the source. This relates to the solid angle, Ω, covered by the detector. The solid angle coverage of a detector can be obtained via integration of an element of solid angle, $\Delta\Omega$ over appropriate angular ranges. $\Delta\Omega$ is defined as shown in figure 4.9; it corresponds to

$$\mathrm{d}\Omega = \frac{\mathrm{d}S_r}{r^2}, \tag{4.4}$$

where $\mathrm{d}S_r$ is a surface element spanning from θ to $\theta + \mathrm{d}\theta$ and from ϕ to $\phi + \mathrm{d}\phi$ for a spherical surface of constant radius, r, which is given by

$$\mathrm{d}S_r = \sin\theta\, \mathrm{d}\theta\, \mathrm{d}\phi. \tag{4.5}$$

Clearly, the complication is that most detectors do not have a curved frontage corresponding to the sphere of radius, r, around the source and so some geometrical

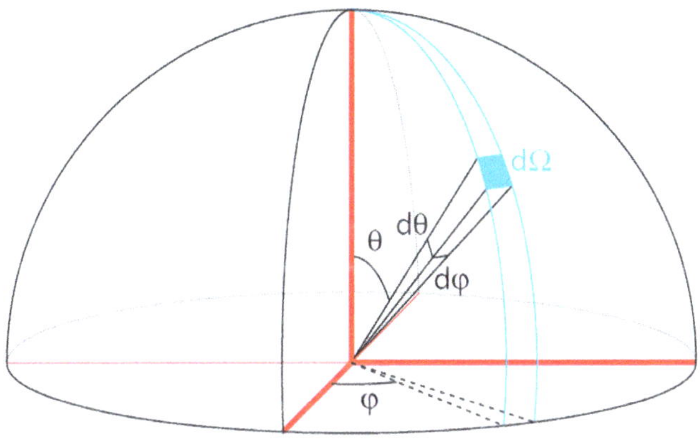

Figure 4.9. Illustration of the concept of solid angle where dΩ represents the element of spherical surface area for a sphere of unit-radius around a point source. (Reprinted from [6] with permission from OSA under Creative Commons License.)

work will be needed to relate the specific detector dimensions to this spherical geometry.

The units of solid angle, Ω, are steradians. Recalling that the surface area of a sphere is $4\pi r^2$, equation (4.4) shows that 4π steradians would correspond to a detector fully and hermetically surrounding the source. It is therefore common to speak of 4π detector arrays which approximates this idealised case.

The solid angle coverage is what relates the intrinsic and absolute efficiencies, i.e.

$$\varepsilon_{\text{abs}} = \varepsilon_{\text{intr.}} \times \Omega/(4\pi). \tag{4.6}$$

4.1.5 Angular coverage and detector geometry

There are many situations in experimental nuclear physics where charged particles/gamma rays/neutrons are emitted from a source in a non-isotropic manner, associated with angular distributions or angular correlations. This can be in the case where different particles/gamma-rays are emitted in coincidence and one is detected setting up an angular correlation with respect to the second one (see section 1.7.1). In nuclear reactions, kinematics are important and angular distributions of particles are observed; the natural axis of reference for such angular distributions is the beam axis of the incoming ion beam used to initiate a given reaction. Extracting the physics associated with these angular distributions will be most easily achieved where the detectors are arranged in an appropriate geometry such as spherical, cylindrical or annular. This might entail an array of individual detectors obeying such a geometry or a position-sensitive detector, e.g. an annular silicon detector. For example, in a spherical array of detectors of the same size, each will subtend the same solid angle (see section 4.1.4) and span an equivalent range of θ, ϕ angles. The situation is more complicated where, for example, gamma rays are emitted from a nucleus at a high relativistic velocity since the gamma rays will be *Lorentz boosted.* Figure 4.10 illustrates the effect on the angular distribution of various gamma-ray

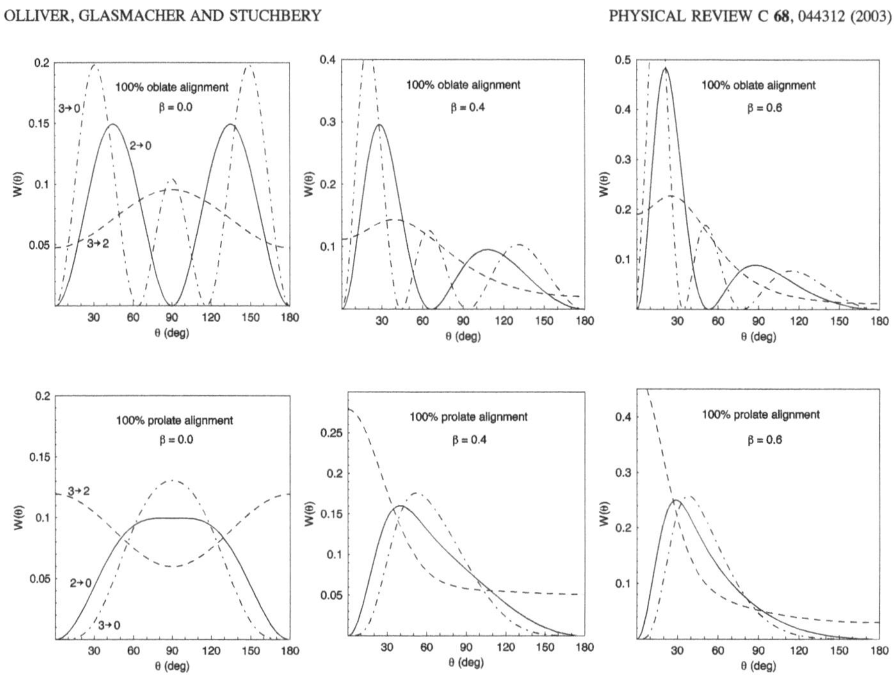

Figure 4.10. The effect of Lorentz boost on calculated angular distributions for 2 → 0, 3 → 0, and 3 → 2 transitions under three different scenarios, $\beta(=\frac{v}{c}) = 0.0$, 0.4, and 0.6. (The top set of angular distributions are for 100% prolate alignment and the bottom set for 100% oblate alignment: this is a detail relating to the m-substate distribution). (Reprinted with permission from [7]. Copyright (2003) by the American Physical Society.)

transitions which are symmetrical about 90° in the rest frame but become strongly forward-focussed at source velocities, β (i.e. v/c) of 0.4 and 0.6. Clearly for high velocity beams, gamma rays are more likely to be detected at forward angles, i.e. close to 0° as opposed to backward angles, close to 180°. In fact, this often motivates design of gamma-ray tracking arrays such as GRETINA and AGATA (see section 6.5.2) where they cover mostly the forward angle range.

4.1.6 Dynamic range

In many studies, the detection system will need to be sensitive to a wide range of particle or gamma-ray energies. We could describe this as the necessary *dynamic range*. The dynamic range could be limited by the thicknesses of the detectors employed, i.e. that they lack sufficient stopping power for high energy particles or gamma rays. The dynamic range might also be limited, e.g. by the photodetectors reading out scintillation light where high energy particles/gamma rays produce too much light which saturates the photosensor and/or leads to non-linearity. Finally, the electronics readout may be unable to cope readily with the dynamic range because the amplitude of some signals exceeds the range of voltage accepted by the data acquisition system. This could be partly corrected by reducing the amplification of the signals but then some signals may end up too small. A *logarithmic preamplifier* (see section 10.2) could be a solution for some special cases, e.g. where GeV heavy ions are implanted in a silicon detector and then followed by the decay of these ions by, e.g. alpha particles with MeV energy.

4.2 Detector design and modelling

As detector systems grow in complexity, it is increasingly common to model apparatus in advance in order to understand its likely performance and how to optimise it to a specific application. The pathway to this is through Monte Carlo simulation where the interaction of different types of ionising radiation is tracked as it transits through the materials which make up the experimental setup as well as the detectors employed. This technique is extremely powerful and can allow many different detector geometries to be explored in a relatively short space of time. Indeed, present day computing power, including access to multiple cores, supports a level of Monte Carlo simulation which would simply not have been possible in the past, where design of detector systems would have relied more heavily on the user's experience and experimental improvement through trial and error.

Conveniently, there are several software packages available which provide the necessary functionality. Some of the more frequently used codes are GEANT4 [8], MCNP and FLUKA [9]. These packages have varying applicability and are the more heavily used by different communities, for example, but not exclusively: GEANT4—nuclear/particle physics, MCNP—nuclear industry, FLUKA—radio-protection. There are also simulation packages that treat specific types of detector, e.g. GARFIELD which can simulate gas-filled detectors and their associated electromagnetic fields. Let us briefly review the two most common Monte Carlo simulation packages, GEANT4 and MCNP.

4.2.1 GEANT4

GEANT4 [8] is a CERN-supported Monte Carlo simulation package which is a very powerful toolkit for modelling the behaviour of radiation detectors to all types of incident ionising radiation. It is widely used and popular with the physics community given it was originally introduced as a means of simulating particle physics experiments and therefore is able to fully treat the zoo of exotic particles that are produced at colliders such as the LHC in CERN. Nevertheless, as a by-product of its origin, it is equally suited to the much lower energy heavy ions, alpha particles, beta particles, gamma rays, and neutrons commonly encountered in experimental nuclear physics and the wider societal applications of this branch of physics. It has also been more widely applied in space science, e.g. in spacecraft design and estimation of dose received by electronics in space applications [10].

GEANT4 is a complex software package and requires programming in C++ which can be a barrier to access for some users. That said, there are many examples supplied on the GEANT4 website and there is a large user community with whom knowledge can be exchanged. However, the output of the Monte Carlo simulations is only as good as the input and the care taken in getting details right. There are many potential pitfalls which the user must be careful of (and the author is well aware of having made such mistakes for himself and seen them in the work of his research team). For example, there is no 'error checking' in the input so it is easy to make mistakes in defining materials, mixing up dimensions, densities etc, which will pass as input but give wildly incorrect results. Moreover, the user needs to have a good understanding of the underlying physics to ensure that all the appropriate types of interaction are turned on for the energy range of interest and that the step size over which particles are tracked is appropriate since these may default to values more appropriate to high energy physics. This may particularly occur when adapting examples from different domains. Particular attention must be paid to the quality of the nuclear data being used in GEANT4 and the extent to which it accurately represents all of the physics processes going on. An example of this would be the treatment of fast neutron interactions with organic scintillators where it was found, in the past, that the interactions with ^{12}C were not fully incorporated, in particular, the tendency for fast neutrons to break up ^{12}C into three alpha particles leading to a characteristic detector signal. This full physics description needed to be added into GEANT4 to resolve discrepancy between experimental measurements and simulation [11].

Some aspects of GEANT4 to note are the following:

- Via extensions and links to appropriate viewers, simulations can be visualised which is an extremely helpful way to get an immediate sense of how a given particle or photon is interacting, e.g. see figure 4.11.
- Routes exist to import detect geometries directly from CAD drawings which simplifies the incorporation of complex set-ups and removes potential for mistakes.
- Extensions of GEANT4 can provide more sophisticated treatment of the working of specific detector types. For example, optical photons generated within a scintillator can be tracked from their production, through reflection and losses at the boundaries to their detection with a given light sensor. This is highly relevant to the simulation and design of, for example, inorganic scintillator detectors (see section 5.1.2). Extensions to GEANT4 can also treat contemporary photosensors such as silicon photomultipliers (see section 5.3.3), e.g. G4SiPM [12].
- By default, GEANT4 knows nothing about the energy resolution of detectors. An appropriate functional form must be added in by hand in order to turn initial simulations into a form comparable with an experimentally-obtained spectrum.
- For highly complex simulations, GEANT4 can be accelerated by accessing the capabilities of multithreading [10].
- There are extensions to GEANT4 which are heavily focussed on medical imaging techniques such as PET, SPECT and CT scanning. An example of this is GATE [13], which allows a rapid entry to simulation in this domain based on standard components.

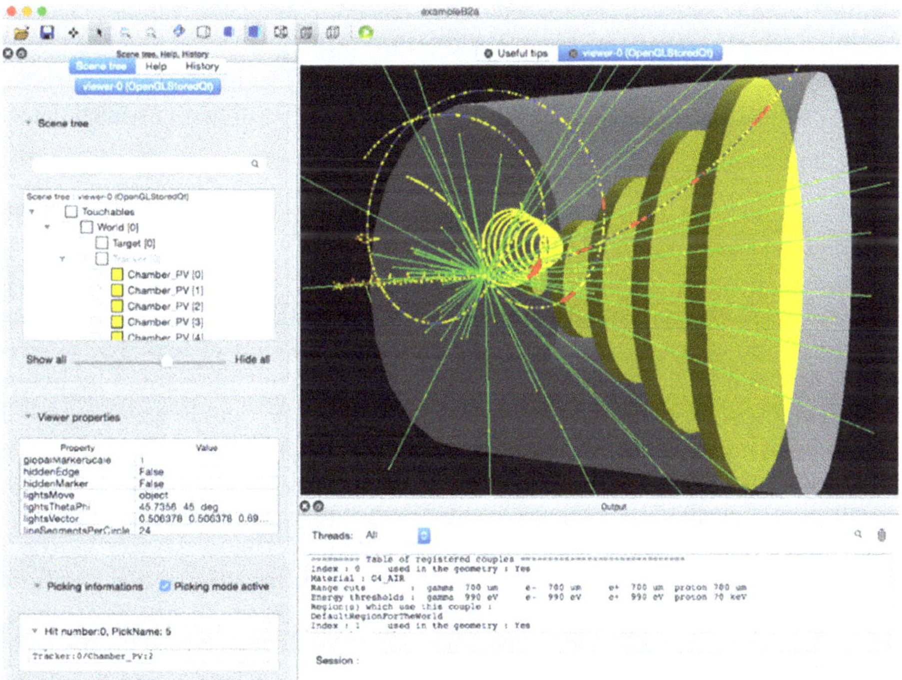

Figure 4.11. Visualisation of GEANT4 simulation geometry using OpenGL viewer. (Reprinted from [10] with the permission of Elsevier under Creative Commons License.)

4.2.2 MCNP

MCNP was originally developed by Los Alamos National Laboratory and its name derives from 'General Monte Carlo N-particle Transport Code'. MCNP is more tailored to those working in nuclear energy and applications. It is widely applied to modelling of fission and future fusion reactors. It is also used often used in design of shielding and medical applications. MCNP is subject to export control, meaning that users based outside the US may not be able to obtain the source code but only the executable-code version.

4.3 Overview of major detector types

Returning to our aim to identify the right detector for the right job, let us review some of the major classes of radiation detector and their basic operating principles. These types of detectors will be returned to later in this book with more specific and detailed discussion of their operation and performance. Critical to the operation of any detector is its readout electronics. This topic warrants its own chapter—chapter 10.

4.3.1 Gas-filled detectors

We begin with gas-filled detectors which comprised some of the earliest detectors used in nuclear physics such as the famous Geiger–Müller tube. Gas-filled detectors are still important, particularly for heavy-ion detection in nuclear physics. The basic operating principle is that ionising radiation transiting a gas-filled volume will ionise the gas atoms/molecules producing electron–ion pairs. The number of electron–ion pairs, N, produced by a particle depositing an energy ΔE within a gas-filled detector is given by:

$$N = \frac{\Delta E}{W}, \tag{4.7}$$

where the W-value is the energy required to produce an electron–ion pair. The W-values are in the approximate range 20–40 eV depending on the gas, and exceed the ionisation potential of the gas by approximately a factor of two. The electron–ion pairs generated will move towards electrodes under the influence of an external electric field. Figure 4.12 illustrates some of the different regimes in which a gas-filled detector can operate. In the low or zero electric field regime, ions and electrons will slowly recombine forming neutral atoms. As the electric field increases, recombination will progressively decrease and eventually become negligible, leading to a flat regime where all charges can be efficiently collected. This is known as the *ion chamber regime*. Beyond the ion chamber regime, as the electric field is further increased, it is possible for *avalanche multiplication* to occur, i.e. the electrons and ions moving in the field are sufficiently energetic that they can ionise further atoms/molecules in the gas. The total charge collected will be proportional to the strength of the electric field, i.e. bias voltage. This is described as a *proportional regime*. The rate of change of the number of charge pairs, N as a function of distance, is given by

$$\frac{dN}{dx} = \alpha N, \tag{4.8}$$

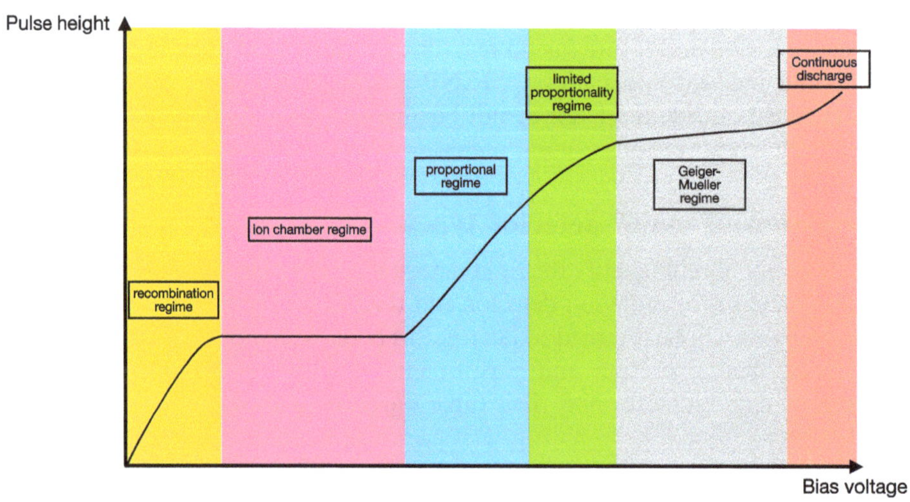

Figure 4.12. Different regimes of operation for a gas ionisation detector shown as a plot of pulse height as a function of bias voltage.

where α is the *first Townsend coefficient*, which is a function of the type of gas, the pressure of the gas, and the electric field intensity [14]. This multiplication of charges can be described by a multiplication factor, M, where

$$M = \frac{N}{N_0}, \tag{4.9}$$

where N_0 is the number of primary charge pairs generated.

Increasing the electric field further, the detector will enter a regime of *limited proportionality* where the cloud of slow moving positively charged ions will shield the external electric field felt by the charge pairs and so proportional response is lost. This is not a useful regime in which to operate a gas-filled detector. At very high electric field, an ionising particle entering the gas volume causes a dramatic cascade of avalanches throughout the volume, generating a large output pulse. This is the *Geiger–Müller* regime and one of the principles behind the familiar *Geiger–Müller tube*. It is an efficient regime for counting but the proportionality to the energy of the ionising radiation is lost.

In summary, therefore, we have several regimes in which a gas-filled detector can operate. The most interesting possibilities for nuclear physics, setting aside the ubiquitous *Geiger–Müller* tube used for radiation monitoring, are the *ion chamber* and *proportional counter* regimes. Both provide a signal proportional to the number of primary charge pairs produced and hence, to ΔE, the energy deposited by an energetic particle passing through the gas volume.

Ion chambers have distinct advantages in that they sit in the flat region shown in figure 4.12 so that small variations in the bias voltage do not greatly effect the charge collected. However, they lack the intrinsic amplification of a proportional counter and so output signals can be very small and depending on the application, potentially difficult to resolve from noise. There are many parameters which can be varied in the design of ion chambers and proportional counters depending on the application: the geometry, arrangement of electrodes, electric field and choice of fill gas. Natural choices for geometry of ion chamber, for example, are parallel-plate electrodes or a cylindrical geometry with an anode wire down the centre. Parallel-plate geometries suffer from the amplitude of the pulse being dependent on the position of interaction in the gas volume. This effect can be handled by the introduction of a *Frisch grid*—a further porous electrode structure within the gas volume through which electrons can pass on their way to the anode.

4.3.2 Scintillator detectors

A scintillation detector comprises two essential components: a scintillating material which produces flashes of light (luminescence) when ionising radiation interacts with it, where the light output is largely proportional to the energy deposited by a gamma ray or charged-particle/neutron interaction in the crystal, and a photodetector—classically, a photomultiplier tube (PMT), which converts the photons into photoelectrons, i.e. an electrical signal which can then be suitably multiplied, i.e. amplified and subsequently digitised[7].

The earliest scintillators used were in the form of a scintillating screen made of materials such as zinc sulfide activated with silver. This was famously used by Geiger and Marsden in their gold foil scattering experiment described in section 1.1.1. These early experiments were limited by the sensitivity of the human eye and practical limits in counting. Moreover, zinc sulfide is only available as a polycrystalline powder and so while a thin scintillating screen may be relatively efficient for short-range particles such as alpha particles, it will be of very little use for direct detection of gamma rays or neutrons. This led on to the development of scintillator materials which can be produced in a large volume, e.g. as blocks or crystals that have the stopping power needed for penetrating radiations such as gamma rays and fast neutrons. Such scintillators fall into two basic categories: (1) organic materials where the scintillation process is an intrinsic property of the molecules concerned and can be observed in different phases or solutions and (2) inorganic crystals where the scintillation process is intrinsically related to the crystalline structure of the material.

Mechanism of organic scintillation

In an organic scintillator, it is appropriate to consider valence electron orbitals at the molecular level. Figure 4.13 shows the molecular orbital structure of a typical organic scintillator. There are two classes of states: the singlet state, of which S_0 is the ground state which has a series of excited singlet states, S_1, S_2 etc, built on top of it with a typical energy spacing of the order of 1 eV. Each of these singlet states has an associated set of molecular vibrational modes with spacing of order 0.1 eV which are denoted e.g. S_{01}, S_{02}, etc. In parallel, there are a set of excited triplet states, labelled T_0, T_1, etc, again with their corresponding set of vibrational modes, T_{01}, T_{02}, etc. The excitation of the scintillator occurs from the ground state up to any of the excited singlet or triplet states or their associated vibrational modes. The singlet excited states immediately relax to the S_1 state on a fast timescale through a non-radiative process known as *internal degradation*. The S_1 state may then decay on a relatively fast time scale of order nanoseconds to the S_0 ground state or its associated vibrational states. This is the primary mechanism for generation of scintillation light or *fluorescence*. A complicating feature of organic scintillators is that there are alternative decay modes present, either through excitation of triplet states which decay rapidly to the T_0 state or through a so-called *inter-system crossing* where the S_1 states decays to the T_0 state. The decay of the T_0 state is intrinsically slow as it has to be mediated by interaction with another molecule excited into a T_0 state[8]. This transition gives rise to a long-lived decay mode or *phosphorescence*. The peak wavelength of the phosphorescence will typically differ from that of the fast fluorescence due to the differing energies of the excited states involved. Organic

[7] For an extended overview of inorganic (and organic) scintillators the reader is directed to the monograph of J B Birks [15] which while dating from the 1960s is still a key reference regarding scintillation counting.

[8] In fact, the reasons for the direct decays being forbidden are down to selection rules for atomic transitions which are similar to the selection rules applying to gamma-ray emission described in section 1.7.1.

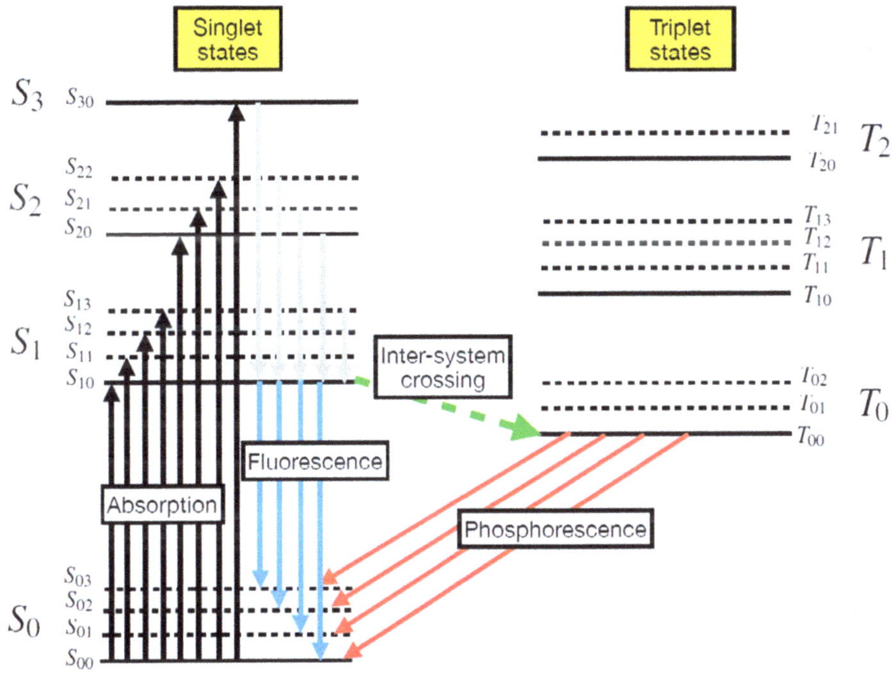

Figure 4.13. The scintillation mechanism in an organic scintillator. The excited states of the organic molecule are shown along with the different modes of excitation and subsequent decay. The mechanism is more fully explained in the accompanying text.

scintillators will be described in more detail in chapter 9 where the focus is on neutron detection as these are a very common material for fast neutron detection.

Mechanism of inorganic scintillation

Inorganic scintillators may be used for gamma-ray or charged particle detection. Inorganic scintillators are the focus of chapter 5 where they are discussed with an emphasis on their suitability for gamma-ray detection. This chapter also describes the operating principles and parameters of photomultiplier tubes (PMTs), and their solid-state replacements such as avalanche photodiodes (APDs) and silicon photomultipliers (SiPMs) (see section 5.3) which are also fully relevant to organic scintillators described above. We return to the topic of inorganic scintillators in chapter 8 where we cover their employment for charged-particle detection, exemplified by the material, CsI(Tl), which has multiple components to its scintillation light with differing decay times that provide sensitivity to the type of particle interacting. Here, we provide an overview of the inorganic scintillation mechanism based on gamma-ray detection.

Gamma rays interact with an inorganic scintillator by one of the typical interaction modes: photoelectric absorption, Compton scattering and pair production (for high energy gamma rays). The primary electrons generated by these processes will excite secondary electrons from the valence band to the conduction

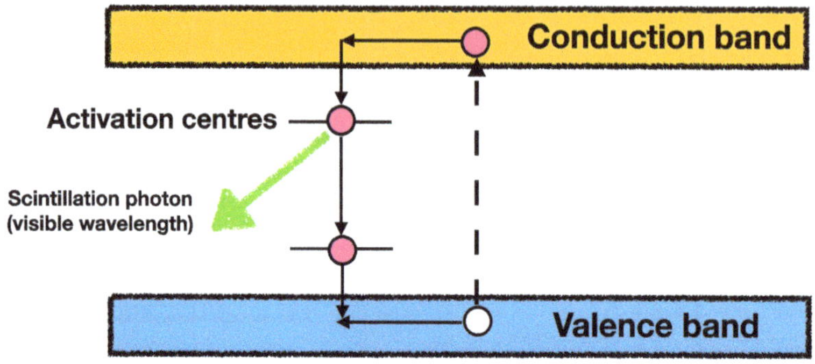

Figure 4.14. Cartoon of the scintillation mechanism in an inorganic crystal e.g. NaI. The addition of activation centres such as thallium to the crystal structure allows an alternative pathway for de-excitation resulting in emission of visible wavelength photons.

band leaving behind a hole in the conduction band (see figure 4.14). When the electrons de-excite, photons will be emitted. If the scintillator is (at least, largely) transparent to this scintillation light emission then it can be collected with a suitable photodetector, classically a photomultiplier tube (described in section 5.3.1) which effectively converts the light emitted into a useful electrical signal. In order for a scintillator to be useful as a *spectroscopic detector*, the light output should be *proportional* to the energy deposited in the crystal—not necessarily linearly proportional—an issue to be discussed in section 5.1.1. Scintillation mechanisms are not particularly efficient; most of the energy deposited in the crystal by gamma rays is converted into phonons and the yield of scintillation photons is typically of the order of 10 000 photons per MeV deposited. Ideally, the light emission of the scintillator should be fast so that high counting rates are achievable and pulses of light are not piled up on top of each other—this is explored further in section 5.1.1. Such direct emission is called *luminescence*. In competition, there will also be much slower light emission called *phosphorescence* or *after glow*.

Some pure inorganic compounds are efficient scintillators but many are poor scintillators such as pure NaI which has a low light emission at room temperature and only becomes an efficient scintillator at liquid nitrogen temperatures. Moreover, it emits scintillation light in the UV part of the spectrum. A key early development in the history of inorganic scintillators was to add activation centres via inclusion of a small fraction of thallium to the crystal structure which forms an alternative de-excitation pathway for the electrons. This faster and more efficient route (see left-hand side of figure 4.14) leads to emission of photons in a visible wavelength (typical peak wavelength of 415 nm for NaI(Tl)) much better matched to the characteristics of typical PMTs and allowing them to be an efficient scintillator at room temperature.

4.3.3 Semiconductor detectors

Semiconductor detectors are the highest-performing radiation detectors in terms of achievable energy resolution. Semiconductors are a special class of material whose properties are intermediate to those of conductors (like metals) and insulators. The key parameter which distinguishes these materials is the *band gap*. The band gap is the separation between the upper occupied energy band for electrons—the so-called *valence band* and the *conduction band*. The evolution of the band gap in different materials is presented in figure 4.15.

In an insulator, the band gap is large (~10 eV) and so it is not possible to thermally excite electrons into the conduction band. It is hard to drive an electrical current in such materials irrespective of the electric field applied because the electrons are essentially immobilised. In a conductor, the opposite is true as the conduction and valence bands are semi-overlapping, meaning that electrons are highly mobile and it is easy to drive an electrical current. Semiconductors fall into an intermediate regime where there is a band gap, of the order of 1 eV; two common materials used in radiation detection are silicon with a band gap of 1.106 eV and germanium with a band gap of 0.67 eV (at 77 K). There are also other semiconductor materials relevant to radiation detection based on compound materials such as gallium arsenide and cadmium zinc telluride (CZT).

In a semiconductor material, the occupancy of an energy level with energy, E, at a given temperature, T, is given by the Boltzmann distribution,

$$f(E) = \frac{1}{1 + e^{\frac{(E-E_F)}{k_B T}}}, \tag{4.10}$$

where k_B is the Boltzmann constant and E_F is the Fermi level. The Fermi level corresponds to the midpoint of the band gap. Consideration of equation (4.10) shows that the existence of the band gap will largely suppress thermal excitation of electrons, although some materials such as germanium, with its smaller band gap, need to be cooled to ensure this suppression. However, when ionising radiation such as alpha or beta particles, or electrons produced by gamma-ray interactions, enters the material and deposits energy, this will lead to excitation of electrons across the band gap, resulting in electrons at the bottom of the conduction band and holes at

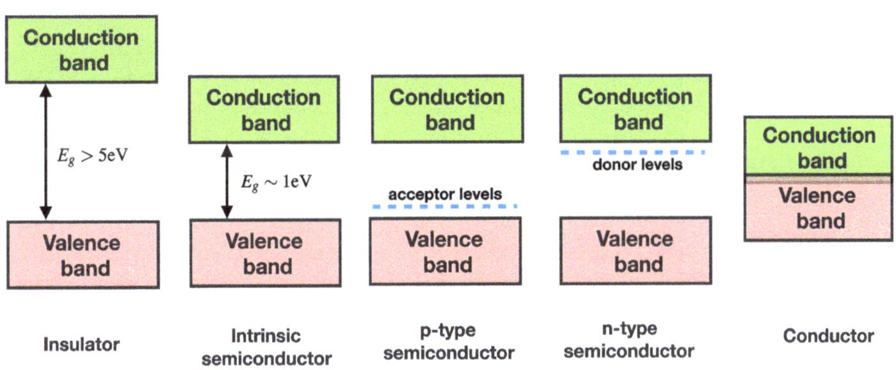

Figure 4.15. Comparison of the band gap in insulators, conductors and semiconductors. In an insulator, the band gap between the valence band and the conduction band is large while in a conductor they overlap. For semi-conductors, the band gap is of order 1 eV.

the top of the valence band. Since the cost in energy in producing electron–hole pairs relates to the value of the band gap, the number of charge carriers produced will be proportional to the energy deposited providing the basis for a semiconductor detector to perform as a *spectroscopic detector*. However, this process is not fully efficient. For example, the energy required to generate an electron–hole pair in germanium is 2.96 eV at 80 K, compared to the band gap of 0.67 eV (at 77 K). The balance of the pair-creation energy goes into, for example, lattice excitations, i.e. phonons. This inefficiency of energy transfer allied to the fact that individual electron–hole creation events are not independent, leads to a degradation of the theoretically achievable energy resolution which in the statistical limit given by

$$\Delta E_{\text{stat}} = 2.35\sqrt{\varepsilon E F}, \tag{4.11}$$

where E is the energy deposited, ε is the energy required to generate an electron–hole pair and F is the Fano factor. The Fano factor represents the deviation from Poisson statistics brought about by the non-independent nature of the electron–hole generation. For example, the Fano factor for germanium is known to be about 0.1 [16].

If an electric field is applied across the semiconductor then rather than recombining, the electrons and holes will migrate in opposite directions. This is the basis of extracting an electrical signal from the detector; the details of this charge collection will be revisited in more detail in chapter 6. The respective velocities of the electrons and holes in an electric field of magnitude, E, are given by:

$$v_e = \mu_e E \tag{4.12}$$

$$v_h = \mu_h E, \tag{4.13}$$

where μ_e is the electron mobility and μ_h is the hole mobility. A semiconductor used as a detector would ideally have high hole and electron mobilities. This is largely realised in silicon and germanium but some semiconductors such as CZT have notably low hole mobilities and special attention has to be applied to this to optimise their performance (see section 6.6).

We will return later in this e-book to specific instances of semiconductor detectors. In chapter 6, we will discuss germanium detectors—the highest-performing class of gamma-ray detector. In chapter 8, we will describe silicon detectors employed for charged-particle detection, principally silicon detectors.

4.4 Map of detector technologies to different applications

In this section, we consider the suitability of the different classes of detectors for detecting different types of ionising radiation in experimental nuclear physics and its wider applications. The technologies discussed should not be seen as exhaustive; there are other types of detectors which might be used in specialist applications.

4.4.1 Alpha particles

We discuss here the example of alpha particle detection but this can be taken as an exemplar of detection methodologies for other light ions, e.g. protons and indeed, heavy ions as well. As discussed in section 1.3, alpha decay is a two-body process resulting in characteristic alpha particle energies for a given isotope, as well as the common occurrence of closely-spaced alpha decay peaks associated with alpha decay fine structure. Figure 4.16 outlines the principal methodologies for alpha particle detection: the main discriminator is the energy resolution required. Silicon detectors, based on semiconductor technology, offer energy resolution as low as 12 keV for 5 MeV alpha particles which would be invaluable in identifying specific alpha particle groups (see section 8.2). Scintillator detectors such as CsI(Tl) (see section 8.1.1) can detect alpha particles but with relatively poor energy resolution and plastic scintillators have worse energy resolution still but can count very fast.

4.4.2 Beta particles and conversion electrons

The strategies for detecting beta particles and conversion electrons are related but distinct. This is because conversion electrons have a discrete energy while beta particles are emitted with a broad distribution of energies up to a beta end-point energy. Figure 4.17 outlines some of the key strategies for electron (positron) detection. For conversion electrons, a spectroscopic detector is important. Here, the relevant technology is silicon detectors (see section 8.2). For lower energy conversion electrons, conventional silicon detectors are appropriate but higher energy electrons will punch through such detectors since they are limited to thicknesses of about 1 mm. A more appropriate technology for higher energy conversion electrons is the lithium-drifted (or Si(Li)) detector (see section 8.4) which can have thicknesses of many millimetres and is often operated cooled to improve energy resolution. For beta particles (electrons or positrons), depending on the end-point energy, a

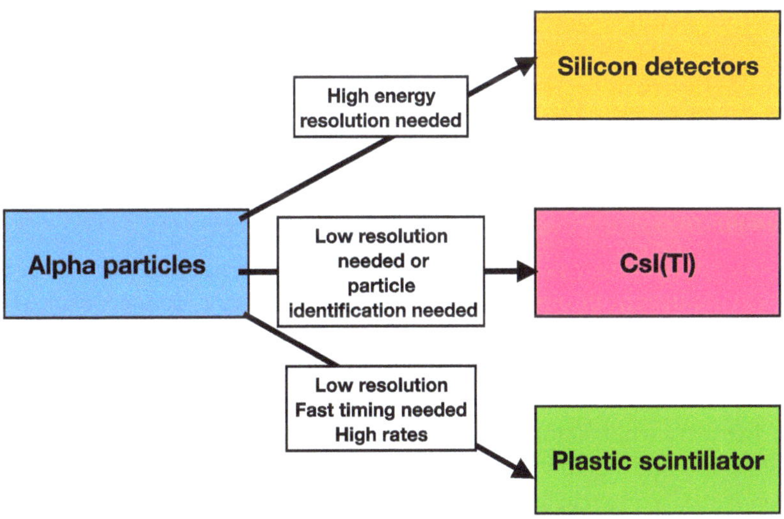

Figure 4.16. A (non-exhaustive) subset of the methodologies for alpha particle detection: the choice between silicon detectors and (in)organic scintillators relates to the demands of the specific application. For example, whether energy resolution is important to resolve peaks in an alpha-particle spectrum or whether high rate capability is needed at the expense of energy resolution.

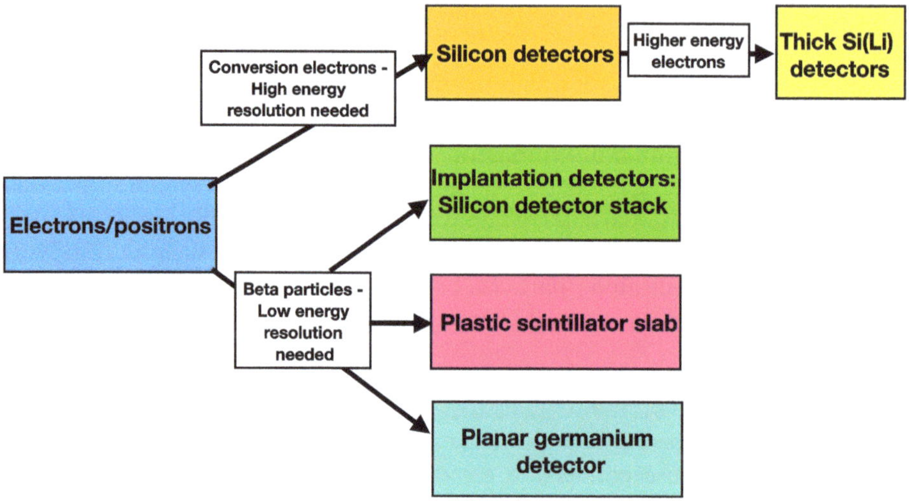

Figure 4.17. Principal methodologies for electron (or positron) detection. The important distinction to be made is between conversion electrons which have a discrete energy and beta particles (electron or positrons) which have a broad energy distribution. For conversion electrons, silicon detectors are suitable for lower energy electrons but since their maximum thickness is of order 1 mm, they are not very suitable for higher energy electrons which will punch through the detector without depositing their full energy. For such cases, cooled Si(Li) detectors are more suitable since they can be produced with much larger thicknesses. Where a full energy measurement is desirable for beta particles, a range of possibilities exist using thick slabs (cm thickness) of material such as plastic scintillator detectors or planar germanium detectors. If the application requires only to establish that a decay occurred then much thinner slabs of material could be used, or, for example, a stack of thin silicon detectors, e.g. see figure 8.38.

reasonable thickness of material will be required to fully stop the beta particles. This corresponds to, e.g. cm's thickness of plastic scintillator or a planar germanium detector. Where simple registration of beta particles is required but it is not necessary to fully stop the particles, other strategies such as a stack of silicon detectors may be appropriate.

4.4.3 Gamma rays

Gamma-ray spectroscopy is a key tool of experimental nuclear physics. It is widely used to provide insight into nuclear structure. Gamma rays connect excited states in a given nucleus and their relative intensity and the associated lifetime for gamma decay directly provides information on aspects such as nuclear collectivity and is used to benchmark nuclear models. Gamma-ray emission shares features with alpha decay in the sense that gamma ray transitions between bound excited states have discrete energies and narrow widths. In fact, the true linewidths are so narrow that they are essentially unmeasurable by any practical detector.

Figure 4.18 sketches the two main strategies for gamma-ray detection: semiconductor detectors, i.e. principally high-purity germanium detectors (but also to a lesser extent, and more commonly for applications, CZT), and scintillator detectors. The main discriminator for this choice of technology is the required energy

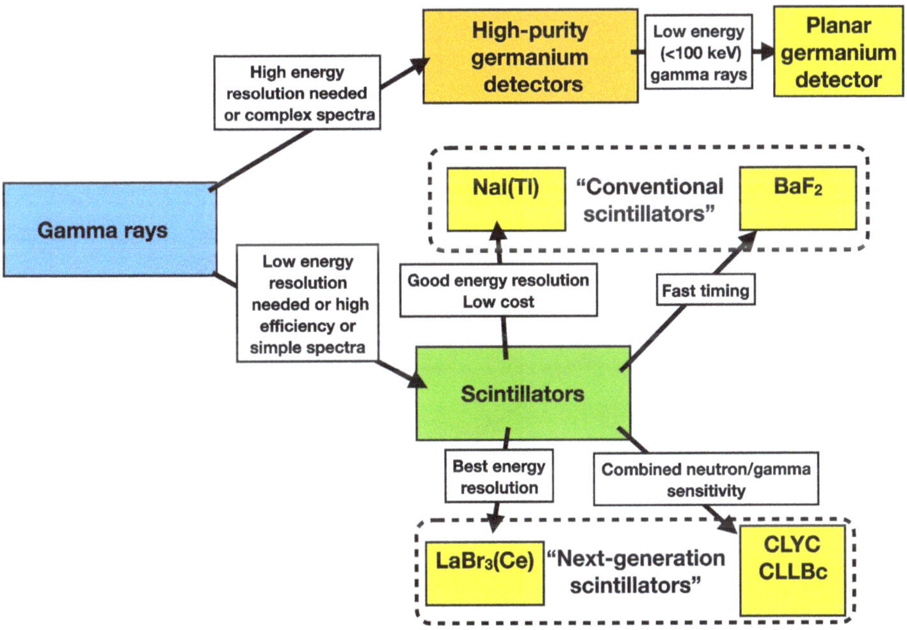

Figure 4.18. Principal methodologies for gamma ray detection: the main discriminator between the option of semiconductor-based, i.e. germanium detector and scintillator detector is energy resolution. However, there is a 'middle way' where next-generation scintillator materials such as $LaBr_3$ and $CeBr_3$ offer much improved energy resolution compared to conventional scintillators such as NaI(Tl). They may also offer additional benefits such as excellent timing resolution and the ability to resolve neutrons and gamma rays through pulse shape discrimination.

resolution and the complexity of the spectra to be studied. High energy resolution mandates semiconductor technology while high efficiency at the expense of energy resolution goes in the direction of scintillator detectors. Historically speaking, the most commonly found scintillation detector was NaI(Tl) due its low cost and decent energy resolution and linearity. In the last twenty years, as partly discussed in section 4.1.1, many new scintillator materials have begun to be used with a particular focus on those with excellent energy and timing resolution, as well as scintillators which are sensitive to both gamma rays and thermal (or fast) neutrons. Scintillator detectors for gamma-ray detection will be outlined in detail in chapter 5, while germanium detectors will be discussed in chapter 6. Since the applications of gamma-ray detection beyond nuclear physics are broad and wide-ranging, we dedicate a chapter (chapter 7) to reviewing some of the many and varied applications across different domains of physics and societal applications where gamma-ray detection plays a prominent role.

4.4.4 Neutrons

For neutrons, detection strategies fall into two camps depending on whether they are *fast* or *thermal* neutrons (see section 2.5, figure 4.19). For fast neutrons, organic scintillators are chiefly used where the ionisation results from proton-recoil of the

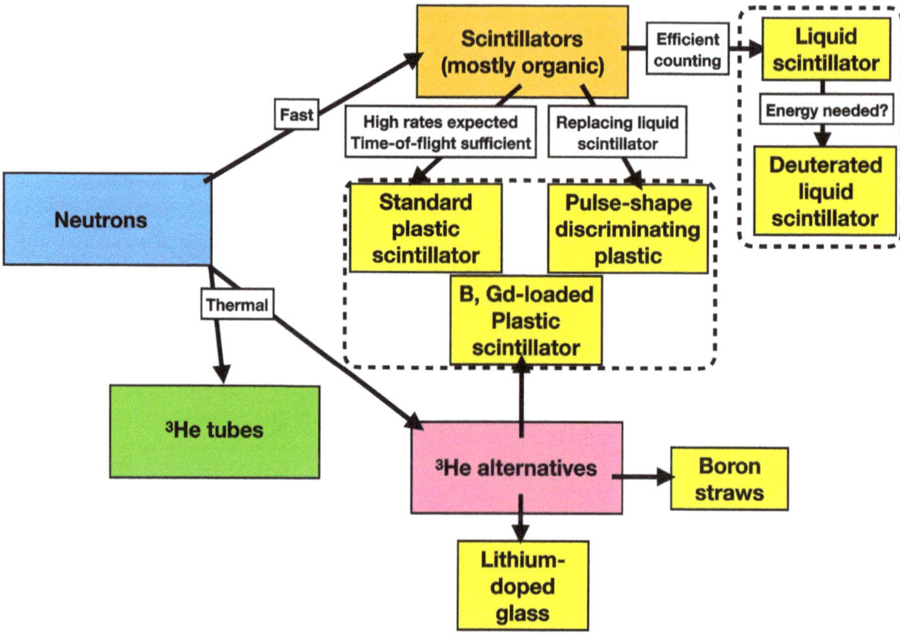

Figure 4.19. Principal methodologies for neutron detection. The main discriminator is the energy of the neutrons i.e. fast or thermal. Counting fast neutrons with high efficiency favours organic liquid scintillators while plastic scintillators are most commonly used for time-of-flight measurements. Thermal neutron detection historically focussed on ^{3}He but alternatives continue to emerge.

hydrogen-rich components in the scintillator. For high efficiency counting of neutrons in a close-packed geometry, liquid scintillator detectors are the most suitable medium since pulse-shape analysis can be used to discriminate neutrons and gamma rays. For time-of-flight measurements which can provide information on neutron energies, plastic scintillator is the material of choice because of its fast time response. For thermal neutrons, the preferred detector was historically a ^{3}He tube due to its high intrinsic efficiency. Rising costs partly attributed to a shortage of ^{3}He (and helium more generally) in recent years have led to a reawakening of interest in alternative detector materials, particularly focussing on those enriched in ^{6}Li or ^{10}B which have notably high thermal neutron capture cross-sections. Chapter 9 provides an overview of the available detector technology for neutron detection and compares and contrasts its performance.

References

[1] Bell S J, Aitken-Smith P, Beeke S, Collins S M, Regan P H and Shearman R 2016 A comparison of emerging gamma detector technologies for airborne radiation monitoring *J. Phys.: Conf. Ser.* **763** 012010

[2] Taniuchi R *et al* 2019 ^{78}Ni revealed as a doubly-magic stronghold against nuclear deformation *Nature* **569** 53–8

[3] Lauritsen T *et al* 2002 Octupole vibration in superdeformed $^{152}Dy_{86}$ *Phys. Rev. Lett.* **89** 282501
[4] Brown M, Gundacker S, Taylor A, Tummeltshammer C, Auffray E, Lecoq P and Papakonstaniou I 2014 Influence of depth of interaction upon the performance of scintillator detectors *PLoS One* **9** e98177
[5] Scoullar P A B, McLean C C and Evans R J 2011 Real time pulse pile-up recovery in a high throughput digital pulse processor *AIP Conf. Proc.* **1412** 270
[6] Rosskamp D, Truffler F, Bolay S and Geiser M 2007 Forward scattering measurement device with a high angular resolution *Opt. Express* **15** 2683
[7] Olliver H, Glasmacher T and Stuchberry A E 2003 Angular distributions of gamma rays with intermediate-energy beams *Phys. Rev.* C **68** 044312
[8] Agostinelli S *et al* 2003 GEANT4 - a simulation toolkit *Nucl. Instrum. Methods Phys. Res.* A **506** 250–303
[9] Battistoni G *et al* 2015 Overview of the FLUKA code *Ann. Nucl. Energy* **82** 10–8
[10] Allison J *et al* 2016 Recent developments in GEANT4 *Nucl. Instrum. Methods* **835** 186–225
[11] Garcia A R *et al* 2017 New physics model in GEANT4 for the simulation of neutron interactions with organic scintillator detectors *Nucl. Instrum. Methods* A **868** 73–81
[12] Niggemann T, Dietz-Laursonn E, Hebbeker T, Künsken A, Lauscher M and Merschmeyer M 2015 G4SiPM: a novel silicon photomultiplier simulation package for Geant4 *Nucl. Instrum. Methods Phys. Res.* A **787** 344–7
[13] Jan S *et al* 2004 GATE: a simulation toolkit for PET and SPECT *Phys. Med. Biol.* **49** 4543–61
[14] Yuan C, Kudryatsev A A and Demidov V I 2018 *Introduction to the Kinetics of Glow Discharges* (San Rafael, CA: Morgan and Claypool) ch 2
[15] Birks J B 1964 *The Theory and Practice of Scintillation Counting International Series of Monographs on Electronics and Instrumentation* vol 27 1st edn (Oxford: Pergamon)
[16] Croft S and Bond D S 1991 A determination of the Fano factor for germanium at 77.4 K from measurements of the energy resolution of a 113 cm^3 HPGe gamma-ray spectrometer taken over the energy range from 14 to 6129 keV *Int. J. Radiat. Appl.* **42** 1009–14

IOP Publishing

Radiation Detection for Nuclear Physics
Methods and industrial applications
David Jenkins

Chapter 5

Scintillator detectors for gamma-ray detection

This chapter provides an overview of one of the most commonly encountered technologies used in gamma-ray detection: inorganic scintillator detectors. These detectors are a workhorse for gamma-ray spectroscopy providing high efficiency and depending on the choice of scintillator, a good energy and timing resolution. This chapter will describe the physics behind the functioning of inorganic scintillators and provide an overview of the key parameters of the most commonly encountered scintillators: they are compared and contrasted in relation to their use in different applications. In addition, the chapter will review photomultiplier tubes (PMTs)—the most commonly employed photodetector as well as more modern alternatives such as silicon photomultipliers (SiPMs).

Concepts: Inorganic scintillators for gamma ray detection; key parameters of inorganic scintillators; choice of scintillator for varying applications; optical readout of scintillators: photomultiplier tubes (PMTs) and silicon photomultipliers (SiPMs); next-generation scintillators, future directions of scintillator research

5.1 Inorganic scintillator detectors

Inorganic scintillator detectors, which were presented in brief in section 4.3.2, are a workhorse of many aspects of experimental nuclear physics and also find application in related physics areas such as hadron physics, particle physics and in space-based astronomy e.g. gamma-ray telescopes. The present chapter focusses on inorganic scintillators widely employed for gamma-ray detection. Let us review some of the key properties of such materials.

5.1.1 Key parameters of scintillators

Energy resolution

Energy resolution is an important parameter since it sets the limits on the complexity of a gamma-ray spectrum, which can be analysed, and the extent to which individual peaks can be resolved. The energy resolution is conventionally expressed for a specific peak as R in % where

$$R = 100\left(\frac{\text{FWHM}}{E_{\text{centroid}}}\right), \tag{5.1}$$

where FWHM is the full-width at half-maximum of a peak and E_{centroid} is the centroid of the peak. As a means of comparison of different scintillators, the energy resolution is most commonly given for the 662 keV peak in the ^{137}Cs spectrum (e.g. see figure 5.1); this is the most simple case as there is only one gamma ray in the spectrum so little possibility of ambiguity or peak overlap. It is important that the spectrum is linearised i.e. calibrated e.g. using a ^{152}Eu spectrum before extracting a FWHM or the value obtained will be erroneous.

The energy resolution of a scintillator coupled to an appropriate light sensor such as a photomultiplier tube (PMT) is obtained by summing the individual contributions in quadrature [1]:

$$\left(\frac{\Delta E}{E}\right)^2 = (\delta_{sc})^2 + (\delta_p)^2 + (\delta_{st})^2 + (\delta_n)^2, \tag{5.2}$$

where δ_{sc} is the intrinsic resolution of the crystal, δ_p is the transfer resolution which relates to the photon generating a photoelectron in the PMT which is then fully multiplied in the PMT as well as the effectiveness of the optical coupling of the

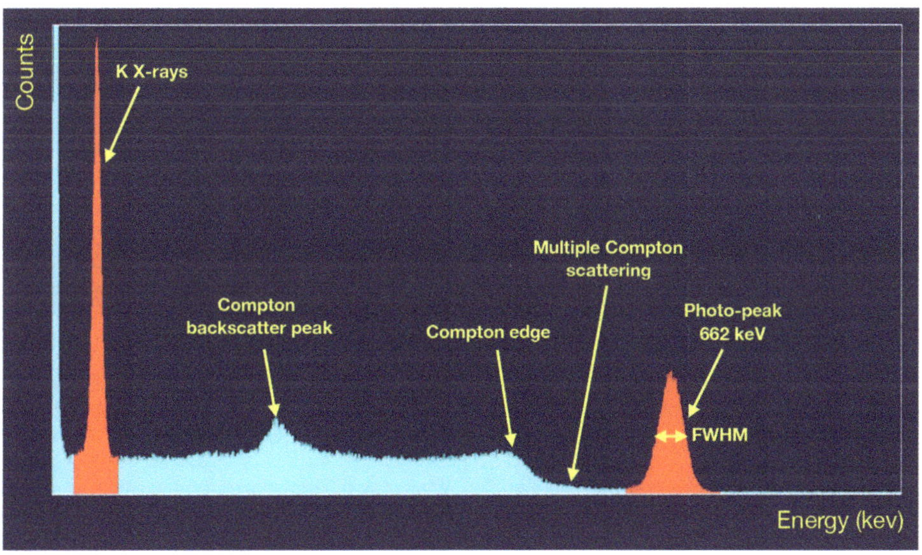

Figure 5.1. Gamma-ray spectrum obtained with a ^{137}Cs source. The detector was a 1″ × 1″ × 2″ cuboid of CsI (Tl) coupled via a lightguide to a 2 ×2 array of 6 mm square SensL C-series SiPMs. The spectrum is labelled with the key features of the photopeak at 662 keV, the Compton edge and the Compton backscatter peak. The K x-ray peaks are associated with internal conversion of the 662 keV transition. The energy resolution of such a detector is around 7% at 662 keV.

scintillator and PMT, δ_{st} is the statistical contribution deriving from the PMT and δ_n is the contribution from dark noise in the photomultiplier and any associated electronic noise. The statistical contribution, δ_{st} can be given by [1]

$$\delta_{st} = 2.35 \times \frac{1}{\sqrt{N}} \times \sqrt{(1+\epsilon)}, \tag{5.3}$$

where N is the number of photoelectrons and ϵ is the variance in the PMT gain which is typically around 0.1–0.2 for a modern PMT. The intrinsic resolution of the crystal, δ_{sc} largely depends on the next two key parameters of a scintillator detector: light output and non-proportionality.

Light output

The scintillation light output of a given scintillator is a dominant contributor to its ultimate energy resolution as a detector. This parameter is typically represented by a value in photons/MeV of incident radiation. This number effectively defines the counting statistics. Strictly speaking, the light yield is energy dependent and is commonly referenced to the light yield at 450 keV. Light output *per se* can be misleading since a scintillator may have a high light output but a poor effective energy resolution because the response of the scintillator is highly non-proportional or it is difficult to get the photons out of the crystal.

Linearity/non-proportionality

An ideal scintillator would generate scintillation light directly proportional to the energy deposited in the crystal by an interacting charged particle or gamma ray. In practice, real scintillators deviate from this linear response and this impacts on the energy resolution [2]. Figure 5.2 provides examples of measured non-proportionality behaviour as a function of incident gamma-ray energy for NaI(Tl) and for the next-generation scintillators, $LaBr_3$(Ce) and $CeBr_3$. The high linearity of the latter scintillators is one of the chief contributors to their improved performance—an issue discussed in more depth in section 5.2.1.

Peak wavelength

Scintillation light output relevant to radiation detection applications typically has a broad spectral distribution in the visible, or near-UV, or near-IR. The distribution can be characterised by a peak wavelength. In order to detect as many photons as possible, the photodetector employed should ideally have a peak sensitivity in accord with the peak wavelength output of the scintillator.

Refractive index

The scintillation photons need to exit the crystal and enter a PMT, e.g. through a quartz-glass window. If the refractive index of the crystal is very high and/or mismatched with the light sensor and its optical window, then light collection will be poor and energy resolution suffers as a consequence. This is a problem which is hard

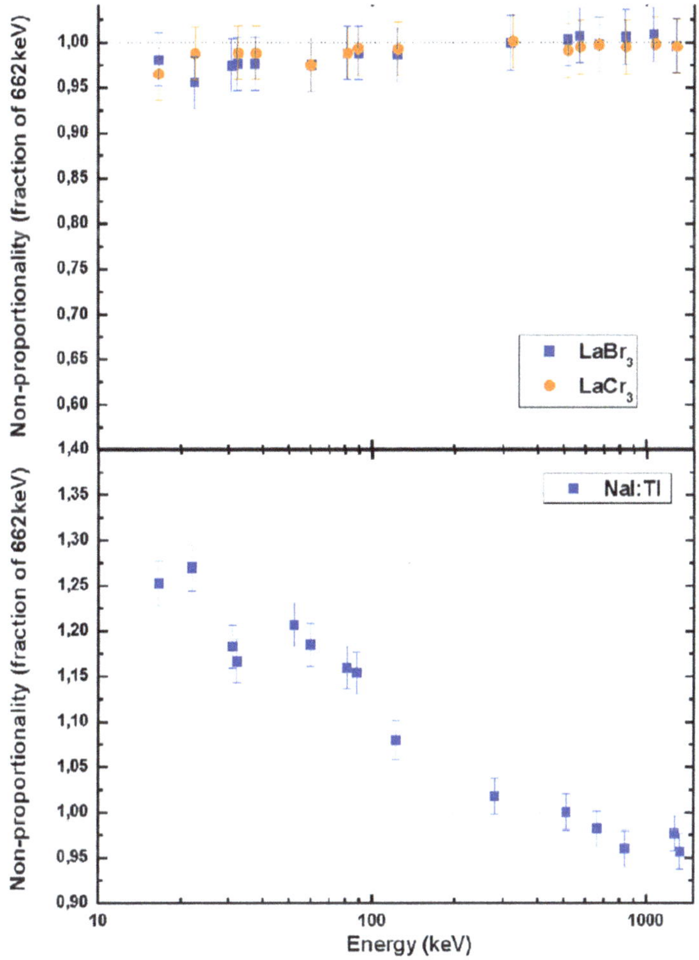

Figure 5.2. Non-proportionality of light yield as a function of incident gamma-ray energy for NaI(Tl) detector (bottom) and $LaBr_3$(Ce) and $CeBr_3$ (top). The plots are referenced to unity for a 662 keV gamma ray from ^{137}Cs source. (Reprinted from [1] with the permission of Elsevier.)

to avoid but some limited work with, e.g. microlens arrays [3] and photonic crystals [4], has shown possible routes to working around it.

Scintillator shape, dimensions and detector efficiency

The dimensions of a scintillator detector play a critical role in its performance and efficiency—see figure 5.3, which presents cartoons of detector response for a putative 1.5 MeV incident gamma ray in detectors of varying dimensions. If a detector is very small, then only a single interaction in the crystal is probable, either photoelectric absorption, Compton scattering or pair production, for the latter process, the subsequent annihilation photons are likely lost from the crystal due to its small dimension (see left panel of figure 5.3)—this gives rise to a dominant Compton

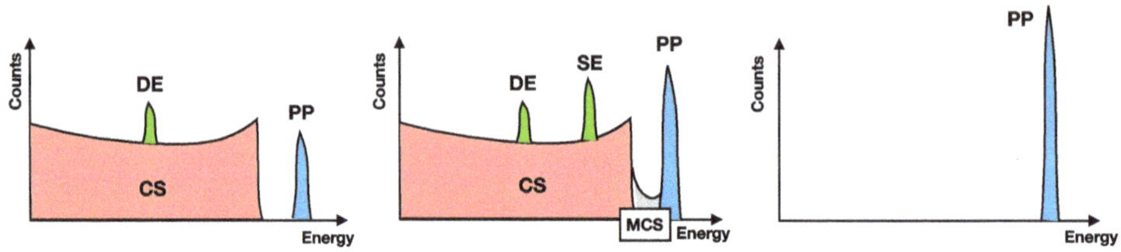

Figure 5.3. Cartoons of scintillator detector gamma-ray spectra acquired from a single-line source with energy above the threshold for pair-production (1.022 MeV)—for the sake of argument, $E_\gamma \sim 2.0$ MeV. The peaks and features in the spectra are labelled as PP—photopeak, CS—Compton scattering, MCS—multiple Compton scattering, SE—single-escape peak and DE—double-escape peak. The left-hand cartoon shows the spectrum obtained from a very small detector where interactions will be predominantly one single interaction due to the dimensions of the crystal. The right-hand cartoon shows the opposite (and physically difficult to realise) extreme of an infinitely large detector. The centre cartoon is for a detector of more usual dimensions found in the laboratory (cf figure 5.1).

scattering feature, a small photopeak and a double-escape peak superimposed on the Compton feature which corresponds to a pair production event where both annihilation photons are lost to the crystal. If the detector is infinitely large (as in the right-hand panel of figure 5.3) then all entering gamma rays will ultimately have all their energy absorbed as Compton-scattered photons and annihilation photons will be subsequently absorbed through multiple Compton scattering and photoelectric absorption. For a more typical scintillator crystal found in the laboratory, such as a 2″ [1] diameter NaI(Tl) crystal, the observed spectrum is intermediate to the two extreme cases discussed above (see centre panel of figure 5.3). Here, the dimension is sufficient for there to be probability of multiple Compton scattering and for annihilation radiation to be subsequently absorbed. This gives rise to features in the spectrum connected with multiple scattering which fills in to some extent the region between the Compton edge and the photopeak. There are also now two escape peaks seen—a double-escape and a single-escape peak. These correspond to either both or just one 511 keV annihilation photon being lost to the crystal.

Scintillators are traditionally produced in cylindrical shapes, which is characteristic of the crystal growth process. However, crystals can subsequently be cut, e.g. into cuboid shapes or indeed tapered along their length, e.g. to allow crystals to better fit together to form a spherical array. Standard crystals are usually available in large sizes and with many options on subsequent machining of the shape. However, many of the more recently-appearing ‘next-generation’ scintillators are only available in small dimensions and are significantly expensive. What can be produced and in what shape/dimension needs to be discussed directly with the crystal manufacturer. Fortunately, the wider scintillator market is so large, encompassing areas such as medical imaging and homeland security, that there are a large number of competing suppliers of the same materials in Europe, North America and Japan, and an increasing number in China. This is quite different to the market for,

[1] The reader should not be surprised to see imperial units being used for scintillator dimensions, i.e. inches, as this seems to be a historical hangover in this particular area.

e.g. silicon and germanium detectors, where there are only a small number of specialist suppliers. It is possible to purchase assembled scintillator detectors with the crystal appropriately canned and coupled to a PMT ready for use. Here, the supplier will warrant the expected performance in terms of energy resolution, timing resolution etc. However, it is equally possible to buy the components, i.e. the crystal, PMT, voltage divider etc, separately and assemble yourself. For non-hygroscopic crystals like CsI (Tl) you can start with the bare crystal and wrap and package it yourself[2]. The reader is encouraged to obtain and play around with these materials themselves as they are relatively inexpensive and a great way to get hands-on experience.

Decay time and multiple decay components

Scintillator decay time (τ_d) is the time for the scintillation light to reduce to $\frac{1}{e}$ of its initial magnitude. This parameter largely depends on the scintillator material but, clearly, it will also vary with crystal dimensions given that it takes time for light to escape the crystal, where some light may reflect multiple times at surfaces before reaching the scintillator face where the light sensor is positioned. Note that this discussion relates to the scintillator crystal only; for a detector formed from a scintillator coupled to a PMT the observed decay time will be a convolution of the scintillator decay time with the time response of the PMT. If the scintillation rise time (τ_t) is fast, i.e. nanoseconds or less, which is commonly the case for inorganic scintillators then the light output rate of the scintillator, $L(t)$ can be expressed as [5]:

$$L(t) \approx \frac{N}{\tau_d} e^{-\frac{t}{\tau_d}}, \tag{5.4}$$

where N is the number of luminescent centres excited, assuming a single characteristic decay time—an assumption to be discussed further below.

The decay time for a given scintillator emission is an important parameter since it defines over what time scale a signal needs to be integrated to ensure maximal light collection. This is significant because even if the same amount of light is collected, the performance of a slower scintillator will be worse since noise will be integrated in the tail of the signal. The decay time also has implications for the highest rate possible with a given detector since if the rate is so high that the detector signal cannot recover to the baseline then this will have implications for spectral quality and energy resolution.

A typical value of decay time for the most commonly encountered inorganic scintillator, NaI(Tl), is 230 ns [6], whereas fast plastic scintillators more appropriate to neutron detection and discussed in chapter 9, can have decay times approaching 1 ns. There are often several different mechanisms for exciting scintillation light in a given material leading to a variety of light emission decay curves. Indeed, some mechanisms are preferentially excited by different types of radiation such as gamma rays and charged particles. This can provide a means of particle identification through pulse shape analysis, as discussed in section 8.1.1.

[2] For hygroscopic crystals, a specialist low-humidity glove box would be needed to handle the crystals.

Afterglow
Many scintillators have long-lived phosphorescence type emission which disturbs measurements since there is a semi-continuous, low-level light output from the crystal.

Stability with temperature
The light output of scintillators can change as a function of temperature which needs to be borne in mind depending on the application under consideration and also in terms of gain-stabilising a detector. Depending on the specific scintillator, the light output can increase or decrease with increasing temperature. In cases where there is more than one component (with a different wavelength centroid) to the scintillation light, the behaviour can be quite different for the two components e.g. BaF_2 where the fast component is semi-independent of temperature and the slow component of the light ouput increases by a factor of two on cooling from room temperature to −25 °C [7]. In a real detector system, the situation is more complicated since both the scintillator and PMT may show a differing trend against temperature and the two combined will fix the temperature dependence of the detector overall. Figure 5.4 shows the measured variation in signal height from a BGO/PMT detector as a function of temperature.

Stability to different environmental conditions
Many inorganic scintillators are hygroscopic, meaning that they react to a greater or lesser extent to moisture in the air. This can readily be countered by hermetically sealing the crystal in an appropriate can or enclosure with optical windows on one or more faces. This will have the effect of attenuating low energy gamma rays and so the material should be minimised where low-energy gamma ray detection is

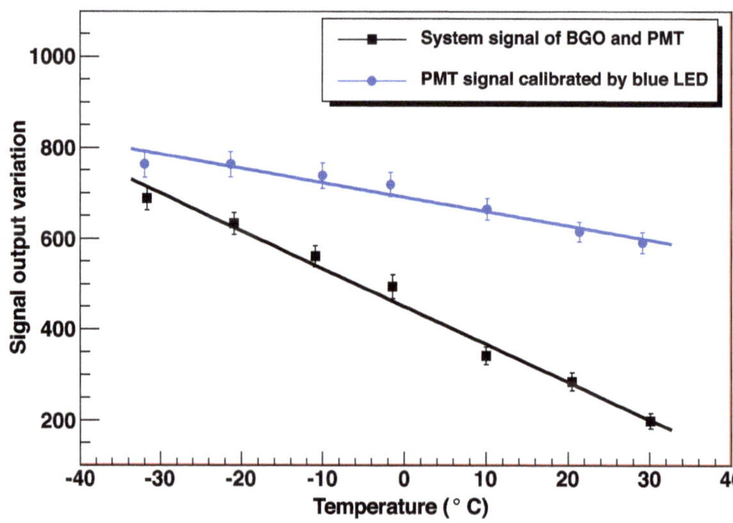

Figure 5.4. Variation in signal height from a BGO/PMT combination as a function of temperature. (Reprinted by permission from Springer Nature [8].)

important. In terms of charged particle detection, the short range of many particles such as alpha particles and heavy-ions would tend to restrict the use of hygroscopic crystals for charged particle detection unless the particles are very high energy e.g. 100 MeV + alpha particles/protons.

Density and atomic number of component elements
In terms of gamma-ray detection, photoelectric absorption is the most useful form of interaction where the full energy of the gamma ray is effectively deposited in the material. Photoelectric absorption for a given material is linearly dependent on density and scales as approximately Z^5 (see section 2.4). This implies that high-Z materials such as inorganic crystals are much more suitable scintillators for gamma-ray spectroscopy than a low-Z material like a plastic scintillator. Indeed, the response of plastic scintillator to gamma rays is generally dominated by Compton scattering. On the other hand, organic liquid scintillators are very suitable for neutron detection as neutron detection relies on proton recoil of hydrogen within the material and the low Z means that sensitivity to gamma rays is minimised (see chapter 9).

Radiation damage
All scintillator materials can suffer from radiation damage, though the dose of radiation needed to cause noticeable damage varies strongly as a function of material [6]. There are several effects of radiation damage including the formation of colour centres which traps some of the scintillation light, and the increase of afterglow in the crystal. Crystals can recover slowly at room temperature. This process can be accelerated by annealing at higher temperatures. In general, the doses needed to damage crystals are so high that they will be irrelevant to most applications in experimental nuclear physics and societal applications. Radiation damage is a relevant consideration, however, in certain specialist applications where high doses might be encountered, for example in nuclear decommissioning.

5.1.2 Typical inorganic scintillators

There are a wide range of inorganic scintillators used for gamma-ray detection with different properties that make them more or less suited to different applications; an excellent review can be found in reference [5]. Here we focus on the properties of a few of the major ones.

Sodium iodide
While pure sodium iodide itself is a scintillator, its properties are not very suitable and it is conventionally doped with thallium to produce a brighter scintillator—NaI(Tl) [9]. NaI(Tl) was historically the most common scintillator used in gamma-ray spectroscopy for nuclear physics and applications; see reference [10] for a detailed review of its properties and applications. This is based on the long experience in its use and it being probably the least expensive of the inorganic scintillators. The light output of NaI(Tl) peaks at 415 nm and it affords a typical

energy resolution of around 7% at 662 keV. NaI(Tl) exhibits reasonably good linearity (see figure 5.2). A key disadvantage of NaI(Tl) is that it is hygroscopic, which makes its employment difficult in most charged-particle detection applications since particles will not penetrate the detector canning. Further work on studying NaI(Tl) has shown that there are two components to its light output: one fast with a decay time of 230 ns, and a slower one with a decay time of around 1 μs. The fast component originates from direct feeding of the (Tl^+) level while the slow component relates to a delayed feeding of this level through a process of 'thermal-assisted' hopping followed by capture and recombination (see figure 5.5). This diversity of mechanisms in, perhaps, this most well-known of scintillators exhibits the complexity of describing the physics underlying the scintillation process.

Caesium iodide

Caesium iodide is itself a scintillator but as with sodium iodide, it is most commonly doped to enhance light output. Common dopings are with sodium—CsI(Na) and thallium—CsI(Tl). The former is hygroscopic while the latter is only slightly hygroscopic, meaning that bare crystals can be employed without canning making it more flexible to use. Moreover, CsI(Tl) has a higher light output at room temperature than CsI(Na). The light output of CsI(Na) peaks at 420 nm, while CsI(Tl) scintillation light peaks at 550 nm. In many respects, CsI(Tl), provides a good, non-hygroscopic alternative to NaI(Tl) but its light output is less linear with deposited energy. CsI(Tl) has two characteristic light emissions with differing decay times [12]. The excitation of these two components varies depending on whether the crystal is excited by gamma rays or different species of charged particles; this application will be discussed in detail in section 8.1.1.

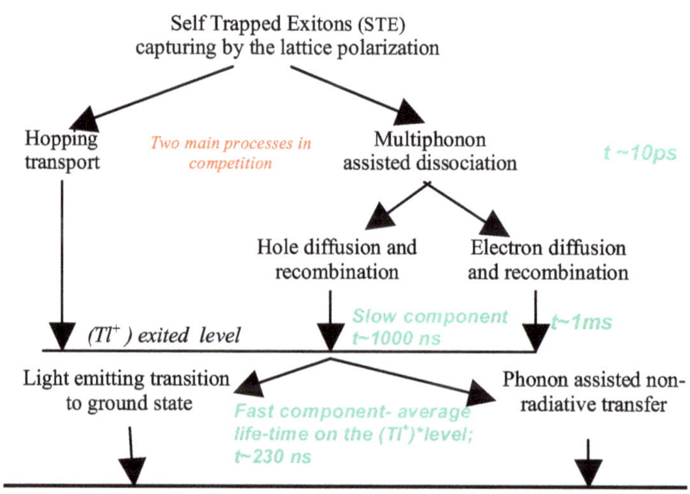

Figure 5.5. The scintillation mechanism in NaI(Tl). There are two main mechanisms for feeding the (Tl^+) level—one is direct and the second introduces a delay leading to a longer decay time. This is the origin of the two components in the NaI(Tl) light emission. (Reprinted from [11] with the permission of Elsevier.)

Barium fluoride—BaF_2

Barium fluoride is a scintillator widely used in applications related to fast timing [13, 14] since BaF_2 has two components to its scintillator light output: a slow component with decay time of 630 ns and a fast component with decay time of about 700 ps. Figure 5.6 shows the physical origin of the two scintillation mechanisms and their associated emission spectra. The fast component compares favourably to the decay characteristics of fast plastic scintillators, but, as can be seen in figure 5.6, it is typically only about 15% of the scintillation light output of BaF_2. The fast component peaks at 220 nm, which is a challenging region for most photodetectors, but it is well separated from the slow component peaking about 300 nm, meaning that use of dedicated photosensors with short wavelength response or appropriate filters can allow the slow component to be suppressed [15]. For fast-timing applications, BaF_2 detectors benefit from temperature stabilisation and cooling.

Bismuth germanate—BGO

Bismuth germanate (BGO) is a scintillator characterised by the high-Z of one of its components, bismuth ($Z = 83$), and its high density (7.13 g cm^{-3}). This means that it is one of the most effective scintillators in terms of absorption of incoming gamma radiation. It is also not hygroscopic which makes its handling more straightforward. However, BGOs light output is low (around 15% of that of NaI(Tl)) and it has a high refractive index (2.15) making its effective coupling to photosensors more difficult. The achievable energy resolution for BGO is therefore rather poor—typically 10% at 662 keV using large-area avalanche photodiode (LAAPD) read-out [16]. BGO finds a major application in nuclear physics as the scintillator of choice for Compton suppression shields placed around high-purity germanium detectors—an issue to be discussed in detail in section 6.5.1.

Cadmium tungstate—$CdWO_4$

Cadmium tungstate is an effective alternative to BGO as a non-hygroscopic material which has high stopping power and a better energy resolution than BGO—7% at

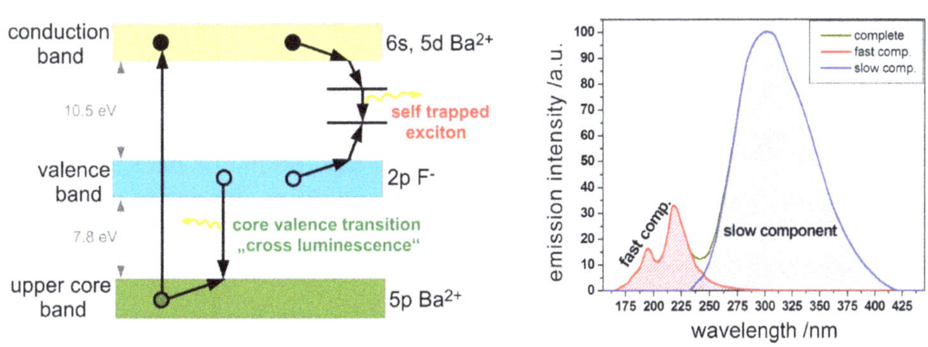

Figure 5.6. (left) The scintillation mechanisms in BaF_2—the fast mechanism corresponds to the core valence transition. (right) The emission spectra for the fast and slow scintillation components from BaF_2. (Reprinted from [15] with permission from IOP Publishing.)

662 keV [17]. This material is more widely employed in high energy physics than in experimental nuclear physics.

LYSO

Lutetium yttrium silicate (LYSO) is a material with a high effective *Z*, high density (7.2 g cm^{-3}), fast decay time and scintillation light yield of around 75% of that NaI (Tl) [18, 19]. LYSO suffers from internal radioactivity due to the long-lived ($t_{1/2} \sim 4 \times 10^{10}$ yr) beta-decaying isotope ^{176}Lu which makes up 2.6% of natural lutetium [20]. The most common application of LYSO is in positron-emission tomography (PET), a medical imaging technique to be described more extensively in section 7.5.1. Here, coincident pairs of annihilation photons are detected to create an image where individual detectors are sticks of LYSO. This coincidence technique largely avoids the issues of the internal radioactivity of the crystals.

5.1.3 Phoswich detectors

A phoswich detector [21] is one where two crystals of different scintillator material are optically coupled together. The two crystals are usually chosen such that they have rather different decay times so the two components can be separated with appropriate analogue electronics or digital pulse-shape analysis. This can be used, e.g. to effectively determine depth of interaction or to most efficiently profit from use of a reduced volume of a more expensive scintillator material (see description of the PARIS calorimeter in section 5.4). They can also be used to discriminate e.g. alpha particles from gamma rays based on depth of interaction [22], or positrons from gamma rays using wavelength discrimination for the light output of two different scintillators [23].

5.2 Recent advances in scintillator technology

While NaI(Tl) remained the standard scintillator of choice for gamma-ray detection for many decades [9], there has been a renaissance in scintillator development over the last twenty years driven by demands, e.g. of gamma-ray astronomy [24, 25] and expansion in the homeland security market notably in the US. Next-generation scintillator detectors developed to meet needs in these parallel fields have also become increasingly important in nuclear physics [26]. Such next-generation detectors significantly bridge the gap between relatively low-cost traditional scintillators such as NaI(Tl) and intrinsically expensive, high resolution detectors such as germanium detectors in that they offer energy resolutions close to 3% at 662keV compared to 7% for a typical NaI(Tl)-based detector. The ideal scintillator would have high light yield, high linearity/non-proportionality, short decay time and excellent energy resolution. In practice, it is difficult to obtain detectors where all these parameters are optimised and the best performing can be an order of magnitude more expensive than traditional scintillators such as NaI(Tl) so their employment needs to be very carefully justified to a particular application and the benefits validated through simulation. An overview of some of the available

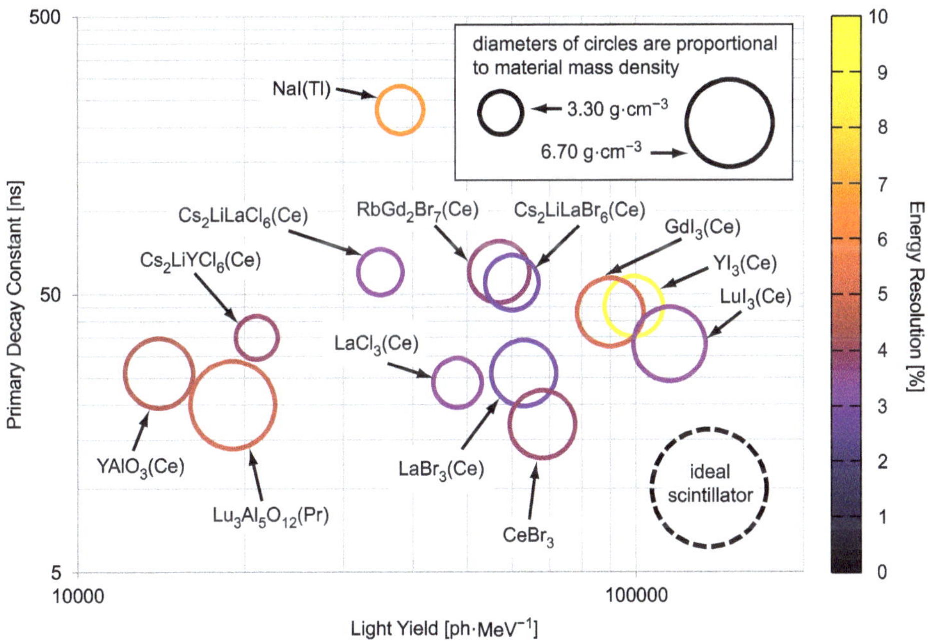

Figure 5.7. Comparison of the light yield, decay time and energy resolution at 662 keV obtained with different scintillators. (Reprinted from [27] with permission from IOP Publishing under Creative Commons License.)

materials and how they compare in terms of light yield, decay constant and energy resolution is given in figure 5.7.

5.2.1 Next-generation scintillators available commercially

Cerium-activated lanthanum bromide—$LaBr_3(Ce)$

The stand-out scintillator at the present time is $LaBr_3$(Ce) which has excellent light output (63 000 photons MeV^{-1}), a fast decay time (~16 ns) and high linearity[3]. Moreover, it is available in large sizes, e.g. as cylinders 3.5″ diameter and 8″ long, e.g. as deployed in the OSCAR array described below (see section 5.4.2). Gamma-ray detectors formed from $LaBr_3$(Ce) coupled to an appropriate PMT may offer energy resolutions better than 3% at 662keV [29, 30]. A clear disadvantage of $LaBr_3$(Ce) in many applications, however, is that it contains internal activity originating from ^{138}La—a naturally-occurring radioactive isotope of lanthanum and ^{227}Ac present due to actinium being a chemical homologue of lanthanum. The presence of these radioactive isotopes leads to alpha, beta and gamma activity within the crystal [31]. This internal activity can disturb measurements depending on the level of sensitivity needed. Low background measurements are generally prohibited. The presence of the activity may not be so important for nuclear physics applications, however, since measurements are generally carried out in coincidence mode, i.e. requiring two or more coincident gamma rays to be registered in an array

[3] There also exist slightly cheaper alternatives to $LaBr_3$(Ce) such as $LaCl_3$(Ce) [28].

of scintillator detectors or for an ancillary detector such as a charged-particle detector to record an event in prompt coincidence. In such cases, the randomly occurring activity can be effectively removed from the gamma-ray data recorded in off-line analysis.

Recently, the performance of $LaBr_3(Ce)$ has been improved through co-doping with Sr to make so-called enhanced lanthanum bromide, $LaBr_3(Ce + Sr)$. The energy resolution at 662 keV improves to 2.2%. This builds on extensive investigations of co-doping of scintillator crystals in recent years where the improved energy resolution is derived from improvements to the scintillator non-proportionality [32]. Figure 5.8 illustrates the improved non-proportionality for $CeBr_3$ crystals when co-doped with different elements.

$LaBr_3(Ce)$ is already finding wide application in experimental nuclear physics (see section 5.4.2) where the excellent energy resolution has been exploited and it has also been shown to be superior to BaF_2 for fast timing applications. $LaBr_3(Ce)$ has also been used in various societal applications such as radionuclide assay [26]. $LaBr_3(Ce)$ detectors are supplied by e.g. Saint-Gobain crystals.

Cerium bromide—$CeBr_3$

A useful alternative to $LaBr_3(Ce)$ is $CeBr_3$ which is also available in large sizes [33, 34]. It has the significant advantage over $LaBr_3(Ce)$ in that it does not contain internal activity [33]. Moreover, the performance of $CeBr_3$ as a scintillator is not significantly worse than $LaBr_3(Ce)$, with a typical energy resolution at 662keV of around 4% [33, 34]. A comparison of the energy resolution of a typical $CeBr_3$ detector to a $LaBr_3(Ce)$ one is given in figure 5.9.

Strontium iodide—SrI_2:Eu

SrI_2:Eu is also a viable next-generation scintillator detector for gamma-ray spectroscopy. High energy resolutions of around 3% at 662 keV have been obtained but for

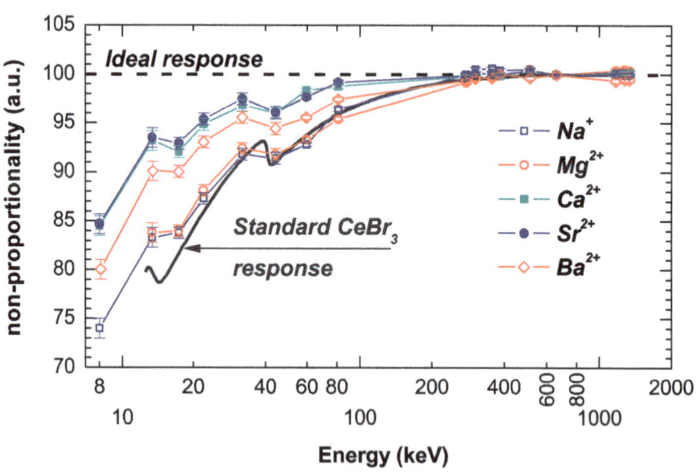

Figure 5.8. Non-proportionality of standard and co-doped $CeBr_3$ samples. (Reprinted from [32] with the permission of Elsevier.)

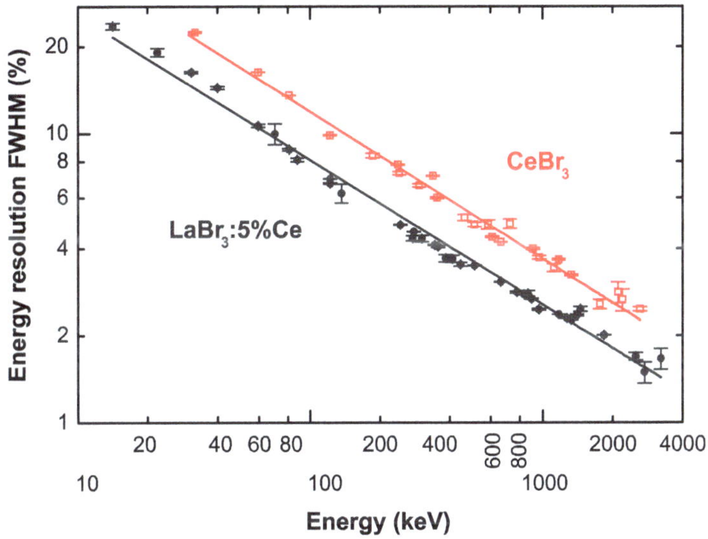

Figure 5.9. Comparison of energy resolution of 2″ × 2″ $CeBr_3$ detector 2″ × 2″ $LaBr_3(Ce)$ detector. (Reprinted from [33] with the permission of Elsevier.)

small samples [35]. Early detectors made from the material were seen to exhibit strong self-absorption of scintillation light leading to strong disimprovement in performance for larger detectors [35, 36]. This has largely been eliminated at the time of writing, nevertheless, SrI_2:Eu exhibits a long decay time—up to 10 μs that can limit its use in high-rate applications.

Elpasolite scintillators

Recently, other scintillators such as CLLB [37] and CLYC [38] have come to market. CLLB has a high light output of around 40 000 photons MeV^{-1} and has energy resolution similar to $CeBr_3$. Both CLLB and CLYC are of interest since they are also sensitive to fast and thermal neutrons. Thermal neutron sensitivity comes about from the presence of the isotope ^{6}Li which has a high thermal neutron capture cross-section and breaks up into ^{3}H + ^{4}He which can produce a signal in the scintillator when they deposit energy in the crystal (see section 9.2.2 for an extended discussion of ^{6}Li-based detectors). Similarly, a fast neutron signal arises from the presence of ^{35}Cl via the ^{35}Cl(n,p) reaction where the outgoing proton deposits energy in the scintillator [39]. Gamma-ray and neutron interactions can be distinguished using pulse shape analysis.

5.2.2 Future prospects

Improvements in scintillator performance are continuously reported [40, 41] and there are extensive efforts worldwide investigating novel materials and techniques. Particularly promising are investigations of ceramic-based detectors which are not hygroscopic and are highly robust [27]. It is worth reflecting, however, that employment in nuclear physics experiments is a fairly niche application and the scintillator market and associated developments are largely driven by the needs of the well-funded homeland security market (see section 7.1) and medical imaging such as positron emission tomography (see section 7.5.1). Nevertheless, many of the objectives of these larger societal programmes are well aligned with the requirements of detectors for physics so the benefits of the industrial-scale development can flow back into nuclear physics. The interested reader is well-advised to keep abreast of advances in parallel fields and applications in order to find new approaches and technology relevant to nuclear physics.

Position sensitivity within the crystal

Light transport within crystals is complex with photons reflecting at surfaces but there is expected to be some correlation between where in the crystal the interaction takes place and where the light is expected to be collected. Recent work has validated this assumption, albeit with relatively limited position resolution for readout of monolithic crystals with segmented photomultiplier tubes [42] and SiPMs [43]. There is much more work to do in this area but the initial work is very promising for experimental nuclear physics since determining the point of interaction effectively sub-divides the detector reducing effects due to Doppler broadening (cf gamma-ray tracking in high-purity germanium detectors in section 6.5.2). This technology could also be adapted to Compton camera applications in medical imaging and homeland security (see section 7.1.2).

3D printing

3D printing is becoming an increasingly important manufacturing technique in industry. The first steps towards 3D printing scintillators have been reported in recent years, including printing of plastic scintillator [44] and of a complex-shaped scintillator of YAG:Ce [45] (see figure 5.10). These 3D printing applications to scintillators are in their infancy but once the technology is mature, it will clearly have high value in experimental nuclear physics and its applications.

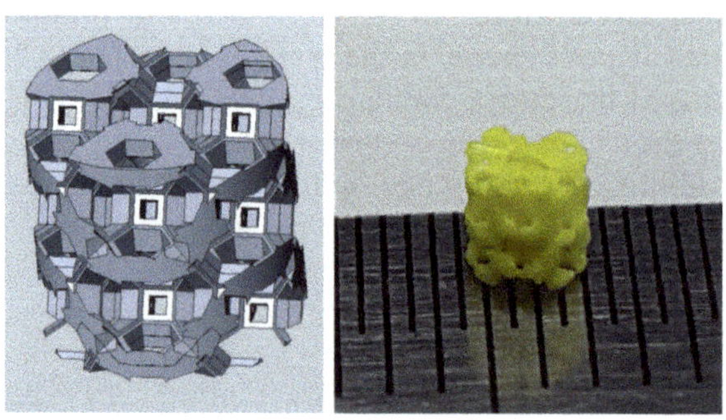

Figure 5.10. A complex shape of YAG:Ce scintillator achieved through 3D printing. (Reprinted from [45] with permission from RSC under Creative Commons License.)

5.3 Photosensors for scintillation light collection

A scintillator crystal is little use without a suitable photosensor attached to it. Remarkably, the first photosensor used in nuclear physics was the human eye where a physicist sat in darkness counting flashes of light on a scintillating screen made of a phosphor such as zinc sulfide. The development of photomultiplier tubes (PMTs) in the 1930s allowed the physicist to rest their eyes and use a device that could convert scintillation light into an electrical signal. PMTs have been a mainstay of nuclear physics ever since but they are not the only means of transforming scintillation light into a useful signal. There are also solid-state devices such as photodiodes, avalanche photodiodes and in recent years, silicon photomultipliers (SiPMs). The latter provide a performance comparable to a PMT with some advantages in terms of their compact size and low voltage operation, but still with some significant disadvantages compared to PMTs. The PMT is by no means a dead technology but SiPM technology is evolving extremely rapidly at the time of writing, driven largely by demands of medical imaging. In this section, we review the conventional PMT technology and its solid state rivals, noting along the way, the advantages and disadvantages of each. The choice of photosensor is crucial to any application involving a scintillator and it may well be different depending on whether the goal is the best energy resolution or, e.g. whether fast timing is the most important feature. The reader is encouraged to review carefully their options in matching to their scintillator of choice.

5.3.1 Photomultiplier tubes

PMTs[4] comprise evacuated tubes with a number of components inside (see figure 5.11). In a PMT, photons transit the optical window at the front and are

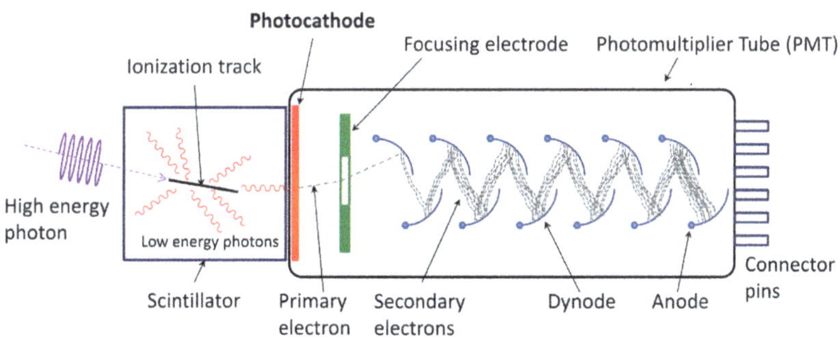

Figure 5.11. The interior of a photomultiplier tube (PMT). The PMT has an optical window at the front leading to a photocathode where primary electrons are generated. These are then accelerated between a succession of dynodes where secondary electrons are generated. The electrons are collected at the end of the tube onto the anode. This image has been obtained by the author from the Wikimedia website where it was made available by Qwerty123uiop under a CC BY-SA 3.0 licence. It is included within this article on that basis. It is attributed to Qwerty123uiop.

[4] PMTs are widely available from many different manufacturers such as Hamamatsu, ET Enterprises, Photek, etc. Extensive details and application notes for given PMTs are found on these manufacturers' webpages which the reader is encouraged to review.

incident on a *photocathode*. The cathode material has a low work function, chosen for the wavelength band corresponding to the photons of interest. The photon generates a photoelectron via the photoelectric effect, which is accelerated and focused using electric fields to impinge on a secondary-emission electrode, or *dynode*, which then emits a few secondary electrons. The multiplication process is repeated typically up to ten times in series to generate a sufficient number of electrons, which are collected at the anode for delivery to the external circuit. The net effect of a photomultiplier is to convert an incoming photon into an electrical signal comprised of 10^5–10^7 electrons depending on the gain of the PMT. Pulse rise times are usually in the few nanosecond range, meaning PMTs are very fast, and depend strongly on the physical dimensions of the PMT: the fastest PMTs are usually physically the smallest and with the fewest dynode stages.

Photocathode

According to which photocathode material is chosen, the spectral range of a PMT can vary widely within the optical, near-IR and near-UV ranges. Most photocathode materials are semiconductors—common materials comprising antimony (Sb) and alkali metals. Semiconductors are more suitable than metals as photocathodes because they have far fewer electrons in their conduction band thus limiting losses due to collisions. The key parameter for a given photocathode is the quantum efficiency, commonly expressed as a percentage for a given peak wavelength, $\eta(\lambda)$. This expresses the probability that a photon with the peak wavelength falling on the photocathode will produce a photoelectron. In recent years with the advent of ultrabright photocathodes it has been possible to raise peak quantum efficiencies from the 20%–25% range to values approaching 40%. Particularly critical is the electric field between the photocathode and the first dynode in the chain, this needs to be tailored so that the collection efficiency for photoelectrons is high and so that despite the range in exit points from the photocathode, the efficiency is uniform and the time spread is limited.

Dynode chain

The amplification in a PMT is achieved using a series of dynodes, typically of order 10, leading to a total gain up to 10^7. The most negative voltage is applied at the cathode end and the positive voltage at the anode end. Successive dynodes have a potential difference between them of order 100 V. Electrons are accelerated across the gap between dynodes so that they arrive on the next dynode with a kinetic energy of around 100 eV. This will lead to the emission of secondary electrons by the dynode. The electron gain from each successive dynode or *secondary emission factor* is given by δ. The total gain of a PMT depends on the applied high voltage step, V, between the dynodes and can be expressed as:

$$G = AV^{kn} \tag{5.5}$$

where $k \approx 0.7$–0.8 (depending on the dynode material), n is the number of dynodes in the chain, and A is a constant (which also depends on n). Since the gain is related to the voltage step raised to a high power, *n*, i.e. the number of dynodes, the applied voltage most be highly stabilised in order to achieve a stable gain.

Voltage divider
A *voltage divider* is a circuit used to ensure that the voltage step applied across the dynode chain is appropriate and that the voltage step is strongly stabilised. This is achieved using a chain of resistors. In order to have high stabilisation of the voltage step, the voltage divider needs to be specially designed to ensure that the so-called *bleeder current*, i.e. the current in the resistor chain, is as large as possible compared to the anode current in the tube. In fact, the variation in gain depends linearly on the ratio of these two currents. This is challenging because the current in the tube can vary strongly depending on event rate. It can be achieved by adding capacitors in parallel to the final stages of the voltage divider.

PMTs need to be driven by a high voltage supply of order 1000 V and capable of supplying mA current. The precise voltage to be applied will depend on the PMT design and the design of the voltage divider; the specification will be supplied by the manufacturer. Historically, PMTs were all driven by a positive high voltage but nowadays as many are driven with negative high voltage. There are advantages and disadvantages to each of the two approaches and naturally, they require a somewhat different design for the voltage divider. In any case, the polarity of the high voltage supply is a key parameter and the user should assure that the correct polarity of high voltage is applied to ensure that no damage results.

Magnetic shielding
Since PMTs function through acceleration of electrons, they are particularly sensitive to stray magnetic fields which alter the electron trajectories. In limited stray fields, PMTs may be effectively shielded using mu-metal or similar material [46]. However, substantial fields cannot be efficiently shielded and PMTs will not perform properly in such conditions.

Ancillary electronics
To operate a single PMT-based scintillator detector in standard analogue mode, several key pieces of electronics, e.g. NIM electronics would be required. Firstly, a well-stabilised, high voltage power supply[5] able to deliver sufficient current, e.g. Ortec 556 high voltage supply or the ISEG NHS. The output of the PMT would go into a preamplifier, either a charged-sensitive preamplifier or a current-sensitive preamplifier for some fast timing applications. The specification of the preamplifier required for inorganic scintillator detectors with PMT light collection is significantly less demanding than for semiconductor detectors as strongly constraining electronic noise is less important. The preamplifier output goes to a shaping amplifier e.g. Ortec 572. A key parameter for the shaping amplifier in terms of an inorganic scintillator detector is the shaping time which must be set appropriately to the decay time of the scintillator in

[5] The author is not recommending specific instrumentation. There are a range of options on the market and their availability has been checked at the time of writing.

question. The output of the shaping amplifier would then go to a multi-channel analyser (MCA). These readout solutions are discussed in more detail in chapter 10.

For those working on nuclear applications or in educational settings where only a single detector, e.g. NaI(Tl) detector, is of interest, then it may be beneficial to obtain an integrated base unit for a PMT which includes the basic electronics such as power supply, preamplifier and shaping amplifier as well as a built-in MCA so that the unit can connect directly to a computer via a USB cable, e.g. [47]. An example of such a system is the digiBASE.

Using PMTs

Clearly, PMTs are photosensitive devices and so they should not be operated, powered on, while exposed to ambient light; this has the potential to damage them since a very high current could result. Even when not in use, PMTs should be stored away from light as the dark current in the tubes will increase. This effect can be reversed by resting the tube for an extended period in darkness.

5.3.2 Solid state light sensors: photodiodes

A photodiode is a semiconductor-based device, comprising a p–n junction, which transforms incident light into an electrical signal; such devices are familiar to us as the photovoltaic elements of solar panels, effectively large area photodiodes. While solar panels operate in *photovoltaic* mode with no applied bias, in nuclear physics applications, they are typically reverse-biased and operated in a *photoconductive* mode; commonly, the photodiodes used have a *PIN structure* leading to the name *PIN diode*. The 'I' in the name refers to intrinsic material which in a typical photodiode comprises very pure n-type silicon with high resistivity (up to 8 kΩ cm). The front of the diode from which photons enter is made from a very thin p^+ layer [48]. Figure 5.12 shows a schematic of a PIN diode and its applied reverse bias.

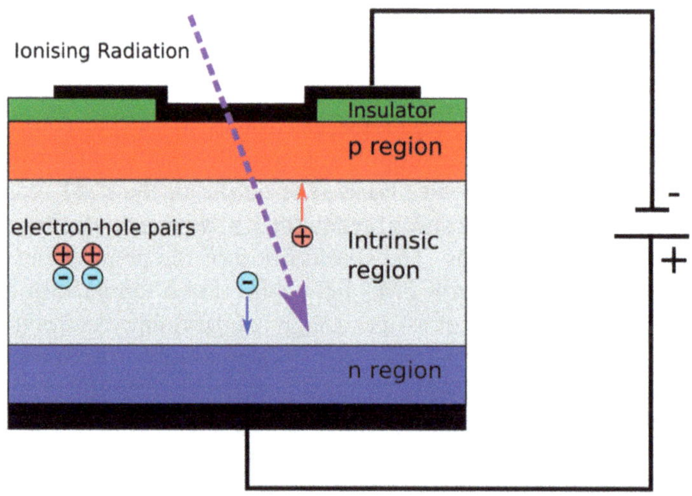

Figure 5.12. Schematic of a PIN diode. (Reprinted from [50] with the permission of Elsevier.)

When light of a compatible wavelength falls on the PIN diode, it generates electron–hole pairs. The holes are swept to the anode and the electrons to the cathode under the action of the applied electric field. The reverse bias increases the width of the depleted region within the sensor, lowering the device capacitance and speeding up charge collection. The response of the photodiode is linear to the light intensity falling on it over a wide dynamic range.

Photodiodes are commonly made from silicon but can also be fabricated from other semiconductor elements or compounds. Silicon has a high quantum efficiency in the range 500–900 nm (see figure 5.13). The quantum efficiency can be much higher than a PMT but unlike a PMT, the photodiode has no internal amplification—it has unit gain, i.e. gain factor is one. This means that energy resolution at low energies is determined by electronic noise not photon counting statistics [48]. Accordingly, special attention has to be paid to the fabrication of the photodiode and the electronics read-out chain to minimise the contribution from electronic noise and ensure a large signal-to-noise ratio. Photodiode read-out is conventionally achieved with a low-noise charge sensitive preamplifier coupled to a shaping amplifier. A related issue concerning the impact of electronic noise relates to the geometry of light collection. It might appear sensible to cover the whole face of a crystal with an array of photodiodes but the electronic noise in the system will grow at nearly the same rate as the increased area of collection—this effect is known as Groom's theorem [49]. The consequence for detection of gamma rays with low energies is that the signal-to-noise and hence energy resolution will barely improve with scaled-up light collection. In summary, photodiodes could be used for collection of scintillation light from scintillators in applications relating to low energy gamma ray detection, but superior performance will likely be obtained with avalanche photodiodes (APDs) and silicon photomultipliers (SiPMs) to be described below, which hold significant advantages. That said, photodiodes can perform better where the signal is large, i.e. high intensity of

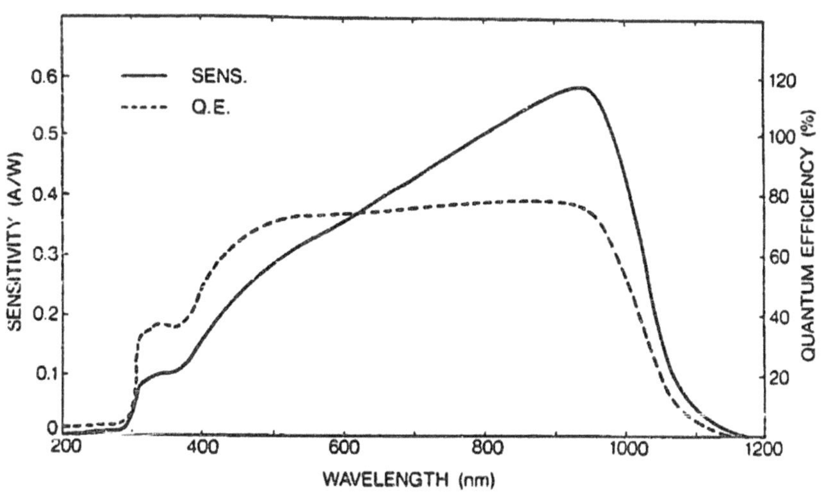

Figure 5.13. Sensitivity and quantum efficiency of a typical silicon photodiode as a function of wavelength. (Reprinted from [48] with the permission of Elsevier.)

scintillation light falling on the photodiode, meaning that they can be more relevant to detection of scintillation light from higher energy (~10 MeV) charged particles—an application to be more fully described in section 8.1.1. Even in this application, SIPMs offer significant advantages as no preamplification is required.

Avalanche photodiodes
Avalanche photodiodes (APDs) [51] are photodiodes with a large reverse bias applied (typically around 100–200 V) which are capable of multiplication. In this regime, the electron–hole pairs generated are accelerated to energies which exceed the bandgap in the material so that they create additional electron–hole pairs which are in turn accelerated. This process continues exponentially creating an *avalanche*. The gain achieved is proportional to the reverse bias voltage. The typical gain for an APD is around a factor of 100. APDs are frequently used for scintillator light collection in different areas of science, e.g. [52] and in the context of experimental nuclear physics; examples include the CALIFA calorimeter for measuring high energy gamma rays and charged particles [53].

5.3.3 Silicon photomultipliers

In the same way that NaI(Tl) dominated the market for scintillator detectors for decades, photomultiplier tubes were essentially the only means for detecting scintillator light in most spectroscopic applications for over fifty years. In recent times, silicon photomultipliers (SiPMs) have been making inroads into the scintillation light detection domain, largely initially in technological applications such as medical imaging (see section 7.5.1) but more recently also in domains such as nuclear physics [54, 55]. SiPMs typically comprise arrays of 10 000s of Geiger-mode avalanche photodiodes (APDs), also known as single photon avalanche diodes (SPADs). As long as the number of photons falling on the SiPM is significantly less than the number of available SPAD pixels then the signal should be essentially linear. This point needs to be considered carefully in applications where high energy gamma rays are to be detected, i.e. the number of scintillation photons is large and may approach or exceed the number of available SPAD pixels. The working principle of an individual SPAD is presented in figure 5.14.

Silicon photomultipliers are unlikely to fully displace conventional photomultipliers but they offer a number of significant advantages which make them useful for a range of applications. For example:

- low operating voltage (~50 V) as compared to photomultiplier tubes which require high voltage (~1 kV);
- robust and compact compared to photomultipliers which are bulky and have an evacuated tube which is intrinsically delicate;
- photon detection efficiency can exceed that of competing PMTs;
- gain is comparable to a PMT i.e. of order 10^6;
- insensitive to magnetic fields—implying suitability for applications like combined PET/MRI (see section 7.5.1);
- in principle, sensitive to single photon events.

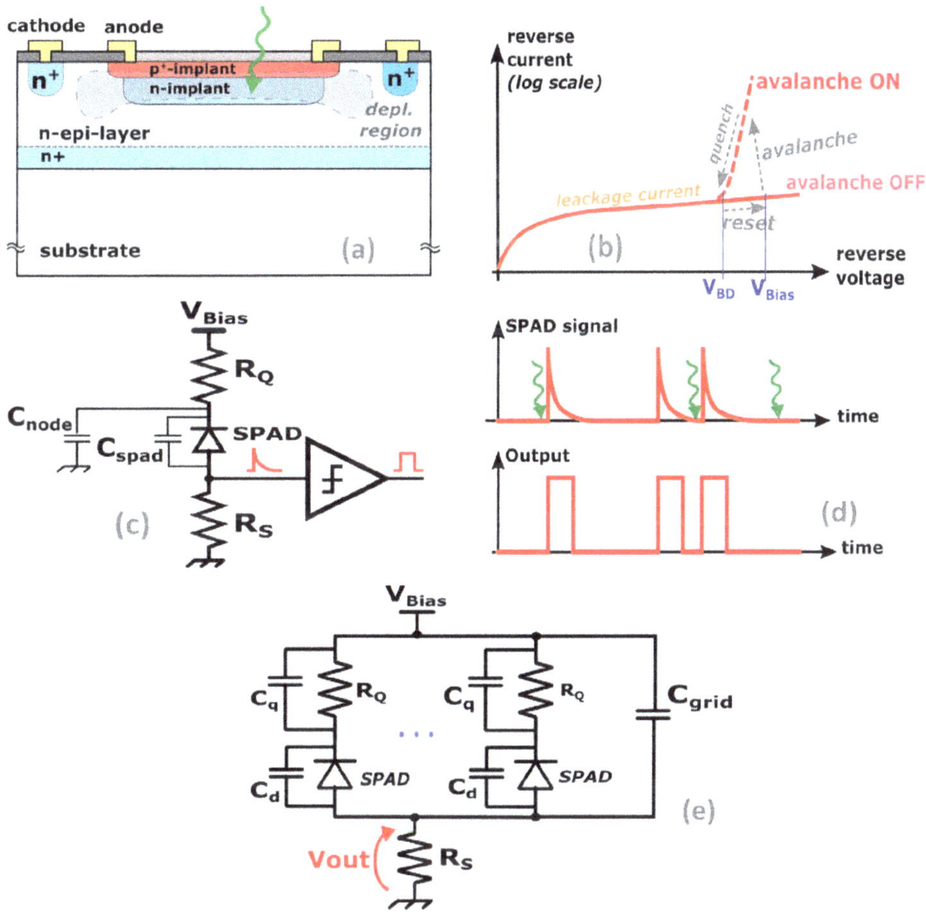

Figure 5.14. The elements of a SiPM—individual SPADs: (a) typical p-on-n silicon SPAD structure, (b) reverse current versus bias voltage, with avalanche, quenching and reset phase, (c) single-SPAD readout circuit, with discriminator to digitise the signal, (d) example of analogue output and digitised signal, and (e) typical analogue SiPM circuit, composed of many SPADs with integrated quenching resistors (R_q), and with sensing resistor (R_s). (Reprinted from [54] with the permission of Elsevier.)

While there are these set of advantages, there are significant issues which affect SiPMs but not conventional PMTs. For example, they exhibit a dark current which becomes worse as a function of temperature. This, for example, degrades achievable energy resolution when coupled to a scintillator and in practice, means that SiPMs are not very effective at elevated temperatures such as above 80 °C. On the other hand, the dark current can be strongly reduced at low temperatures achieved through active cooling and they have been employed even at cryogenic temperatures [56]. Of more practical concern is the significant change in SiPM gain as a function of temperature. This can be readily corrected using a feedback circuit which varies the bias voltage as a function of temperature; suitable power supplies on chips matched to common SiPM types are now readily available, e.g. the CAEN DT5485P. SiPM technology is evolving rapidly at the time of writing. For example, the development of through silicon vias in the last few years has allowed much closer packing of SiPMs in arrays (see figure 5.15). This produces marked improvement in collection of scintillation light. Figure 5.16 shows spectra obtained for a 2″ cubic $LaBr_3$ scintillator coupled to a SensL C-series and a J-series array. The energy resolution at 662 keV improves from 3.8% to 3.2% when using the J-series array.

Ancillary electronics and optimisation of an SiPM

The electronics chain required for SiPM readout is very similar to that for a conventional PMT, although it is often possible to dispense with the preamplifier stage. Again, shaping time is a key parameter which needs optimisation. It is common to explore the effects of varying the bias voltage above the breakdown point to locate the point where energy resolution is optimised.

Figure 5.15. Comparison of an 8x8 array of 6 mm square SensL C-series (left) and J-series (right) SiPMS. The J-series uses through silicon vias allowing closer packing. (Photo credit: the author.)

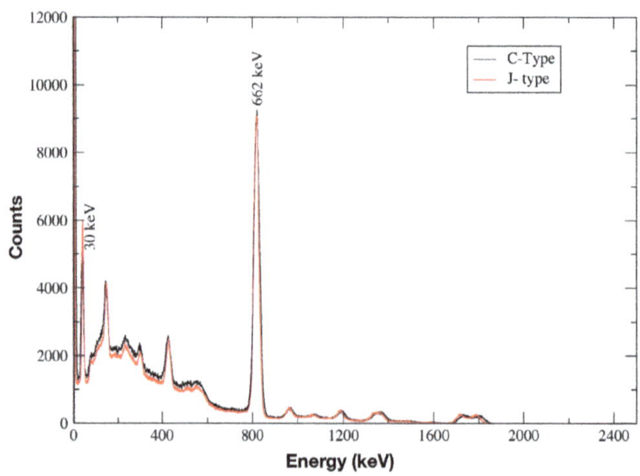

Figure 5.16. Spectra obtained coupling a 2″ cubic $LaBr_3$ scintillator to a SensL C-series (black) and a J-series (red) array. The energy resolution improves from 3.8% to 3.2% at 662 keV.

5.3.4 Simulation of scintillators and SiPMs

GEANT4 (described in more detail in section 4.2.1) is an excellent simulation tool for evaluating the performance of scintillator detectors. GEANT4 is capable of studying the generation and transport of the optical photons resulting from the scintillation process (see example in figure 5.17) and extensions such as GODDeSS exist for complex geometries and optical fibre read-out [57]. Such simulations assist in matching the geometry of the scintillator to the photodetector employed and also in design of light guides. In addition, there is an extension to GEANT4 called GEANT4SiPM [58] which handles the physics associated with SiPM readout.

5.4 Scintillator detector arrays

In experimental nuclear physics, scintillators are often deployed as large arrays or calorimeters (meaning a system designed to catch all gamma rays emitted from a nuclear reaction or following beta decay). Key parameters in array design, e.g. for a spherical arrangement of detectors, are the radius and timing resolution of the detectors employed since time-of-flight can potentially be exploited to discriminate gamma rays from neutrons.

5.4.1 Total absorption spectrometers

Total absorption spectrometers (TAS) are used for study of nuclear beta decay—examples can be found in references [60–62]. When the beta decay proceeds to excited states in a daughter nucleus, gamma rays will be subsequently emitted. A TAS is designed to fully capture all these gamma rays as far as possible. This can be used to infer key details of the beta decay process and the fraction of the beta decays which reach specific excited states. Figure 5.18 provides a standard example of a TAS—the Lucrecia system used at CERN—which employed NaI(Tl) detectors to surround the beta-decaying sample. Figure 5.19 is a concept design for an ambitious 128 element $LaBr_3(Ce)$ array to form a TAS system that could be used at future radioactive beam facilities.

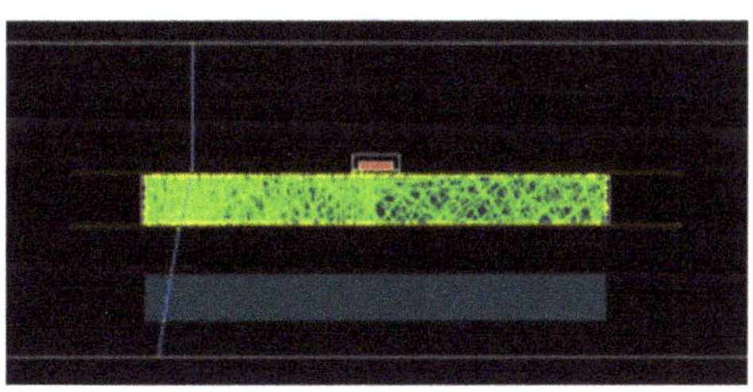

Figure 5.17. Visualisation of a GEANT4 simulation a beta particle (blue line) interacting with a plastic scintillator. The trajectories of the optical photons generated are shown in green. The SiPM collecting the scintillation light is shown in red. (Reprinted from [59] with permission from IOP Publishing.)

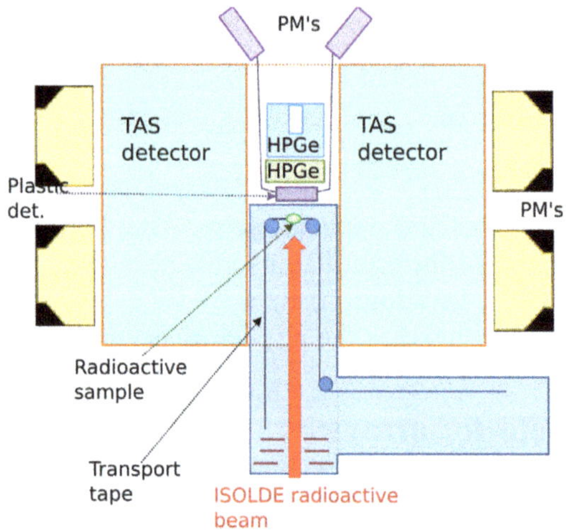

Figure 5.18. Design of Lucrecia total absorption spectrometer used at ISOLDE-CERN. A low energy beam of a radioactive isotope from the ISOLDE facility at CERN is implanted on a moving tape system. A fast plastic scintillator detector detects the beta particles to act as a trigger. Gamma rays are detected both in high-purity germanium detectors and in the TAS which comprises NaI(Tl) scintillators surrounding the whole system. (Reprinted from [60] with permission from IOP Publishing.)

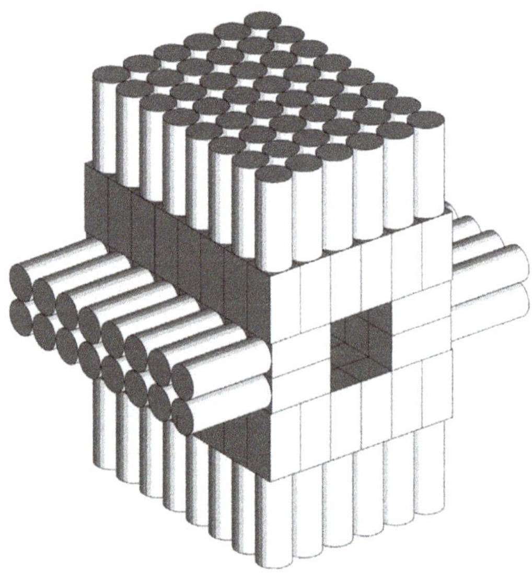

Figure 5.19. An ambitious design for a TAS system comprising 128 individual LaBr 3(Ce) detectors each of 5.5 × 5.5 × 11.0 cm readout with 2″ diameter photomultiplier tubes. (Reprinted from [62] with the permission of Elsevier.)

5.4.2 Next-generation scintillator arrays

There are several detector arrays constructed in recent years which exploit next-generation scintillator crystals.

OSCAR

OSCAR is an array of 30 large volume $LaBr_3$(Ce) detectors (see figure 5.20)—each detector is cylindrical with a 3.5″ diameter and 8″ long. This amounts to almost 200 kg of $LaBr_3$(Ce) material. The array is operated in conjunction with the cyclotron facility at the University of Oslo where proton beams (up to 35 MeV) as well as deuteron, ^{3}He and ^{4}He beams can be delivered. The array is principally used for studies of level density in nuclei where benefits are seen in the improved energy resolution and in the excellent timing resolution which allows gamma rays and neutrons to be discriminated on the basis of time of flight.

NANA

The NAtional Nuclear Array, NANA, is a twelve-element $LaBr_3$(Ce) scintillation detector array, based at the UK's National Physical Laboratory (NPL). It is designed to act as a primary standard [26, 63, 64].

FATIMA

The FATIMA array is a fast-timing detector system designed for the future DESPEC facility at FAIR [65]. It comprises 36 $LaBr_3$(Ce) detectors. Each detector has a cylindrical crystal of 1.5″ diameter and 2″ long. Each crystal is coupled to a Hamamatsu R9779 photomultiplier tube. The detectors are readout using a dedicated VME-based data acquisition system used to support the fast timing measurements.

$LaBr_3$(Ce) performs very well as a fast timing detector due to its excellent timing and energy resolution. Lifetimes of nuclear excited states down to less than 10 ps can be measured using such an array using the centroid shift method [65]. Figure 5.21

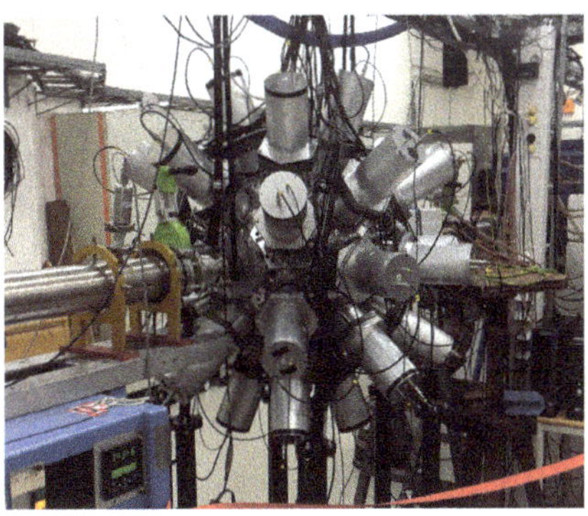

Figure 5.20. The OSCAR array of 30 large $LaBr_3$ (Ce) detectors at the University of Oslo. (Photo credit: the author.).

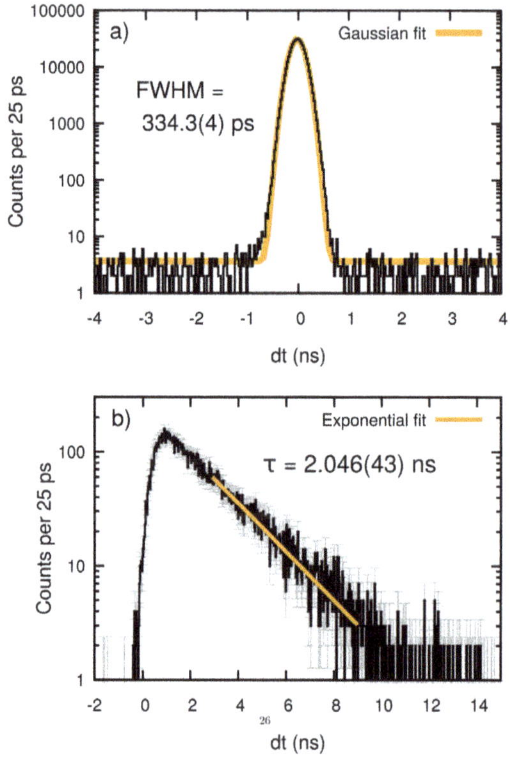

Figure 5.21. Time spectra obtained with the FATIMA array of 34 $LaBr_3$(Ce) detectors: (a) time distribution between the detection of the gamma rays in the decay of ^{60}Co—these gamma rays are in prompt time coincidence so the coincident timing resolution may be fitted as shown (CTR ≈ 334 ps). (b) Delayed coincident time distribution of the 122 keV–1408keV gamma ray cascade in ^{152}Sm, from a ^{152}Eu source. The measured time distribution can be fitted to obtain the lifetime of the 2^+ state in ^{152}Sm corresponding closely to the literature value of 2.020(20) ns. (Reprinted from [65] with the permission of Elsevier.)

illustrates the coincident timing resolution (CTR)[6] of the detector array, which is on average around 320 ps.

The FATIMA array is highly modular and can be readily deployed in different configurations and in conjunction with different detector systems such as high-purity germanium detector (HPGe) arrays. Figure 5.22 illustrates some of the varying configurations of the array.

PARIS

PARIS is a calorimeter based on the phoswich concept. The individual phoswich detectors comprise 2″ cubic $LaBr_3$(Ce) or $CeBr_3$ scintillators coupled directly to 2″ × 2″ × 6″ cuboid NaI(Tl) scintillators. The scintillation light is collected with a single photomultiplier tube mounted on the rear. Pulse shape analysis based on the very different decay times of the $LaBr_3$(Ce)/$CeBr_3$ and NaI(Tl) can be used to

[6] See section 4.1.2 for discussion of the meaning of CTR.

Figure 5.22. Different experimental setups incorporating the FATIMA array: (a) in a hemisphere to couple to half of the Gammasphere HPGe detector array at Argonne National Laboratory (b) at University of Surrey in three rings (c) mounted around a HPGe clover detector (d) at IPN Orsay as part of the ν-ball array with HPGe detectors (e) at IPN Orsay as part of the STELLA apparatus (see section 8.2.4) (f) at RIKEN where they are placed in the spaces between HPGe cluster detectors (g) and (h) at GSI in its intended configuration as part of the DESPEC decay station. (Reprinted from [65] with the permission of Elsevier.)

discriminate interactions in the two portions of the detector. The rationale behind the phoswich design is that the $LaBr_3$(Ce)/$CeBr_3$ portion can provide good energy resolution for gamma rays up to 5 MeV, while the thick NaI(Tl) crystal has a high efficiency for stopping gamma rays allowing sensitivity to gamma rays of 10s of MeV.

References

[1] Moszynski M, Syntfeld-Kazuch A, Swiderski L, Grodzicka M, Iwanowska J, Sibcynski P and Szcezesniak T 2016 Energy resolution of scintillators *Nucl. Instrum. Methods Phys. Res.* A **805** 25–35

[2] Wang Z, Williams R T, Grim J Q, Gao F and Kerisit S 2013 Kinetic Monte Carlo simulations of excitation density dependent scintillation in CsI and CsI(Tl) *Phys. Status Solidi* B **250** 1532–40

[3] Chen X *et al* 2018 Light extraction enhancement and directional control of scintillator by using microlens arrays *Opt. Express* **26** 23132

[4] Salomoni M, Pots R, Auffray E and Lecoq P 2018 Enhancing light extraction of inorganic scintillators using photonic crystals *Crystals* **8** 78

[5] McGregor D S 2018 Materials for gamma-ray spectrometers: inorganic scintillators *Annu. Rev. Mater. Res.* **48** 245–77

[6] Zhu R Y 1998 Radiation damage in scintillating crystals *Nucl. Instrum. Methods* A **413** 297–311

[7] Nanal V, Back B B and Hofman D J 1997 Temperature dependence of BaF_2 scintillation *Nucl. Instrum. Methods* A **389** 430–6
[8] Wang P, Zhang Y, Xu Z and Wang X 2014 Study on the temperature dependence of BGO light yield *Sci. China: Phys. Mech. Astron.* **57** 1898–901
[9] Hofstadter R 1949 The detection of gamma-rays with thallium-activated sodium iodide crystals *Phys. Rev.* **75** 796–810
[10] Heath R L, Hofstadter R and Hughes E B 1979 Inorganic scintillators: a review of techniques and applications *Nucl. Instrum. Methods* **162** 431–76
[11] Ianakiev K D, Alexandrov B S, Littlewood P B and Browne M C 2009 Temperature behavior of NaI(Tl) scintillation detectors *Nucl. Instrum. Methods* A **607** 432–8
[12] Grassmann H, Lorenz E and Moser H-G 1985 Properties of CsI(Tl)—renaissance of an old scintillation material *Nucl. Instrum. Methods* A **228** 323–6
[13] Mach H, Gill R L and Moszynski M 1989 A method for picosecond lifetime measurements for neutron-rich nuclei *Nucl. Instrum. Methods* A **280** 49–72
[14] Fraile L M 2017 Fast-timing spectroscopy at ISOLDE *J. Phys.* G **44** 094004
[15] Diehl S *et al* 2015 Readout concepts for the suppression of the slow component of BaF_2 for the upgrade of the TAPS spectrometer at ELSA *J. Phys. Conf. Ser.* **587** 012044
[16] Moszynski M *et al* 2004 Intrinsic energy resolution and light yield nonproportionality of BGO *IEEE Trans. Nucl. Sci.* **51** 1074–9
[17] Bardelli L *et al* 2006 Further study of $CdWO_4$ crystal scintillators as detectors for high sensitivity 2β experiments: Scintillation properties and pulse-shape discrimination *Nucl. Instrum. Methods* A **569** 743–53
[18] Melcher C L and Schweitzer J S 1992 A promising new scintillator: cerium-doped lutetium oxyorthosilicate *Nucl. Instrum. Methods* A **314** 212–4
[19] Park C H, Lee A, Kim R and Moon J H 2014 Evaluation of the detection efficiency of LYSO scintillator in the fiber-optic radiation sensor *Sci. Technol. Nucl. Install.* **2014** 248403
[20] Alva-Sanchez H, Zepeda-Barrios A, Diaz-Martinez V D, Murrieta-Rodriguez T, Martinez-Davalos A and Rodriguez-Villafuerte M 2018 Understanding the intrinsic radioactivity energy spectrum from ^{176}Lu in LYSO/LSO scintillation crystals *Sci. Rep.* **8** 17310
[21] Wilkinson D H 1952 The phoswich—a multiple phosphor *Rev. Sci. Instrum.* **23** 414–7
[22] Rajabi Moghadam S, Feghhi S A H and Safari M J 2015 A phoswich detector for simultaneous alpha-gamma spectroscopy *Nucl. Instrum. Methods* A **799** 59–63
[23] Ullah M N, Park C, Pratiwi E, Kim C, Choi H and Yeom J-Y 2019 A new positron-gamma discriminating phoswich detector based on wavelength discrimination (WLD) *Nucl. Instrum. Methods* A **946** 162361
[24] Gostojic A *et al* 2016 Characterisation of $LaBr_3$:Ce and $CeBr_3$ calorimeter modules for 3D imaging in gamma ray astronomy *Nucl. Instrum. Methods* A **832** 24–42
[25] McConnell M L, Bloser P F, Legere J and Ryan J M 2016 Applications for new scintillator technologies in gamma ray astronomy *J. Phys.: Conf. Ser.* **763** 012008
[26] Regan P H *et al* 2016 Applications of $LaBr_3$(Ce) gamma-ray spectrometer arrays for nuclear spectroscopy and radionuclide assay *J. Phys.: Conf. Ser.* **763** 012004
[27] de Faoite D, Hanlon L, Roberts O, Ulyanov A, McBreen S, Tobin I and Stanton K T 2015 Development of glass-ceramic scintillators for gamma-ray astronomy *J. Phys.: Conf. Ser.* **620** 012002
[28] van Loef E V D, Dorenbos P, van Eijk C W E, Kraemer K and Guedel H U 2001 High-energy-resolution scintillator: Ce^{3+} activated $LaCl_3$ *Appl. Phys. Lett.* **77** 1467–8

[29] van Loef E V D, Dorenbos P, van Elijk C W E, Kraemer K and Guedel H U 2001 High-energy-resolution scintillator: Ce^{3+} activated $LaBr_3$ *Appl. Phys. Lett.* **79** 1573–5
[30] Quarati F G A *et al* 2011 High energy gamma-ray spectroscopy with $LaBr_3$ scintillation detectors *Nucl. Instrum. Methods* A **629** 157–69
[31] Camp A, Vargas A and Fernandez-Vega J M 2016 Determination of $LaBr_3$(Ce) internal background using a HPGe detector and Monte Carlo simulations *Appl. Radiat. Isot.* **109** 512–7
[32] Quarati F G A, Alekhin M S, Krämer K W and Dorenbos P 2014 Co-doping of $CeBr_3$ scintillator detectors for energy resolution enhancement *Nucl. Instrum. Methods* A **735** 655–8
[33] Quarati F Q A, Dorenbos P, van der Biezen J, Owens A, Selle M, Parthier L and Schotanus P 2013 Scintillation and detection characteristics of high-sensitivity $CeBr_3$ gamma-ray spectrometers *Nucl. Instrum. Methods* A **729** 596–604
[34] Drozdowski W, Dorenbos P, Bos A J J, Bizarri G, Owens A and Quarati F Q A 2008 $CeBr_3$ scintillator development for possible use in space missions *IEEE Trans. Nucl. Sci.* **55** 1391–6
[35] van Loef E V, Wilson C M, Cherepy N J, Hull G, Payne S A, Choong W-S, Moses W W and Shah K S 2009 Crystal growth and scintillation properties of strontium iodide scintillators *IEEE Trans. Nucl. Sci.* **56** 869–72
[36] Mitchell L J and Phlips B 2016 Characterization of strontium iodide scintillators with silicon photomultipliers *Nucl. Instrum. Methods* A **820** 95–101
[37] Shirwadkar U, Glodo J, van Loef E V, Hawrami R, Mukhopadhyay S, Churilov A, Higgins W M and Shah K S 2011 Scintillation properties of $Cs_2LiLaBr_6$ (CLLB) crystals with varying Ce^{3+} concentration *Nucl. Instrum. Methods* A **652** 268–70
[38] Smith M B, Achtzehn T, Andrews H R, Clifford E T H, Ing H and Kovaltchouk V D 2013 Fast neutron spectroscopy using Cs_2LiYCl_6:Ce (CLYC) scintillator *IEEE Trans. Nucl. Sci.* **60** 855–9
[39] Mentana A, Camera F, Giaz A, Blasi N, Brambilla S, Ceruti S, Gini L, Groppi F, Manenti S and Million B 2016 Measurement of fast neutron detection efficiency with ^{6}Li and ^{7}Li enriched CLYC scintillators *J. Phys.: Conf. Ser.* **763** 012006
[40] Derenzo S E, Bourret-Courshesne E, Bizarri G and Canning A 2016 Bright and ultra-fast scintillation from a semiconductor? *Nucl. Instrum. Methods* A **805** 36–40
[41] Dujardin C, Auffray E, Bourret-Courshesne E, Dorenbos P, Lecoq P, Nikl M, Vasil'ev A N, Yoshikawa A and Zhu R-Y 2018 Needs, trends and advances in inorganic scintillators *IEEE Trans. Nucl. Sci.* **65** 1977
[42] Giaz A *et al* 2015 Investigation on gamma-ray position sensitivity at 662 keV in a spectroscopic 3" × 3" LaBr3:Ce scintillator *Nucl. Instrum. Methods* A **772** 103–11
[43] Ulyanov A, Morris O, Roberts O J, Tobin I, Hanlon L, McBreen S, Murphy D, Helms N and Shortt B 2017 Localisation of gamma-ray interaction points in thick monolithic $CeBr_3$ and $LaBr_3$:Ce scintillators *Nucl. Instrum. Methods* A **844** 81–9
[44] Son J, Kim D G, Lee S, Park J, Kim Y, Schaarschmidt T and Kim Y K 2018 Improved 3D printing plastic scintillator fabrication *J. Korean Phys. Soc.* **73** 887–92
[45] Dosovitskiy G A, Karpyuk P V, Evdokimov P V, Kuznetsova D E, Mechinsky V A, Borisevich A E, Fedorov A A, Putlayev V I, Dosovitskiy A E and Korjik M V 2017 First 3D-printed complex inorganic polycrystalline scintillator *CrystEngComm* **19** 4260
[46] Faradzhaev R M, Trofimov Y A, Lupar E E and Yurov V N 2016 Performances investigation and material selection of PMT magnetic shields for the space experiments with GRIS and PING-M instruments *J. Phys. Conf. Ser.* **675** 042008

[47] Xue T, Zhu J, Gong G, Wei L and Li J 2018 The prototype design of integrated base for PMT with pulse digitalization and readout electronics *J. Instrum.* **13** T06007
[48] Suffert M 1992 Silicon photodiode readout of scintillators and associated electronics *Nucl. Instrum. Methods* A **322** 523–8
[49] Groom D E 1984 Silicon photodiode detection of bismuth germanate scintillation light *Nucl. Instrum. Methods* **219** 141–8
[50] Turkington G, Gamage K A A and Graham J 2018 Beta detection of strontium-90 and potential for direct *in situ* beta detection for nuclear decommissioning applications *Nucl. Instrum. Methods* A **911** 55–65
[51] Campbell J C 2016 Recent advances in avalanche photodiodes *J. Lightwave Technol.* **34** 278–85
[52] Kataoka J *et al* 2005 Recent progress of avalanche photodiodes in high-resolution x-rays and gamma-rays detection *Nucl. Instrum. Methods* A **541** 398–404
[53] Knyaez A *et al* 2019 Properties of the CsI(Tl) detector elements of the CALIFA detector *Nucl. Instrum. Methods* A **940** 393–404
[54] Acerbi F and Gundacker S 2019 Understanding and simulating SiPMs *Nucl. Instrum. Methods* A **926** 16–35
[55] Simon F 2019 Silicon photomultipliers in particle and nuclear physics *Nucl. Instrum. Methods* A **926** 85–100
[56] Cervi T, Bonesini M, Falcone A, Menegoli A, Raselli G L, Rossella M, Simonetta M and Torti M 2016 Characterization of SiPM for cryogenic applications *Nucl. Instrum. Methods* A **824** 89–91
[57] Dietz-Laursonn E, Hebbeker T, Künsken A, Merschmeyer M, Nieswand S and Niggemann T 2017 GODDeSS: a Geant4 extension for easy modelling of optical detector components *J. Instrum.* **12** P04026
[58] Niggemann T, Dietz-Laursonn E, Hebbeker T, Künsken A, Lauscher M and Merschmeyer M 2015 G4SiPM: A novel silicon photomultiplier simulation package for Geant4 *Nucl. Instrum. Methods Phys. Res.* A **787** 344–7
[59] Korpachev S and Chadeeva M 2017 Geant4 simulation of optical photon transport in scintillator tile with direct readout by silicon photomultiplier *J. Phys. Conf. Ser.* **798** 012218
[60] Rubio B *et al* 2017 Beta decay studies with total absorption spectroscopy and the Lucrecia spectrometer at ISOLDE *J. Phys.* G **44** 084004
[61] Karny M *et al* 2016 Modular total absorption spectrometer *Nucl. Instrum. Methods* A **836** 83–90
[62] Tain J L *et al* 2015 A decay total absorption spectrometer for DESPEC at FAIR *Nucl. Instrum. Methods* A **803** 36–46
[63] Lorusso G *et al* 2016 Development of the NPL gamma-ray spectrometer NANA for traceable nuclear decay and structure studies *Appl. Radiat. Isot.* **109** 507–11
[64] Collins S M, Shearman R, Keightley J D and Regan P H 2018 Investigation of gamma-gamma coincidence using the National Nuclear Array (NANA) as a primary standard *Appl. Radiat. Isot.* **134** 290–6
[65] Rudigier M *et al* 2020 FATIMA—FAst TIMing Array for DESPEC at FAIR *Nucl. Instrum. Methods* A **969** 163967

Chapter 6

Semiconductor detectors for gamma-ray detection

This chapter provides an overview of semiconductor detectors for gamma-ray detection, focusing heavily on germanium detectors which are the highest performing class of detector for gamma-ray spectroscopy. They offer superb energy resolution well-matched to the demands of state-of-the-art nuclear physics experiments. They also find employment in industrial and societal applications where their high energy resolution can be used to relate the fingerprint of the gamma-ray spectrum to the associated radioisotope. High performance comes with a high cost both financially but also in maintenance overheads since they need to be cooled to liquid nitrogen temperatures in order to operate. This chapter will describe the essential physics behind germanium detectors and the principal classes of detector widely found in the laboratory and industry. It goes on to detail the advantages and disadvantages of different detector designs for different applications. A review is given of germanium detector arrays, widely found in experimental nuclear physics, and the issues informing their design. Finally, a brief overview is given of gamma-ray tracking detectors—the state-of-the-art in this field, which offer unprecedented performance for high-resolution spectroscopy but whose high intrinsic cost is likely to preclude them appearing in industrial or other societal applications any time soon. The chapter is completed with a brief overview of room-temperature semiconductor detectors (RTSDs) for gamma-ray detection.

Concepts: Germanium detectors; physics of germanium detectors; hyperpure germanium detectors; germanium-detector design; arrays of germanium detectors; gamma-ray tracking arrays; CZT detectors

6.1 Germanium detectors—an overview

Germanium detectors represent the premier type of detector for gamma-ray spectroscopy. Germanium detectors are semiconductor-based detectors; silicon

doi:10.1088/978-0-7503-1428-2ch6

detectors, which are also semiconductor-based, will be described later in section 8.2. Silicon detectors excel at charged-particle detection but are generally too thin to have any useful efficiency for gamma-ray detection with the possible exception of lithium-drifted silicon detectors, known as Si(Li) detectors, which can be used for x-ray detection (or low energy gamma ray detection <100 keV). The limited thickness in which Si(Li) detectors can be fabricated (of order millimetres) makes them completely unsuitable for gamma rays in the more typical energy range ~ MeV which is found in experimental nuclear physics and its associated applications (see section 1.7.1). By contrast, the principal feature of germanium detectors that makes them pre-eminent in gamma-ray detection is that they can be fabricated from crystals of significant dimension, i.e. of order 10 cm consistent with the long range of gamma rays in matter due to the discontinuous nature of their interactions (as described in section 2.4). Moreover, they have a high gamma-ray absorption compared to, e.g. silicon, since germanium has a relatively high Z ($Z = 32$) and high density (5.32 g cm^{-3}). Germanium detectors afford excellent energy resolution, typically of the order of 2 keV at 1332 keV, which is significantly better than the best performing scintillator detector (see figure 6.1). This high performance comes at a price, however, in terms of their high cost relative to standard scintillator detectors of similar volume. An excellent review of modern germanium detectors and their uses can be found in reference [1].

The key disadvantage of germanium is that it will not operate as an effective semiconductor detector at room temperature. This is a feature of the low band gap in germanium corresponding to 0.67 eV (at 77 K), which means that too many electron–hole pairs would be thermally excited at room temperature. Instead, germanium detectors have to be operated at liquid nitrogen temperatures. There are some

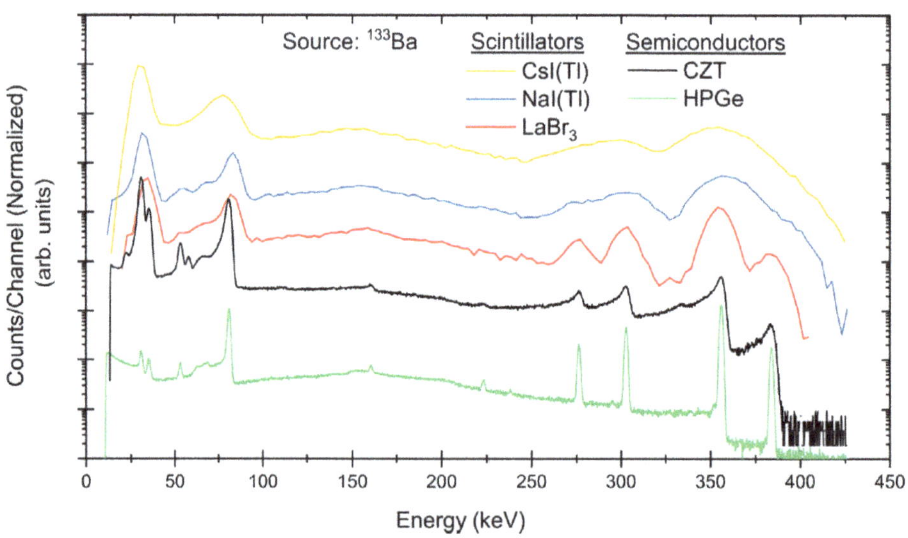

Figure 6.1. Comparison of gamma-ray spectra obtained with a ^{133}Ba source for two semiconductor detectors: high purity germanium (HPGe) and CZT with three scintillator detectors: CsI(Tl), NaI(Tl) and high-performing $LaBr_3$(Ce). (Reprinted from [2] with the permission of AIP Publishing.)

room-temperature semiconductor detectors (RTSDs) such as CZT and thallium bromide [3] but the size in which they can be produced is much smaller than typical germanium detectors. They have significantly poorer energy resolution and suffer additional issues with poor charge carrier mobility that limit their performance (see figure 6.1). RTSDs will be reviewed at the end of this chapter, in section 6.6.

Germanium detectors have an in-built cryostat so that they can operate at liquid nitrogen temperatures. In industrial settings, where one or two detectors are employed, it is realistic to make use of more recently available pumped cooling systems which avoid the need for the user to supply the detector with liquid nitrogen. This is advantageous in non-laboratory settings where liquid nitrogen is not likely to be readily to hand. However, the cost of the pumped-cooling system may be comparable to the price of the detector itself. Where large arrays of germanium detectors are used, for example, in nuclear physics experiments, liquid nitrogen cooling is nearly always provided externally, using an *autofill system* where liquid nitrogen is periodically pumped into a Dewar attached to the detector. This can entail a significant infrastructure overhead for a laboratory as well as significant risk both from oxygen deficiency in a confined space as well as damage to the detectors if they warm up while a high-voltage bias is applied. Modern detectors typically employ a temperature-controlled bias shutdown circuit to avoid this risk of damage. However, human monitoring and intervention is considered mandatory to ensure that detectors continue to be effectively cooled.

6.2 Hyperpure germanium detectors

The present generation of germanium detectors are called hyperpure germanium (HPGe) detectors[1]. The term 'hyperpure' refers to the fact that the germanium crystal has typically less than 1 in 10^{12} impurity concentration achieved through *zone refining* [5]. A hyperpure germanium detector functions as a diode and can be formed from either *p-type* or *n-type* germanium which we define below.

6.2.1 Detector fabrication

An absolutely pure semiconductor would have a balance of thermally-excited electron–hole pairs. In practice, a germanium crystal will not be absolutely pure and will have a low level of impurities in it. The nature of these impurities defines an important property of the material. If the excess impurities are, for example, associated with a three-valent atom such as boron then there will be a net deficit in the number of electrons in the crystal structure and this amounts to an effective hole. This is known as an *acceptor impurity* and germanium with such excess impurity is known as *p-type* germanium. By contrast, the presence of impurities

[1] The first germanium detectors available were known as Ge(Li) detectors since they were lithium-drifted [4]. The presence of the lithium is to compensate for the impurity of the germanium crystal. Such detectors have to be kept permanently cooled to liquid nitrogen temperature since if they are thermally cycled, the lithium will become mobile and damage the detector. Moreover, the lithium drifting process is time-consuming and can only produce compensated Ge crystals of small volume. These limitations and the fact that they are no longer manufactured mean that they are unlikely to be encountered in a modern laboratory.

associated with five-valent atoms such as phosphorus leads to a net excess of electrons. Either *p-type* or *n-type* germanium can be used to fabricate a germanium detector. There are distinct advantages and disadvantages to the choice of material, e.g. in improved low-energy efficiency or in improved resistance to neutron damage. These issues will be covered in more detail later on in this chapter. The bare *p-type* or *n-type* germanium is processed into a germanium detector with either a $p^+n^-n^+$ or a $p^+p^-n^+$ structure, usually by forming a n+ contact by lithium diffusion and a p+ contact by boron implantation. The geometries for the contacts vary greatly depending on the detector design. The detector is then metallised commonly using sputtered aluminium.

Modern germanium detectors are encapsulated in a vacuum enclosure which can serve to limit damage from impurities falling on to the germanium crystal surface and help to facilitate annealing. The charge-sensitive preamplifier(s) are built into the detector cryostat but separated from the vacuum enclosure so they can be readily accessed for repair or replacement. This positioning within the cryostat allows their component field effect transistors (FETs) to also be cooled, which strongly reduces noise. Encapsulating a detector in a material such as a thin aluminium case will naturally lead to some absorption of gamma rays, particularly low-energy gamma rays. However, typical thicknesses of the canning are only around 0.5 mm. In many situations where low-energy detection is important, a germanium detector often has its front face covered by a thin entrance window of low-*Z* material. Historically, a thin beryllium window was widely used but this is deprecated due to its high toxicity and the associated risks in breaking the window[2]. It is now more common to find a carbon-fibre window which has the required properties of thinness and strength. Carbon fibre can also be a better replacement for some applications as it has much lower internal radioactivity compared to beryllium or other end-cap materials.

6.2.2 Depletion depth

Applying a positive high voltage to the n side of the junction in a germanium detector, what is known as a *reverse bias*, will draw electrons away and create a deeper depleted region within the crystal. Germanium detectors are usually operated fully depleted to maximise the active detector volume. The depth of the depleted region within a germanium crystal can be evaluated by solving the Poisson equation, $\nabla\psi = -\frac{\rho}{\varepsilon}$ where ψ is the potential at a given point, ε is the permittivity of the material, and ρ is the charge density profile. The permittivity can be expressed in terms of the permittivity of free space, $\varepsilon = \varepsilon_r\varepsilon_0$ where ε_r is the dielectric constant or relative permittivity. For germanium, $\varepsilon_r \approx 16$, while for silicon $\varepsilon_r \approx 12$.

For a planar detector, which is a simple geometry to solve for, the depletion depth is given by

[2] The presence of a beryllium window should be a matter of concern when disposing or recycling old germanium detectors.

$$d \approx \sqrt{\frac{2\varepsilon V_0}{eN}}, \tag{6.1}$$

where V_0 is the reverse bias voltage, e is the electron charge and N is the net impurity concentration (of either donor or acceptor impurities depending on which dominates in the material). The depletion depth will, hence, vary smoothly with the applied reverse bias as the other parameters in the equation are fixed. We note that much of this underlying physics is fully applicable to silicon detectors which typically have a thin planar geometry and will be covered in detail in section 8.2. By contrast, most germanium detectors in use in laboratories and industry have a coaxial geometry, i.e. a cylinder with a cylindrical hole bored into it from the back. Most detectors are not true coaxial detectors because the hole is not bored all the way through, but for a true coaxial geometry, i.e. a cylindrical crystal of radius, r_2, with a cylindrical hole bored right through it with radius, r_1, the depletion voltage can be calculated as

$$V_d = \frac{1}{2\varepsilon eN}\left[r_1^2 \ln\left(\frac{r_2}{r_1}\right) - \frac{1}{2}\left(r_2^2 - r_1^2\right)\right]. \tag{6.2}$$

Putting numbers for typical dimensions of germanium detectors into equation (6.2), using the dielectric constant, $\varepsilon = 16$ for germanium, will show that depletion voltages are of order of 1000s of Volts. They generally do not exceed 5 kV to avoid breakdown and sparking.

6.2.3 Principle of operation

Gamma rays interact with germanium crystals by the standard modes of interaction: photoelectric absorption, Compton scattering and pair production as described in section 2.4. Ejected electrons from photoelectric absorption and Compton scattered electrons will generate electron–hole pairs as they lose energy in the crystal. The average energy needed to generate an electron–hole pair in germanium is 2.96 eV at 80 K, meaning that a high number of electron–hole pairs are produced per unit of energy of ionising radiation since typical photoelectric/Compton electron energies are in the range 10 keV–1 MeV. The electric field set up in the germanium crystal by the application of the reverse bias will induce motion of the charge carriers parallel to the direction of the electric field. Germanium has the convenient property of having very similar mobilities for electrons and holes at liquid nitrogen temperature of 3.6×10^4 and 4.2×10^4 cm^2 V^{-1} s^{-1}, respectively.

6.2.4 Pulse shapes

A key feature of germanium detectors which has been powerfully exploited in recent years is the information contained in the pulse shapes emerging from the detector's integrated charge-sensitive preamplifier(s). This feature, depending on the design and geometry of the detector can give an insight into the position of interaction of a gamma ray within the crystal. It is the advent of digital electronics which has made these pulse shapes accessible. Implementation of pulse-shape analysis will be described in more detail in section 6.5.2 where we review the topic of gamma-ray tracking but let us first look at why detector pulse shapes should vary. The pulse shapes vary because the path of charge carriers is dependent on the structure of the electrodes and the electric field distribution within the extended crystal. We illustrate this with respect to the specific example of a BEGe detector whose geometry is presented in figure 6.13. This detector has n^+ electrodes on all surfaces apart from a small central p^+ contact on the rear. Consider shining a highly-collimated beam of 60 keV gamma rays from a ^{241}Am into the detector at different radial positions. Such gamma rays will interact predominantly by photoelectric absorption within a short distance (~mm) of the front face. In this example, the pulse shape from the detector will depend almost entirely on the motion of the holes generated in the crystal towards the p^+ contact. This seems at first glance, counter-intuitive because the electrons will reach the much nearer n^+ electrode on the front face first but the holes will traverse almost the entire electric field region within the crystal. During their motion, the electrons and holes will induce image charges on the two electrodes. Their motion can be considered in terms of them travelling independently within a fixed applied field [6]. This allows the trajectory of the electrons and holes to be calculated. The shape of the preamplifier signal is defined by the *weighting potential* within the crystal [7]. The weighting potential, $\phi_i(\mathbf{x})$ is defined in relation to the induced charge Q_{ind} in electrode i for a unit of charge placed at position $\mathbf{x}$ in the detector:

$$Q_{\text{ind,i}} = -q\phi_i(\mathbf{x}). \tag{6.3}$$

The weighting potential can be evaluated using the Ramo–Shockley theorem [8, 9], assuming the electrodes are grounded and solving the Laplace equation. This leads to the conclusion that the time-dependent induced charge on the electrode is given by [6]:

$$Q_{\text{ind,i}}(t) = -\sum_q q\phi_i(\mathbf{x}_q(t)). \tag{6.4}$$

The observed pulse shapes correspond to the solution of equation (6.4).

Figure 6.2 illustrates the average pulse shapes obtained at different radial distances in the example BEGe detector (see figure 6.13) and an immediate connection is seen between the pulse shape and position of interaction which can be exploited with appropriate digital electronics. A few features should be noted. Firstly, the presentation of this figure is somewhat idealised since what is presented are the averaged pulse shapes gathered from an ensemble of 1000s of events. Individual pulse shapes will be distributed around this average and this will impose some limitation, i.e. a position resolution which can be achieved experimentally. Second, the example presented is relatively straightforward as it is a single crystal detector. Many modern germanium detectors, however, particularly those used in experimental nuclear physics, are electrically segmented. Transient signals may be recorded in segments adjoining the segment where the gamma-ray interaction occurs [10]. This additional information may be used to refine the determination of the position of interaction (see section 6.5.2).

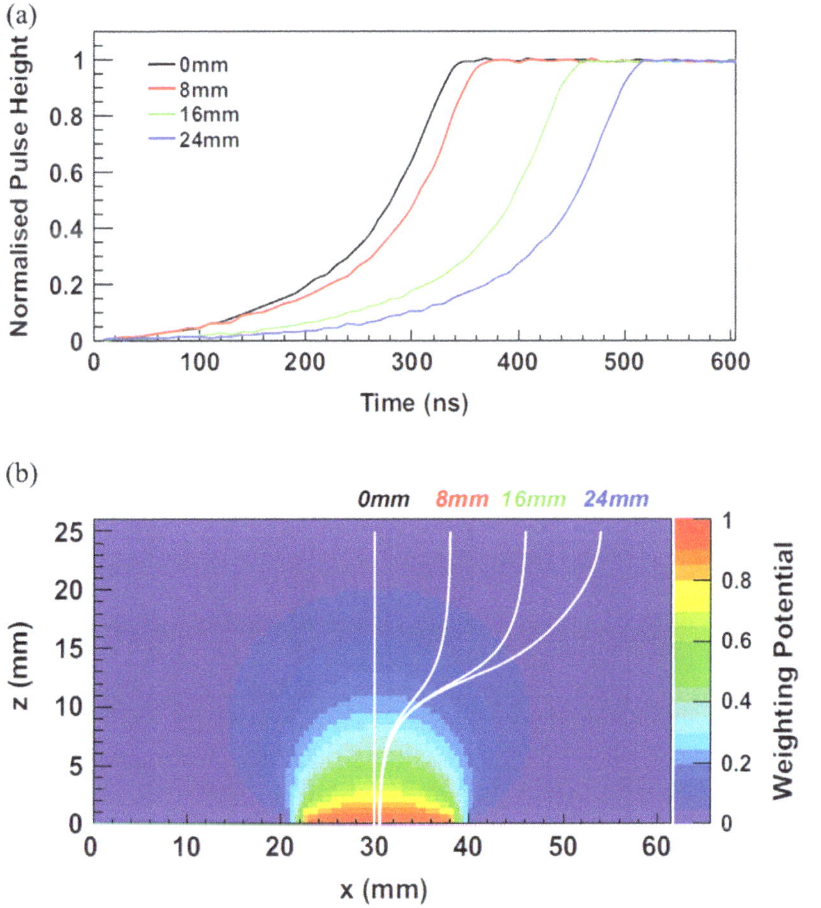

Figure 6.2. (Top) normalised average pulse shapes in a BEGe detector generated at radial distances of 0, 8, 16 and 24 cm with a ^{241}Am source incident on the detector surface. (Bottom) simulated trajectories for holes assuming an interaction depth of 1 mm, overlaid on the calculated weighting potential for the crystal and electrode structure. (Reprinted from [7] with the permission of Elsevier.)

6.3 Key parameters for germanium detectors

Paralleling what was covered for inorganic scintillator detectors in section 5.1.1, let us review some of the key parameters of germanium detectors—some of which have a one-to-one correspondence with inorganic scintillators and some of which are unique.

6.3.1 Energy resolution

Energy resolution is the key parameter in terms of the performance of germanium detectors as high energy resolution allows dense spectra and overlapping peaks to be resolved. It is important to distinguish intrinsic energy resolution from achievable energy resolution in an experiment; the latter can be worse than the intrinsic resolution where the gamma rays are, for example, emitted from a moving source and Doppler broadening occurs.

The intrinsic energy resolution of germanium detectors can be parametrised as

$$\mathrm{FWHM} = \sqrt{a^2 + bE}\,. \tag{6.5}$$

Figure 6.3 shows the measured energy resolution for a BEGe detector as a function of incident gamma-ray energy. The energy resolution in the statistical limit from consideration of the number of electron–hole pairs generated was presented in equation (4.11). However, so far we have not discussed ancillary contributions to the energy resolution from charge trapping and electronic noise. Charge trapping is fairly small in germanium because electrons and holes have high mobility. Most modern HPGe detectors have designs which aim to limit the electronic noise as far as possible using specialist components such as an integrated charge-sensitive preamplifier with a cooled field-effect transistor (FET). The electronic noise in the germanium detector is largely independent of the energy deposited in it, and, hence, the contribution from noise becomes more important where less energy is deposited and fewer charge carriers are generated. This means that energy resolution is significantly worse at low energies. A key parameter in terms of energy resolution is the capacitance of the detector. The capacitance of a coaxial geometry is generally lower than that of a planar geometry leading to better energy resolution [1].

To extract the functional form for the energy resolution of a germanium detector, a curve can be fitted to the data obtained from source spectra using the functional form of equation (6.5). This fitting is appropriate as the gamma-ray photopeaks

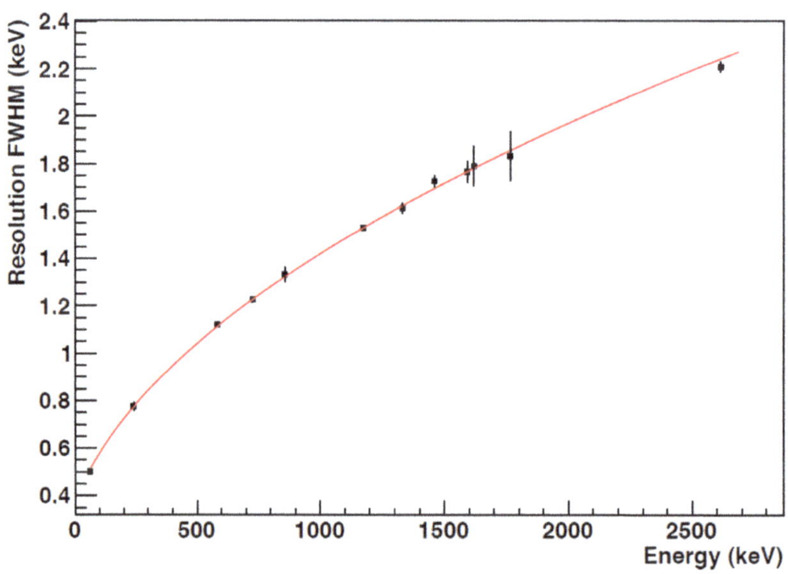

Figure 6.3. Energy resolution (FWHM) as a function of energy for a BEGe detector. The black points are values for FWHM obtained from source spectra. The red curve is a fit to the data using a function of the form given in equation (6.5). (Reprinted from [11] with the permission of IOP Publishing.)

should have a Gaussian shape since the width of the de-exciting nuclear energy level is extremely small. Some points to note on fitting source data are that the inclusion of x-ray peaks and the annihilation (511 keV) photopeak should generally be avoided. This is because the widths of atomic levels are not so small, especially for high-Z atoms. The x-ray line shapes will be a convolution of the Lorentzian energy distribution corresponding to the atomic level and the Gaussian peak-shape arising from the detector which is described by a *Voigt function*. The use of the 511 keV annihilation gamma ray from e.g. a ^{22}Na source should also be avoided as this peak may be Doppler-broadened due to the probability that the positron is not fully at rest at the point of annihilation.

In a practical experiment, where gamma rays are emitted from a moving source such as an accelerated beam, the gamma rays are Doppler-shifted (see section 3.3.1). The photopeaks in a gamma-ray spectrum reconstructed from rings of detectors at different angles will be broadened due to the uncertainty in the angle of emission. Where detectors are electrically segmented or use pulse-shape analysis to determine the point of first interaction (see section 6.5.2), this effect can be minimised. Figure 6.4 shows an example of this effect for the MINIBALL germanium detector array [12] where the point of first interaction is successively localised from the overall cluster detector level, to the individual capsules that make it up, to the sub-capsule level using pulse-shape analysis.

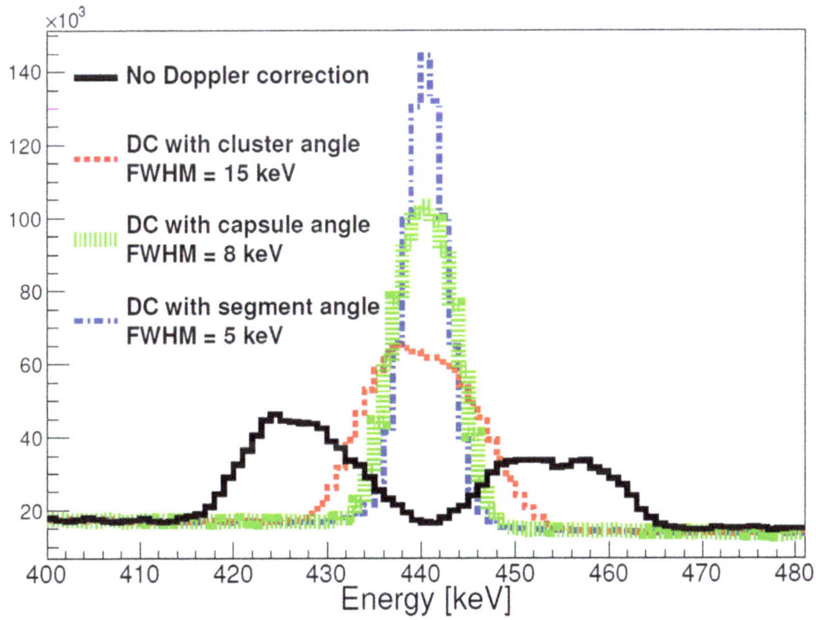

Figure 6.4. Gamma-ray spectrum from the ^{22}Ne(d,n)^{23}Na reaction taken using the MINIBALL germanium detector array. The ^{23}Na was not detected. The energy resolution improves as the point of interaction is successively localised from the overall detector volume (cluster) to the individual crystals which make it up (capsule) to the segment in the crystal obtained from pulse-shape analysis. (Reprinted from [12] with the permission of Elsevier under Creative Commons License.)

6.3.2 Timing resolution

Timing resolution is an important feature of germanium detectors since, within experimental nuclear physics but also some applications, such detectors are operated in a coincidence mode within an array of detectors (see section 6.4.2) or in coincidence with events recorded in other ancillary[3] detectors (e.g. charged particle counters—see section 8.1.1, fast neutron counters—see section 9.1.1). A good timing resolution will strongly reduce experimental backgrounds, e.g. gamma rays emitted semi-continuously from long-lived beta decay where elements of the apparatus have been activated or radioactive beams have been scattered within the reaction chamber.

The timing resolution of germanium detectors is relatively poor compared to some fast scintillators (of order 1–10 ns depending on the size and geometry of the detector). Part of the reason for that can be seen in section 6.2.4 where pulse shapes clearly vary strongly as a function of the position of interaction within the crystal. Specifically, they vary in their rise time, where rise time is conventionally defined as the time for a pulse to rise from 10% to 90% of its full height. Traditional analogue electronics systems would define the time response using a timing-filter amplifier (TFA) and a constant fraction discriminator (CFD)—see section 10.2.5. Access to pulse-shape data allows various algorithms for defining event timing to be trialled. Figure 6.5 presents two coincidence systems—one analogue and one digital—for measuring germanium detector timing resolution with respect to a fast detector, in this case, a plastic scintillator detector (see section 9.1.2 for more on plastic scintillators) [13]. A ^{60}Co source is used for the measurement since it emits a cascade of two gamma rays of 1172 and 1332 keV in *prompt coincidence*. By prompt

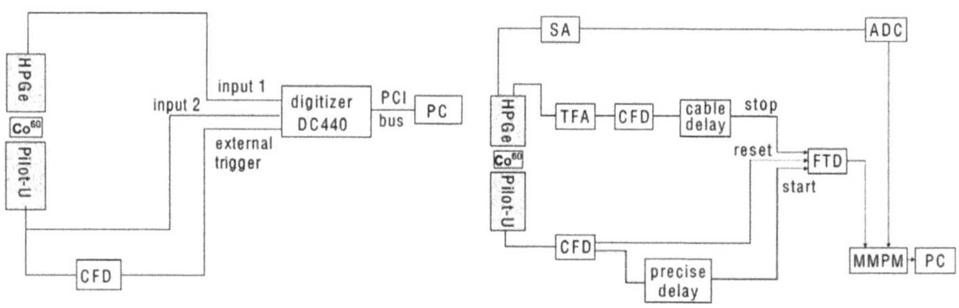

Figure 6.5. Two systems for measuring the timing resolution of germanium detectors with respect to a fast plastic scintillator detector (marked as Pilot-U) using a ^{60}Co source: (left) a digital acquisition system where pulse shapes are analysed. (right) An analogue system for timing resolution measurements. To define some acronyms: TFA = timing-filter amplifier, CFD = constant-fraction discriminator, ADC = analogue-to-digital converter, FTD = fast timing discriminator. (Reprinted from [13] with the permission of Elsevier.)

[3] It is perhaps debateable as to which class of detector is the ancillary but with the advent of large, expensive, germanium detector arrays much of the apparatus they are operated in conjunction with seems to be conventionally described as *ancillary*.

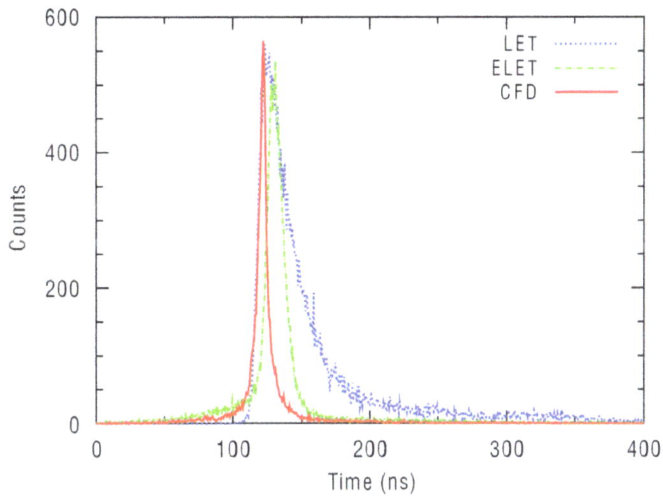

Figure 6.6. Timing resolutions measured for a large-volume HPGe detector with respect to a fast plastic scintillator using three different digital algorithms acting on the pulse shape data obtained: CFD (constant fraction discriminator), LET (leading-edge threshold) and ELET (extrapolated leading-edge threshold). In this example, the FWHM for the timing resolution varies from 6.2 ns to 18 ns depending on the strategy employed. (Reprinted from [13] with the permission of Elsevier.)

coincidence, we mean that the lifetime of the intermediate excited state in the cascade is so short (in fact, $t_{\frac{1}{2}} = 0.735$ ps) that it is tiny with respect to practical detector timing resolutions. Typical timing resolutions for a large volume germanium detector are shown in figure 6.6 where different digital algorithms were applied to the data obtained using the setup shown on the left-hand side of figure 6.5.

6.3.3 Peak-to-total

Peak-to-total is one of the key parameters of a germanium detector. It is defined as the ratio of the number of counts in the photopeak (full energy peak) to the total number of counts in the germanium detector spectrum. Peak-to-total is an energy-dependent parameter. Since the photopeak is generally the most useful part of the spectral information and a high Compton background will interfere with photopeaks of lower energies, a high peak-to-total is important in most applications driving the demand for a larger germanium crystal volume.

Peak-to-total will clearly scale with the dimensions of the germanium crystal since the larger the crystal, the higher the probability an incoming gamma ray will be fully absorbed either in one step through photoelectric absorption or in multiple steps associated with Compton scattering (see the parallel discussion of this effect for inorganic scintillator detectors in section 5.1.1). Figure 6.7 shows how the peak-to-total evolves as a function of gamma-ray energy as a function of germanium crystal volume. Clearly, peak-to-total is largest for low energies where photoelectric absorption is the dominant interaction.

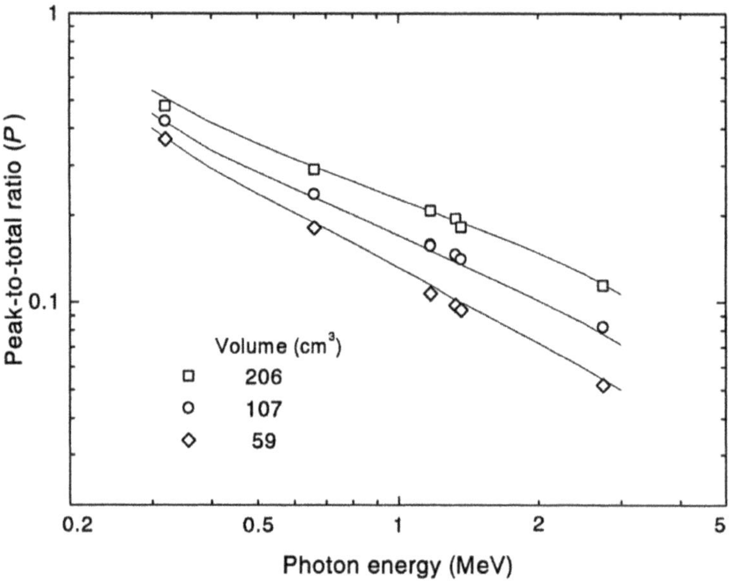

Figure 6.7. Peak-to-total as a function of gamma-ray energy for three germanium detectors with differing volumes (59 cm^3, 107 cm^3 and 206 cm^3). The curves are calculations for the peak-to-total. (Reprinted from [14] with the permission of Elsevier.)

6.3.4 Linearity/calibration

Modern germanium detectors are extremely linear in response but they need to be calibrated, i.e. relating the output signal in terms of ADC channel with gamma ray energies associated with a calibrated source where the energies are known to high precision (figure 6.8). Calibration is typically carried out with standard calibration sources such as ^{152}Eu or a combination e.g. ^{152}Eu/^{133}Ba. To extend to higher energies, a ^{56}Co source is useful as it emits gamma rays with energies above 3 MeV (see discussion of gamma-ray sources in section 3.1.3). To achieve a reliable calibration at higher energies is a somewhat specialist application and requires access to an accelerator to initiate nuclear reactions such as proton-capture reactions to generate high-energy gamma rays.

6.3.5 Efficiency

There are two aspects to efficiency of a germanium detector: the *intrinsic* efficiency normally quoted for an energy of 1.332 MeV corresponding to one of the two well-known gamma-ray decay lines of ^{60}Co, and the *functional form* of the efficiency as a function of energy. Historically, the efficiency of coaxial germanium detectors was given as a % relative to the efficiency of a 3″ × 3″ NaI(Tl) detector for detection of the 1.332 MeV gamma ray (from ^{60}Co decay) at a distance of 25 cm.[4] The largest germanium detectors produced have relative efficiencies exceeding 200% [15].

[4] This can cause confusion as the efficiency may exceed 100% but this is, of course, a relative efficiency.

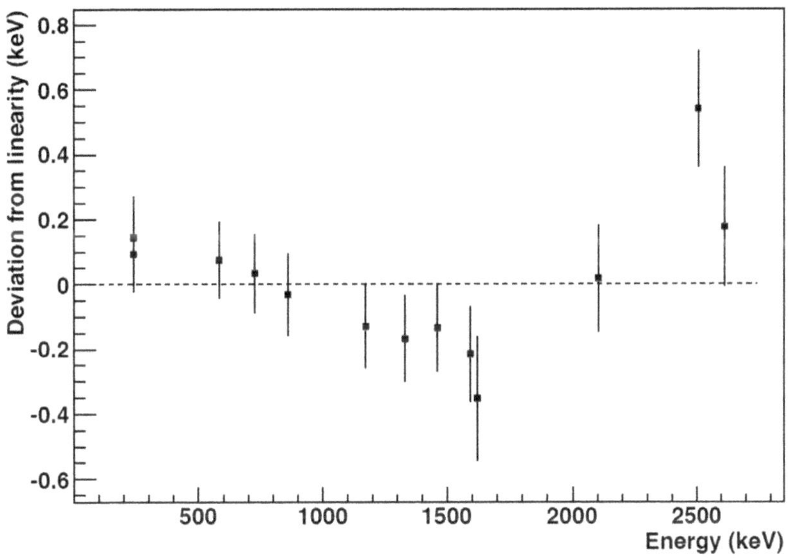

Figure 6.8. An example of measured deviations from linearity for a hyperpure germanium detector. (Reprinted from [11] with the permission of IOP Publishing.)

A characteristic functional form for the full-energy peak efficiency of a coaxial germanium detector as a function of energy is given by:

$$F(E) = aE^b \exp\left\{-\left(\frac{E}{c}\right)^d\right\}, \tag{6.6}$$

where a, b, c and d are constants. These constants are fitted to experimental data. The procedure is to obtain the photopeak area (e.g. fitting a suitable Gaussian or skewed Gaussian peak shape) for a wide range of gamma ray peaks from standard sources: typically this would include ^{152}Eu [16] and ^{133}Ba to cover a broad energy range. The relative intensity of these peaks is known to high precision, i.e. γ rays emitted per decay, and tabulated in nuclear data tables. If several different sources are used but placed at the same position for the measurement, then the results can be normalised together correcting for the present activity of the source based on its initial certified activity on a given date.

Fixing germanium detector efficiency with high precision, i.e. sub-percent, is a challenging undertaking and rarely undertaken in practice. Once it is desirable to enter this high precision regime then various effects must be corrected for, including the probability of *true summing*. True summing is where two coincident gamma rays in a decay cascade enter the detector at the same time, which has a non-negligible probability especially for sources where many gamma rays are emitted in coincidence, e.g. from a ^{152}Eu source [17, 18]. Hardy *et al* have pioneered high-precision efficiency calibration to support their high-precision decay measurements [19]; figure 6.9 provides an example of this work. To achieve such precision requires

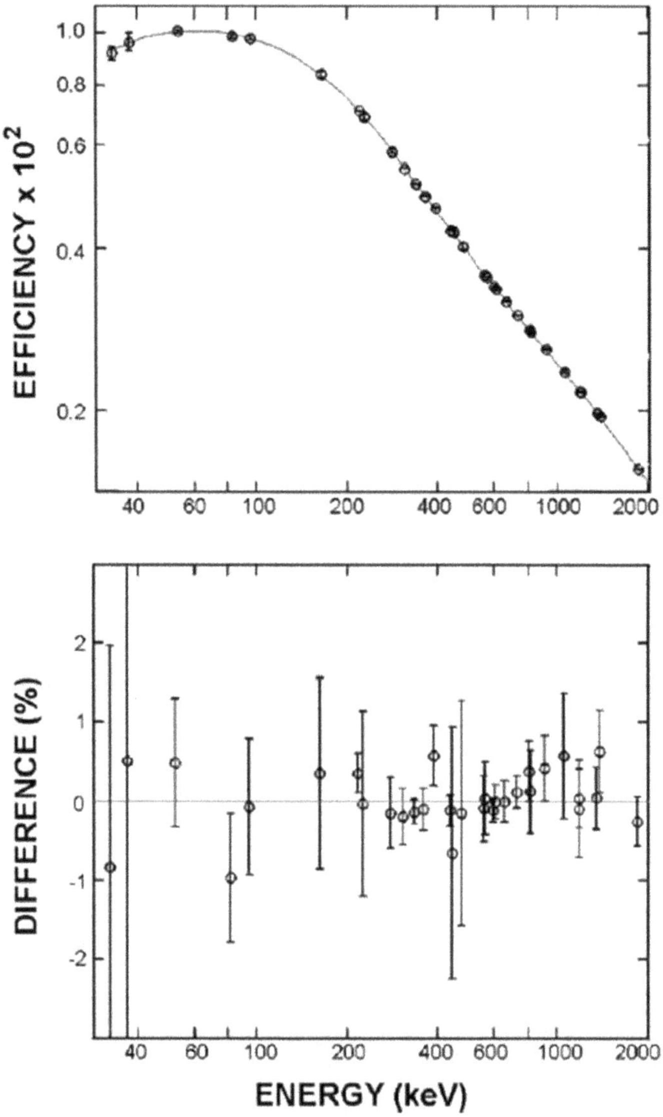

Figure 6.9. (Upper panel) precision efficiency curve for a coaxial HPGe detector with an active volume of 280 cm^3 as a function of gamma-ray energy (in keV). (Lower panel) the percentage deviation from Monte Carlo simulations. (Reprinted from [19] with the permission of Elsevier.)

high quality nuclear data for the calibration as well as detailed detector dimensions for the supporting Monte Carlo simulation.

6.3.6 Neutron damage

During nuclear physics experiments but potentially also some industrial applications, fast neutrons may be present. Fast neutrons are damaging to germanium detectors in a number of ways: neutrons can disrupt the crystal structure of the detectors by displacing germanium atoms from their position. If the neutrons are sufficiently energetic, they can also initiate nuclear reactions with germanium, e.g. (n,2n), leading to transmutation of the detector material. The presence of a fast neutron field can be readily observed in a germanium detector spectrum from the presence of peaks associated with inelastic excitation of the various germanium isotopes e.g. ^{72}Ge (n, $n'\gamma$) (see figure 6.10). Such peaks have an unusual shape with a high-energy tail (see figure 6.10). This is because, depending on the individual kinematics, the germanium nucleus recoils when undergoing scattering by a fast neutron. The recoiling germanium nucleus also creates an energy signal in the germanium detector of order 10s of keV and this is summed with the gamma-ray emitted from the germanium isotope's excited state as the two events are effectively simultaneous.

The effect of neutron damage is to degrade the performance of the detectors, notably, introducing low-energy tailing on the observed photopeaks. Such damage can be largely reversed by annealing the detectors, i.e. raising the temperature to a point where the crystals anneal, i.e. resetting the crystal structure [20]. Typical

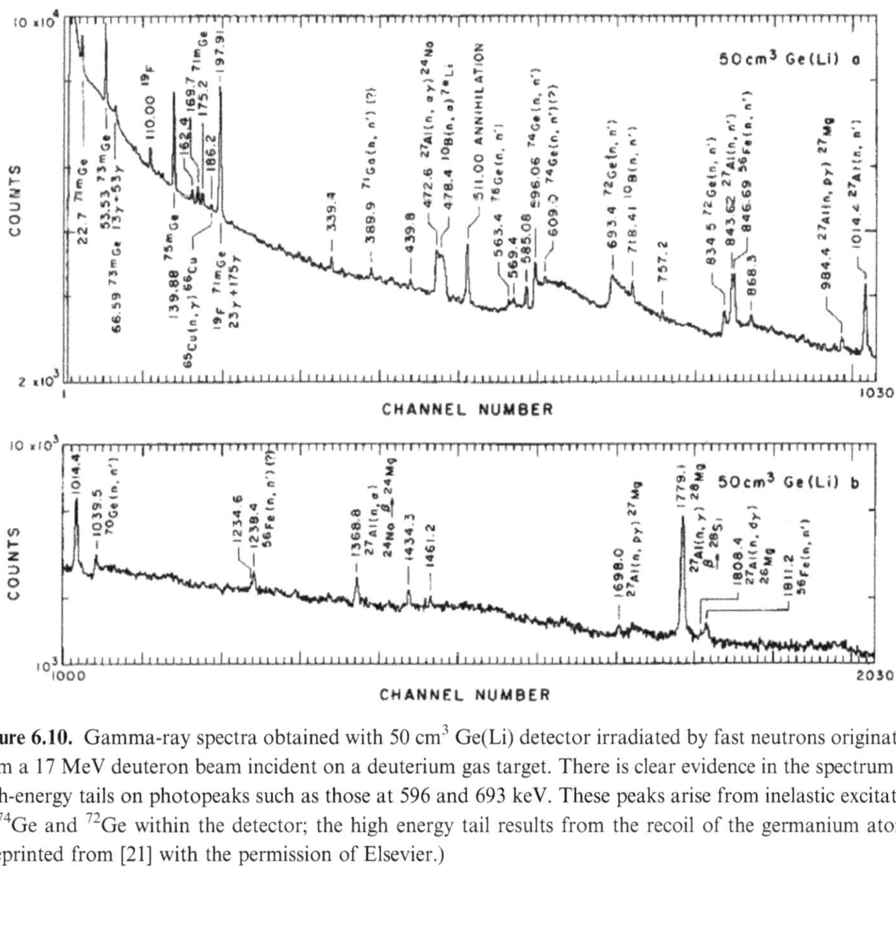

Figure 6.10. Gamma-ray spectra obtained with 50 cm^3 Ge(Li) detector irradiated by fast neutrons originating from a 17 MeV deuteron beam incident on a deuterium gas target. There is clear evidence in the spectrum for high-energy tails on photopeaks such as those at 596 and 693 keV. These peaks arise from inelastic excitation of ^{74}Ge and ^{72}Ge within the detector; the high energy tail results from the recoil of the germanium atoms. (Reprinted from [21] with the permission of Elsevier.)

annealing cycles take a few hours at temperatures around 100 °C. The time between annealing will reduce as a function of increasing fast neutron fluence. Depending on the type of experiments to be conducted with an array, the effects of neutron damage can therefore impose limitations in scheduling an experimental campaign. It is known that, in general, *n-type* germanium detectors are less affected by radiation damage than *p-type* detectors which largely relates to the direction in which the charge carriers move within the crystal and the fact that hole trapping is more important than electron trapping. Moreover, the annealing cycle for *n-type* germanium detectors is often significantly shorter.

6.4 Principal classes of germanium detector

6.4.1 Types of hyperpure germanium detector

Hyperpure germanium detectors are available in several broad classes each frequently seen with different levels of electrical segmentation. They exhibit a range of different shapes from planar to coaxial, and a range of more sophisticated geometries (see figure 6.11). Let us explore some of the more common geometries:

- *Coaxial* detectors are the most commonly found type of detector (see figure 6.12 for the geometry of a typical n-type coaxial detector). This cylindrical geometry typically has a low capacitance leading to a good signal-to-noise and hence, better energy resolution. However, the electric field can be less uniform in this geometry, particularly at the closed end and this can degrade both timing resolution as charge carriers need to travel further, and energy resolution since the charge carriers may encounter more trapping centres on their path to the electrode.
- *Planar* detectors are named after their geometry, i.e. typically, a flat slab of germanium, which may or may not be segmented into strips. Such detectors have also been used in vacuum to detect beta particles [22]. Given the geometry, it is relatively straightforward to have a uniform electric field across a planar detector as opposed to the difficulties discussed above for coaxial detectors. However, limitations on the high voltage (up to 5 kV)

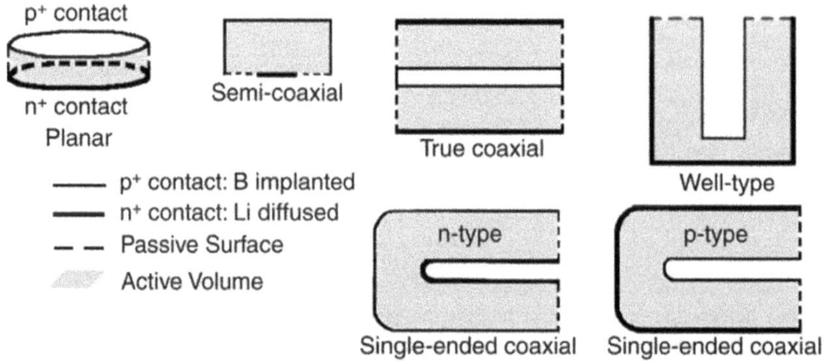

Figure 6.11. Typical shapes of germanium detectors. (Reprinted from [27] with the permission of Elsevier.)

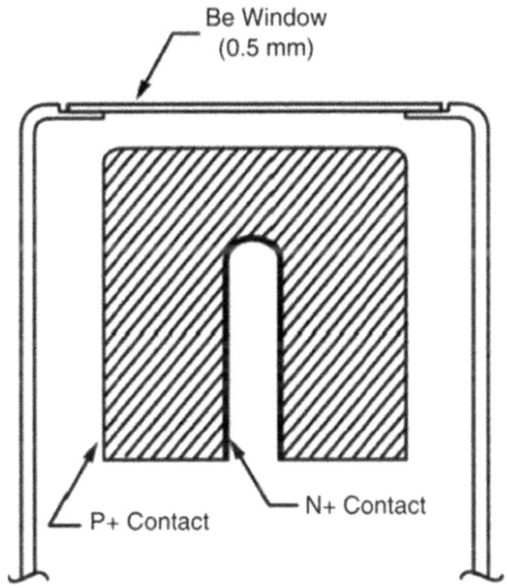

Figure 6.12. Layout of a typical n-type coaxial germanium detector. (Reprinted from [27] with the permission of Elsevier.)

which can be applied due to high leakage currents and sparking impose a practical limitation on planar detector thicknesses of around 2 cm.

- *BEGe* detectors are a relatively new design of germanium detector intended for the energy range 3 keV to 3 MeV. BEGe detectors have a specially designed electrode structure which leads to improved energy resolution at energies below 100 keV [7, 23, 24]. Figure 6.13 illustrates the electrode design: the *p-type* germanium crystal is surrounded by lithium-diffused n^+ contacts on all faces apart from a small region at the centre of the bottom of the detector where there is a boron-implanted p^+ contact.
- *Clover* detectors are so-called because they have four crystals encapsulated together. An example of a clover detector design is the GRIFFIN clover used extensively in beta-decay studies at TRIUMF is shown in figure 6.14. Spectra can be obtained using the crystals individually or the energy recorded in individual crystals can be combined in a process called *add back* which effectively takes account of the fact that an individual gamma ray is likely to Compton scatter and deposit energy in multiple crystals or in pair production, annihilation photons may escape the initial crystal in which the interaction took place. A variety of strategies for carrying out the add-back procedure have been discussed [25, 26].

An energy-dependent add-back factor can be defined as

$$F(E_\gamma) = \frac{N_{AB}}{N}, \tag{6.7}$$

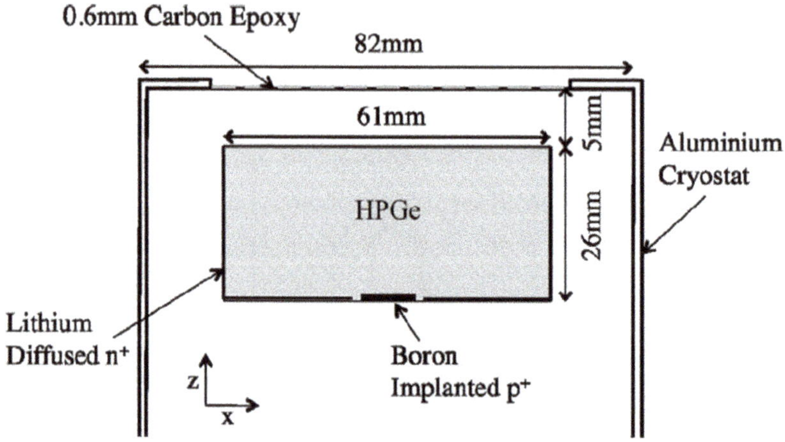

Figure 6.13. Interior arrangment of a BE2825 BEGe detector. The *p-type* germanium crystal is a cylinder with cross-sectional area of 28 cm^2 and thickness of 26 mm. The crystal is surrounded on all faces by a lithium-drifted n$^+$ contact with the exception of small area in the centre of the bottom of the crystal where there is a boron-implanted p$^+$ contact. The entrance window is made from a thickness of 0.6 mm of carbon epoxy which allows transmission of gamma rays down to 10 keV. (Reprinted from [7] with the permission of Elsevier.)

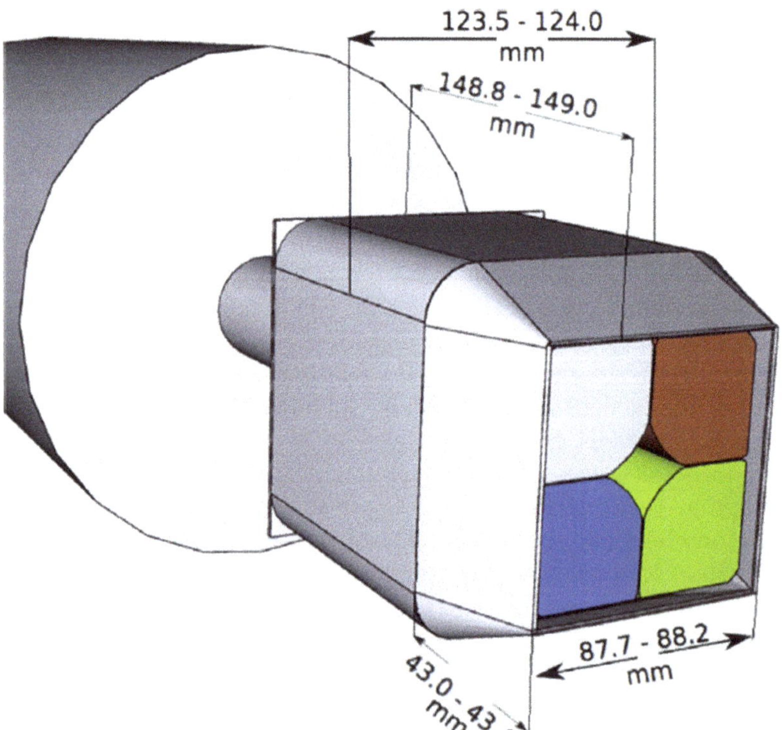

Figure 6.14. 3D model of GRIFFIN clover detector. (Reprinted from [28] with the permission of Elsevier.)

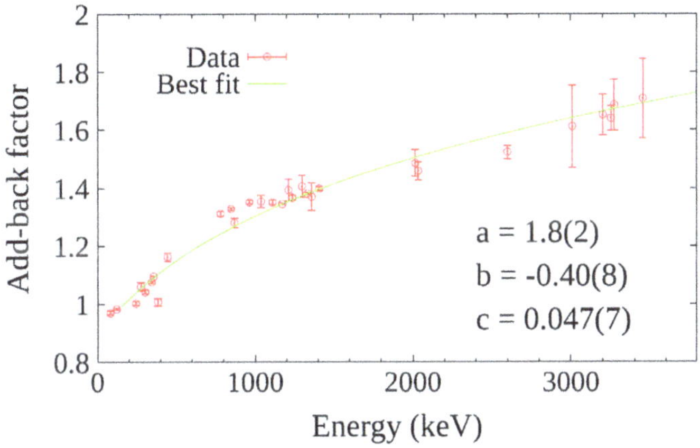

Figure 6.15. Clover add back factor as a function of energy for a GRIFFIN clover detector. (Reprinted from [28] with the permission of Elsevier.)

where N_{AB} is the number of counts after add-back and N is the number of counts before add back. The energy-dependent add-back factor is plotted for a GRIFFIN clover detector in figure 6.15. Clearly, there is limited benefit in add-back at low energies where photoelectric absorption dominates but as Compton scattering and latterly pair production become important with increasing energy, the improvement from add-back is considerable.

The geometry of the clover detector is a natural one for linear polarisation measurements where the polarisation-dependence of the Klein–Nishina formula (see equations (2.23) and (2.24)) can be exploited to infer the electric or magnetic character of gamma-ray transitions. Figure 6.16 illustrates how the difference between parallel and perpendicular scattering between adjacent crystals within a clover detector can be used to determine electric or magnetic character.

6.4.2 Germanium detector arrays

In the field of experimental nuclear physics, germanium detectors are frequently deployed in arrays comprising multiple individual detectors—maybe as many as 100 operated in conjunction [30]. The most obvious array design is to surround the source of gamma-ray emission with a spherical array with the individual germanium crystal facing towards the centre (see discussion of array geometries in section 4.1.5). This is often described as a 4π spectrometer where 4π refers to the solid-angle coverage. Ideally, the detectors are designed so as to be able to be close packed. A good example of a close-to 4π array is Gammasphere, comprising an array of up to 100 high-purity germanium detectors (see figure 6.17), presently situated at Argonne National Laboratory. The advantage of such a spherical spectrometer with identical detectors is that it simplifies angular distribution/angular correlation analysis since the efficiency and solid angle coverage of the detectors is uniform. However, other arrangements of germanium detectors have been employed, e.g. hybrids of different

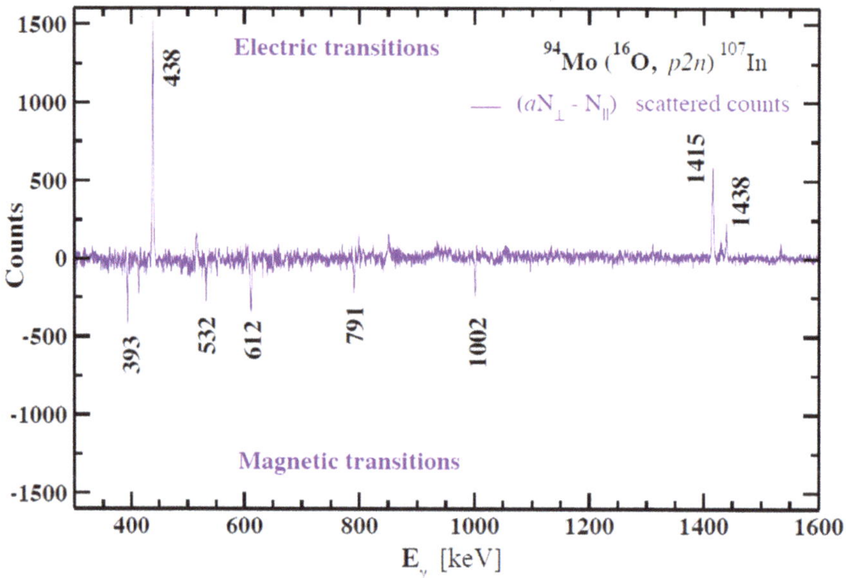

Figure 6.16. Difference spectrum of perpendicular and parallel Compton scattering in an INGA clover detector. In such a spectrum, electric transitions have positive counts and magnetic transitions have negative counts. The gamma rays are from excited states of ^{107}In, populated in the ^{94}Mo(^{16}O,p2n) reaction. (Reprinted from [29] with the permission of IOP Publishing under Creative Commons License.)

Figure 6.17. The Gammasphere array of 100 high-purity germanium detectors situated at Argonne National Laboratory. (Photo credit: the author.)

types of detector where sufficient quantity of identical detectors was not available. This has the benefit of enhanced coincidence efficiency but in cases where the detectors are not the same, angular correlation analysis etc, may prove more cumbersome. Some further examples of germanium detector arrays include TIGRESS [31] based at the TRIUMF laboratory in Vancouver, MINIBALL [12] principally based in the ISOLDE facility at CERN, and INGA [29]—the Indian National Gamma-ray Array which moves between different accelerator facilities in India.

The improvements in experimental sensitivity accompanied by the evolution of germanium detector arrays are illustrated in figure 6.18 with respect to successive studies of excited states of ^{156}Dy carried out over the period 1963–98. The limit in terms of relative intensity of the gamma rays which can be observed improved by four orders of magnitude over that time. Improvements of this kind drove the field of *high-spin spectroscopy*.

In a high-spin spectroscopy experiment, the multiplicity of gamma rays emitted per event, M_γ, may be as high as 25–35. With a 4π detector array, there is significant probability to detect many of these individual gamma-rays in prompt time coincidence. Each event has a corresponding coincidence *fold*, f, corresponding to how many gamma rays are detected. These events can be subsequently unpacked, e.g. a single event with $f = 4$ contains four γ-γ-γ coincidence sub-events ($f = 3$) and six γ–γ sub-events ($f = 2$) [30]. The availability of high-fold coincidence data allows the possibility of placing multiple, or high-fold, selection criteria (known as *gates*) on γ-ray transitions in order to enhance specific features of interest. As higher-fold coincidences are specified, the level of statistics in the resulting coincidence spectrum reduces but the peak-to-background is enhanced. An example of this effect is shown in figure 6.19; it is evident that as the analysis is continued to very high fold, more and more spectral features emerge. This is the true power of such germanium detector arrays for high spin spectroscopy.

It is useful to define an observational limit, α_0 for such high-fold coincidence measurements to understand the limits in spectral improvement where α_0 is the minimum intensity of a gamma ray which can be detected. This is conventionally done in an empirical manner in terms of the *resolving power* of an array, R, which is defined as [27, 30]:

$$R = 0.76\frac{SE_\gamma}{\Delta E_\gamma}P/T. \tag{6.8}$$

where SE_γ is the average separation of γ-ray peaks in a coincident cascade which depends on the nuclear structure properties of the system under study, ΔE_γ is the FWHM of a gamma-ray peak, and P/T is the peak-to-total of the individual germanium detectors. In other words, the larger the peak-to-total and the better the energy resolution, the higher the resolving power.

For a spectrum of fold, f, the peak-to-background ratio, i.e. number of counts in the photopeak divided by the counts in the background underneath the peak, can be expressed in terms of the observation limit, α_0, and resolving power, R, as

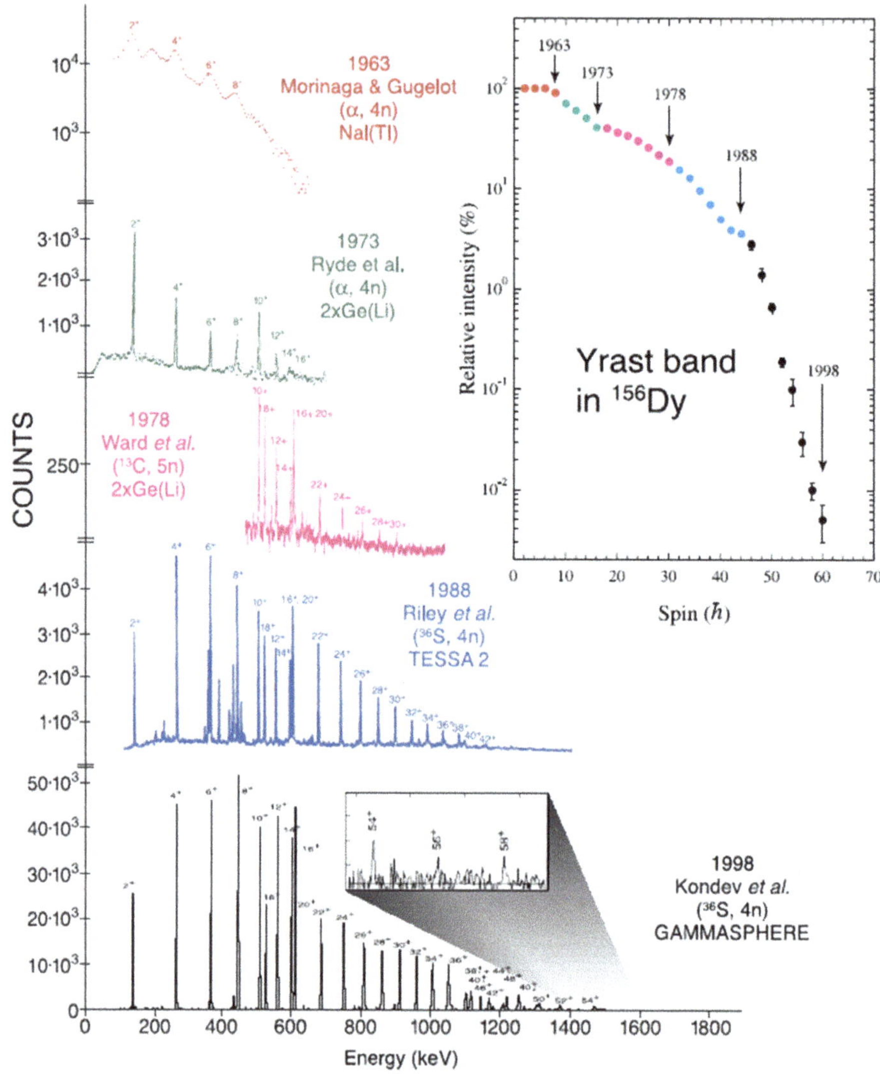

Figure 6.18. Evolution in experimental sensitivity in gamma-ray spectroscopy. The figure shows the improvement in spectroscopic data for studies of the nucleus, ^{156}Dy. The five different spectra published from 1963 (top) to 1998 (bottom) are each labelled with the reaction populating the nucleus and the detectors used. The projectile used: ^{4}He, ^{13}C or ^{36}Ar is significant since heavier ion beams can impart more angular momentum to the compound system leading to the population of high spin states and hence a higher multiplicity cascade of emitted gamma rays. The earliest spectrum (top) is taken with sodium iodide detectors. The improvement in resolution in the next two spectra from 1973 and 1978 which use a pair of Ge(Li) detectors is very striking. In the bottom two spectra, a germanium detector array, TESSA 2 or Gammasphere were used. The level of statistics and spectral quality is markedly improved. The inset shows a different way of presenting the improvement in detector performance in terms of the relative intensity of the gamma rays in the spectrum as a function of spin. The advent of Gammasphere is seen to produce a step-change improvement of two orders of magnitude in this relative intensity scale. (Reprinted from [32] with the permission of IOP Publishing under Creative Commons License.)

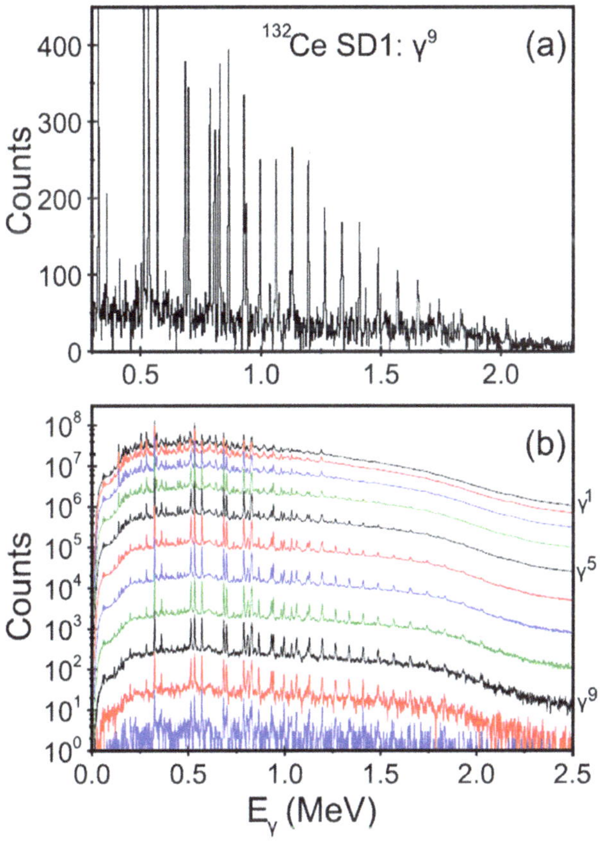

Figure 6.19. (a) Gamma-ray spectrum in coincidence with eight gating transitions in a 'highly-deformed' band in ^{132}Ce. (b) Various spectra for the band as the coincidence fold, f, for gamma rays is increased from 1 to 11. (Reprinted from [32] with the permission of IOP Publishing under Creative Commons License.)

$$\left(\frac{N_p}{N_b}\right)_f = \alpha_0(0.76R)^f, \tag{6.9}$$

where $\left(\frac{N_p}{N_b}\right)_f$ is the final peak-to-background ratio in the spectrum obtained when $f - 1$ selection criteria have been applied. Equation (6.9) shows how strongly peak-to-background improves as a function of the coincidence fold, f, in which the data is analysed.

6.5 Improving germanium detector performance

6.5.1 Compton suppression

Gamma rays interact discontinuously with matter by one of the three main modes discussed in section 2.4: photoelectric absorption, Compton scattering and pair production. As a function of increasing energy, Compton scattering becomes increasingly important and pair production becomes possible above 1 MeV. In these two processes, there is some probability the scattered photon or the annihilation photons following pair production escape the germanium detector. In this case, the energy recorded in the germanium detector will not reflect the full energy of the initial gamma ray. Such incomplete events will add to the experimental background and may prove bothersome when examining γ–γ or higher-fold coincidences with germanium detector arrays. A means of mitigating this effect is to surround the back and sides of the germanium detector with a Compton-suppression shield. Typically, such a shield is formed from scintillator crystal. The shield serves solely as a veto for scattered or escaping gamma rays. An event-by-event suppression of incomplete events in the germanium detector can then be achieved which strongly improves the peak-to-total. Since the shield is merely a veto, the scintillator crystal used does not necessarily need a high energy resolution but does need to have a high efficiency for detecting scattered gamma rays. Bismuth germanate (BGO) (see section 5.1.2) is conventionally used in building Compton-suppression shields (although alternate materials such as CsI(Tl) have also been employed [33]). BGO has relatively poor energy resolution, partly due to low light emission but also due to its high refractive index which hinders optical coupling. It does, however, have high stopping power (due to the high Z of its component elements) and is not hygroscopic which makes it easy to handle. The high stopping power of BGO makes it possible to minimise the width of the Compton suppression shields allowing neighbouring germanium detectors to be closely packed. For example, the Gammasphere array described in section 6.4.2, makes excellent use of BGO suppression to improve its peak-to-total [34]. Figure 6.20 illustrates the layout of BGO suppression shields around the Gammasphere germanium detectors. The advantages of this arrangement of suppression shields are shown in figure 6.21. A shield made of a dense, high-Z material such as 'hevimet' alloy[5] is often placed as a collar on the front of a germanium detector to prevent direct illumination of the BGO shield. A more modern design of BGO suppression shields is shown in figure 6.22 where BGO shields have been retrofitted to MINIBALL triple-cluster germanium detectors.

6.5.2 Gamma-ray tracking

The inclusion of Compton-suppression shields in an array such as Gammasphere strongly improves its spectroscopic performance. However, the BGO shields comprise around half the available solid angle and so limit the efficiency of such an array. This effect becomes compounded when there are a high-multiplicity of gamma rays to be detected. The ultimate performance for a germanium detector array can be achieved by dispensing with the Compton-suppression shields entirely and having a contiguous arrangement of germanium detectors. Gamma-ray interactions can then be reconstructed fully within the germanium crystal using *gamma-ray tracking* (see figure 6.23 for an illustration of the principles behind this). Gamma-ray tracking is based on pulse shape analysis which can establish the position of interaction and energy deposited on

[5] Hevimet is typically comprised of 90% tungsten, 4% copper and 6% nickel.

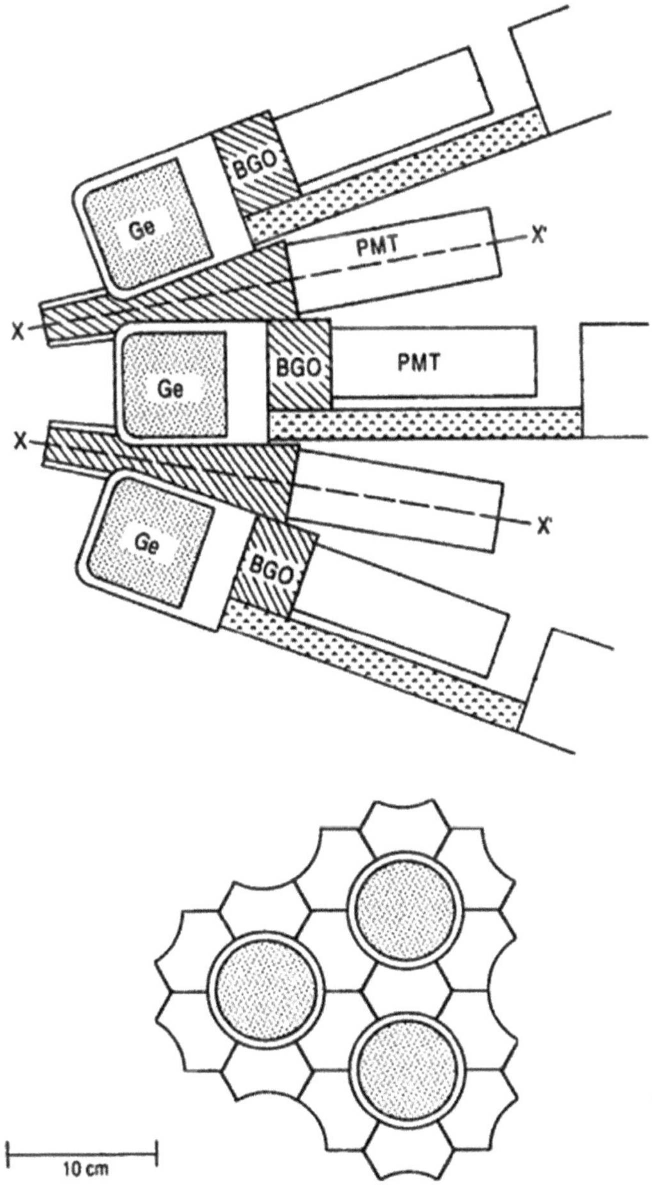

Figure 6.20. Schematic of the Gammasphere high-purity germanium detectors with a honeycomb arrangement of BGO suppression shields and a BGO back plug. (Reprinted from [34] with the permission of Elsevier.)

an event-by-event basis. Pulse shape analysis was already outlined in relation to an individual detector in section 6.2.4 but with a gamma-ray tracking array, it is extended to the array as a whole since the interactions of individual gamma rays by, e.g. Compton scattering, may result in interactions in multiple individual detectors. Significant electronics and computational infrastructure is needed to realise the gamma-ray tracking concept backed up by extensive GEANT4 simulation to fully

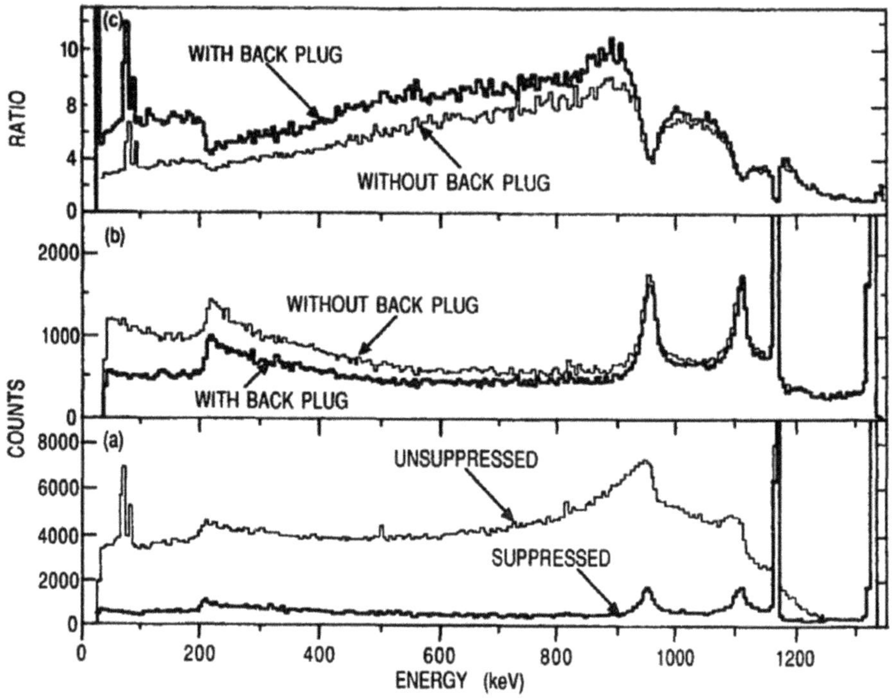

Figure 6.21. The improvements in performance of Gammasphere operated in conjunction its BGO suppression shield: (a) comparison of the unsuppressed and suppressed (vetoed by BGO shield) germanium detector spectrum for a ^{60}Co source. The extent to which the Compton background is reduced is evident. (b) The detailed improvement on including the BGO back plug for Compton suppressions. (c) The ratio of suppression with and without the back plug. (Reprinted from [34] with the permission of Elsevier.)

model the physics processes entailed. Gamma-ray tracking may identify the point of first interaction of a gamma ray to a precision of around two millimetres in x and y and z. This position localisation is of limited benefit for gamma rays emitted from a stationary source but achieves high significance in the case where the gamma rays are emitted from a nucleus moving at relativistic velocities. The position of first interaction can be used in the Doppler correction to restore high energy resolution. Moreover, the ability to reconstruct the details of the Compton scattering allows access to the polarisation of gamma rays (see discussion in section 2.4.2) which is a means of establishing their multipolarity, e.g. distinguishing E1 and M1 transitions.

Following well over a decade of development effort, two gamma-ray tracking arrays have been realised worldwide: one in the US called GRETINA [36]—an acronym for Gamma-Ray Energy Tracking In-Beam Nuclear Array—and a European project called AGATA [37]. Here, we focus the discussion on the first example, GRETINA. The geometry of GRETINA is conceived in terms of a geodesic dome such as famously employed by the architect, Buckminster Fuller. With such a geometry, it is possible to tile a sphere using a sequence of irregularly shaped hexagons and pentagons [38]. For the given radius of the sphere, the size of the individual GRETINA modules is chosen to be as large as possible to limit dead

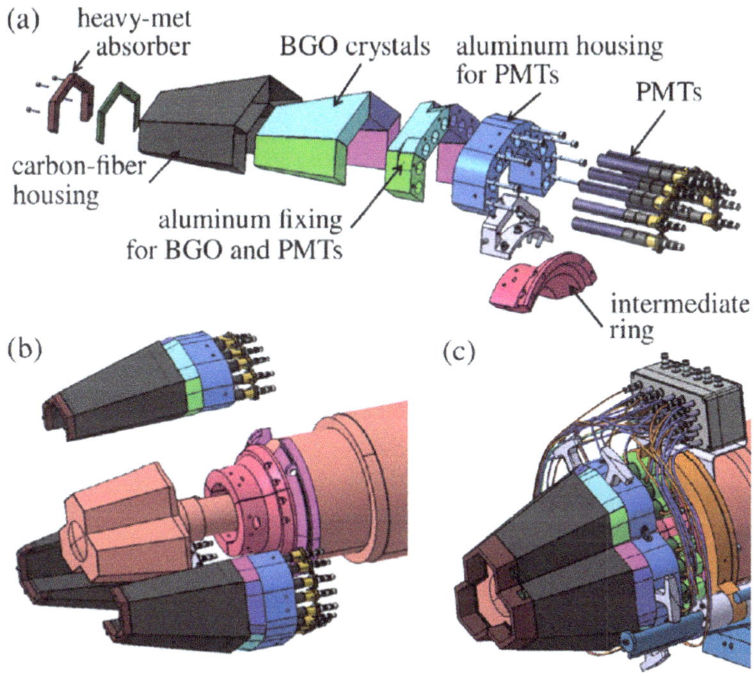

Figure 6.22. Design of BGO shields for MINIBALL triple-cluster germanium detectors. The shields are formed in three parts which bolt together. Each third contains four differently-shaped BGO crystals. Scintillation light collection is achieved with an array of small PMTs. (Reprinted from [35] with the permission of Elsevier.)

space between individual detectors. However, given that in a practical experiment, access is needed for beampipes etc, the full array design leaves the pentagonal elements empty and comprises only the hexagonal shapes. The elegance of such a design is evident in figure 6.24 which shows an example of five GRETINA modules under test. Each GRETINA module is made up of four individual germanium crystals with an irregular hexagonal shape: two of type-A and two of type-B. The arrangement is shown in figure 6.25. The front faces are tapered to conform to the spherical geometry. Each crystal is then effectively subdivided through segmentation of the electrical outer contact. Figure 6.25 shows how this segmentation is applied, with the hexagonal front face subdivided into six equal-sized triangles and the crystal segmented into rings in depth also into six but with variable depth of the individual segments (8, 14, 16, 18 and 20 mm deep, respectively, for rings starting from the front face). Each of the segments provides an individual signal, which taken together with the central contact signal leads to a total of 37 signals per germanium crystal. The average resolution in position for the point of first interaction in a GRETINA detector is 1.9 mm [38].

An example of the improvement of gamma-ray tracking detectors over conventional detector arrays is illustrated figure 6.26 where the performance of GRETINA

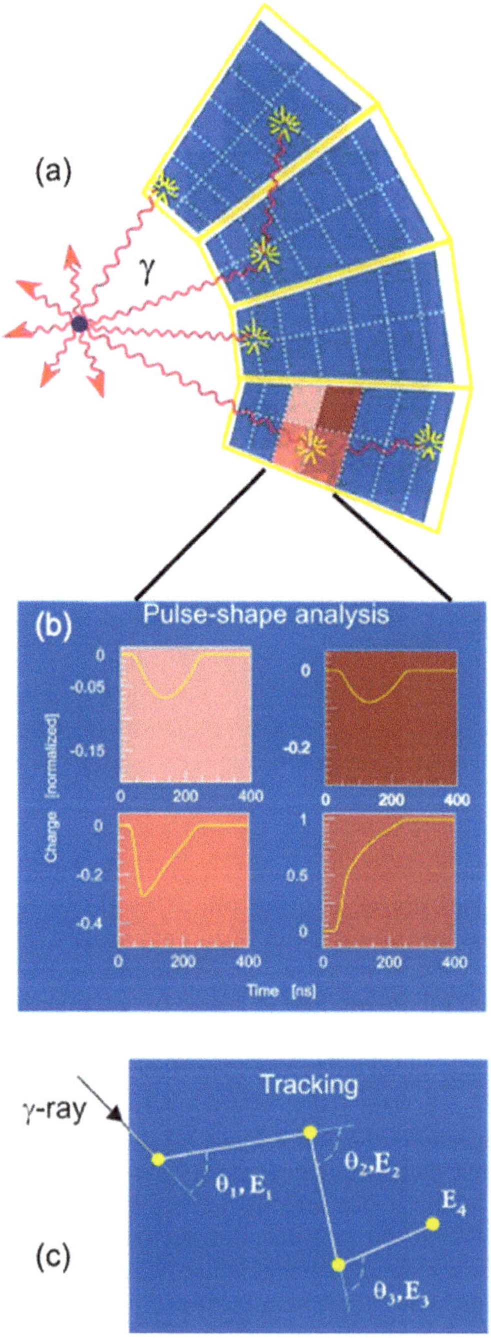

Figure 6.23. Principles of gamma-ray tracking: (a) a detector array is made from a contiguous shell of high-purity germanium detectors each of which is segmented. (b) Pulse-shape analysis on signals recorded in different segments allows the point of interaction to be obtained to a high precision in x and y and z. (c) from the energies recorded and the inferred Compton scattering angles, the initial decay energy and incident angle can be obtained. (Reprinted from [32] with the permission of IOP Publishing under Creative Commons License.)

Figure 6.24. A photograph of the GRETINA gamma-ray tracking array comprising here, five GRETINA modules totalling 20 individual germanium crystals. The close packing of the germanium crystals and their elegant geometrical shape is evident from the photograph. (Reprinted from [38] with the permission of Elsevier.)

is compared for the same experimental study with that obtained using the earlier SeGA array at Michigan State University (MSU).

Gamma-ray tracking detectors and their associated infrastructure are extremely expensive, meaning that neither of the two such systems worldwide—GRETINA and AGATA—has so far been able to advance to a full 4π solid angle configuration. Achieving a 3π or 4π geometry, would, in the future, show spectacular improvement in spectroscopic performance and efficiency [40]. Figure 6.27 shows how replacing an existing NaI(Tl)-based scintillator array with a 4π AGATA would provide a dramatic step-change in spectral quality. The right-hand side of figure 6.27 also shows the possibilities for studying the lifetimes of excited states in relativistic Coulomb excitation where the gamma-ray peak exhibits a lineshape related to the decaying level lifetime.

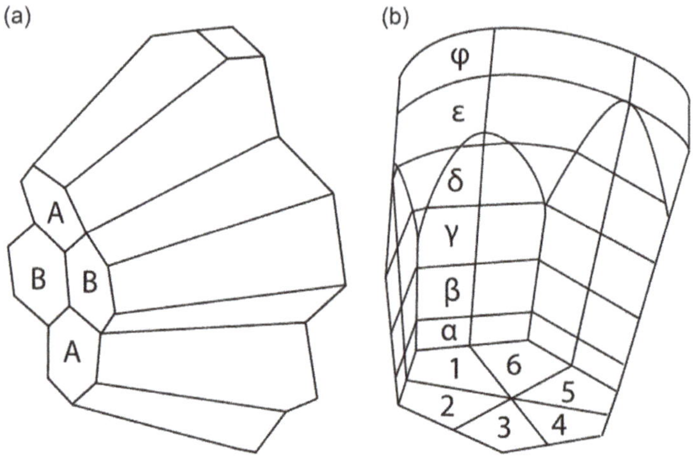

Figure 6.25. A diagram of the internal structure of a GRETINA module. The left-hand diagram illustrates how each module comprises four crystals of two slightly different designs: type-A and type-B. The right-hand diagram shows how the detector is effectively subdivided in length and depth through electrical segmentation of the outer contacts. (Reprinted from [38] with the permission of Elsevier.)

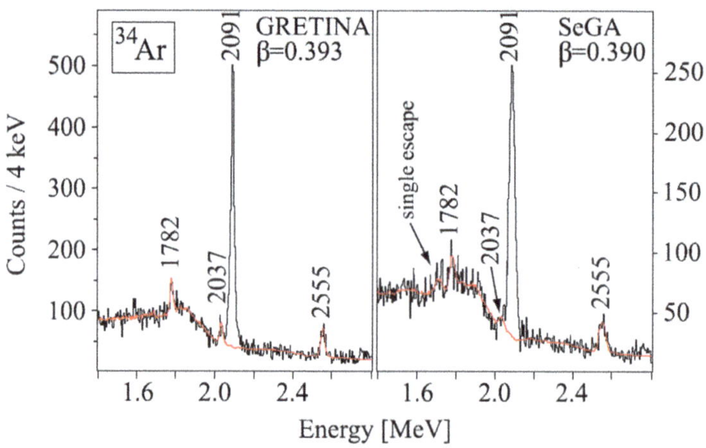

Figure 6.26. A comparison of a gamma-ray spectrum associated with decay of excited states in a radioactive beam of ^{34}Ar travelling at around 39% of the speed of light with the gamma-ray tracking array, GRETINA (left) and an earlier conventional germanium detector array, SeGA (right). The improvement in energy resolution is clear and some gamma-ray peaks which were obscured in the SeGA spectrum are now evident in the GRETINA spectrum. (Reprinted from [39] with the permission of Elsevier.)

6.6 Room temperature semiconductor detectors for gamma rays

For completeness, it is worth mentioning that in addition to germanium which operates cooled to liquid nitrogen temperatures, there are some semiconductors suitable for gamma-ray detection which operate at room temperature (see reference [2] for a recent review. These include semiconductor compounds such as cadmium telluride (CdTe), cadmium zinc telluride (CZT) and thallium bromide (TlBr) [3].

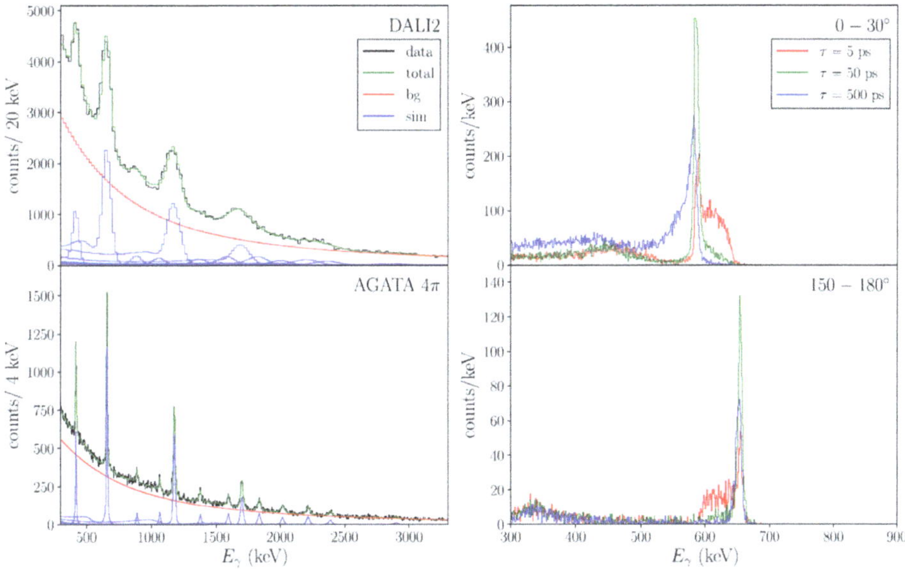

Figure 6.27. (Top left) experimental γ-ray spectrum of ^{135}Te measured with the DALI2 spectrometer (a large array of NaI(Tl) detectors) following the one-neutron knockout from ^{136}Te projectiles at 165 MeV/u on a carbon target at RIKEN. (Bottom left) simulated spectrum assuming the same reaction parameters and background shape, but using AGATA for γ-ray detection. (Right) simulated line shapes of the 618-keV $2_1^+ \to 0_1^+$ transition in ^{132}Cd emitted in forward (top) and backward (bottom) direction following relativistic Coulomb excitation at an energy of 150 MeV/u on a gold target. Lifetimes of 5 ps (red), 50 ps (green), and 500 ps (blue) were assumed for the 2_1^+ state. (Reprinted from [40] with the permission of Elsevier under Creative Commons License.)

Such detectors are not widely employed in experimental nuclear physics but they are finding widespread application in areas such as homeland security [2], nuclear decommissioning and medical imaging [41, 42]. In principle, they are very good gamma-ray detectors from the point of view of absorption as they comprise much higher Z materials, i.e. Cd ($Z = 48$) and Te ($Z = 52$) compared to Ge ($Z = 32$); recall that the probability of photoelectric absorption scales approximately as Z^5 (see section 2.4.1). However, the available CdTe and CZT detectors are strongly limited in size, e.g. a 10 mm cube is a typical crystal dimension for a CZT detector. Moreover, materials such as CZT suffer strongly from the effects of charge trapping, notably poor hole mobility (see figure 6.28). This can impact both on obtainable energy resolution and lead to low energy tails on photopeaks (see example CZT spectrum in figure 6.1). Figure 6.29 illustrates the effects of the charge trapping in a CZT detector which can lead to varying energy resolution/spectral quality depending on the point of interaction in the crystal [42]. Here, the pulse height spectra are seen to depend on the measured drift time of the charge carriers. Fortunately, there are solutions available to address these issues such as co-planar grid technology [43, 44]. Here, the anode is pixellated in x and y and measured drift time effectively pixellates the detector in z. The events can then be corrected on a pixel-by-pixel basis to correct out a lot of the issues to do with poor charge mobility. Figure 6.30 shows the benefits of such a correction to spectral quality and observed energy resolution.

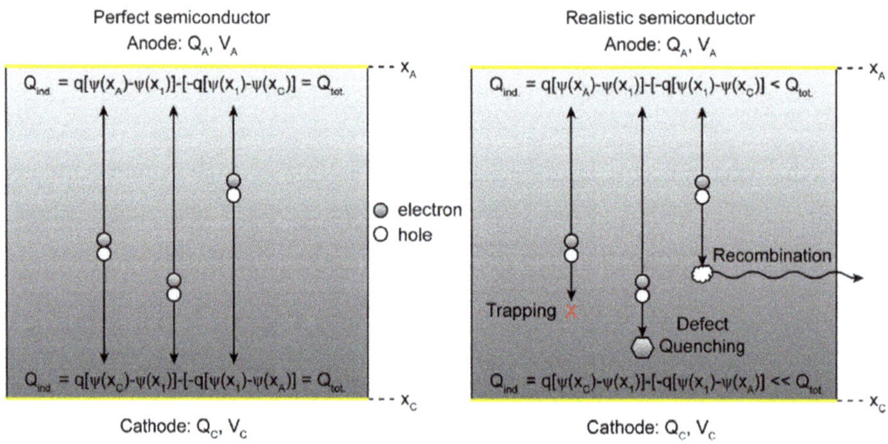

Figure 6.28. As electrons and holes transit throughout a semiconductor, they induce charge on the collecting electrode proportional to the difference in weighting potential between the site of their generation and the site of their termination. In a perfect semiconductor (left), the amount of induced charge is equal to the total amount generated in ionisation. In a realistic semiconductor (right), effects that quench the charge before it reaches the electrode diminish the total induced charge from the interaction. In the example shown, holes are lost through the processes of trapping, defect quenching and recombination. These effects are particularly severe in CZT which has a very low hole mobility. (Reprinted from [2] with the permission of AIP Publishing.)

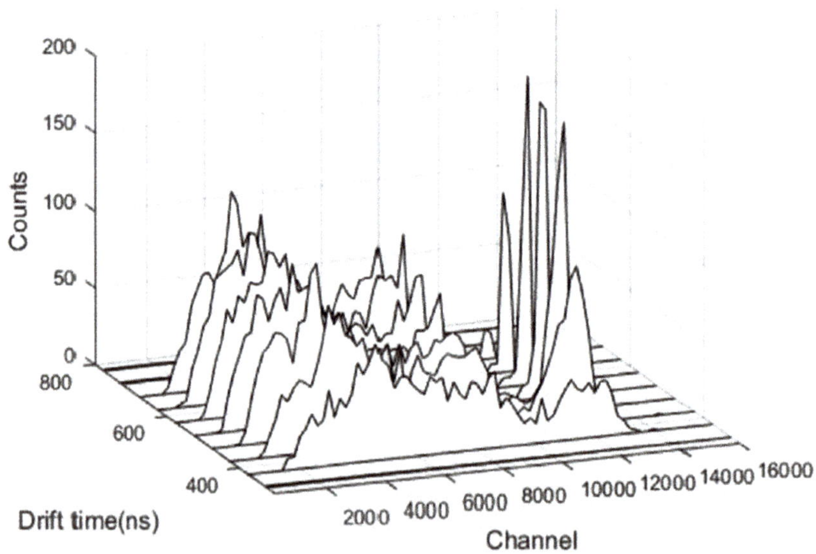

Figure 6.29. Pulse height spectra from a ^{137}Cs source as a function of drift time, i.e. time difference between anode and cathode signal, recorded with a 20 × 20 × 5 mm CZT detector. The detector has an 8 × 8 pixellated anode and a common cathode. The observed variation in spectra is due to the poor charge carrier mobility in CZT. (Reprinted from [42] with the permission from the Korean Nuclear Society.)

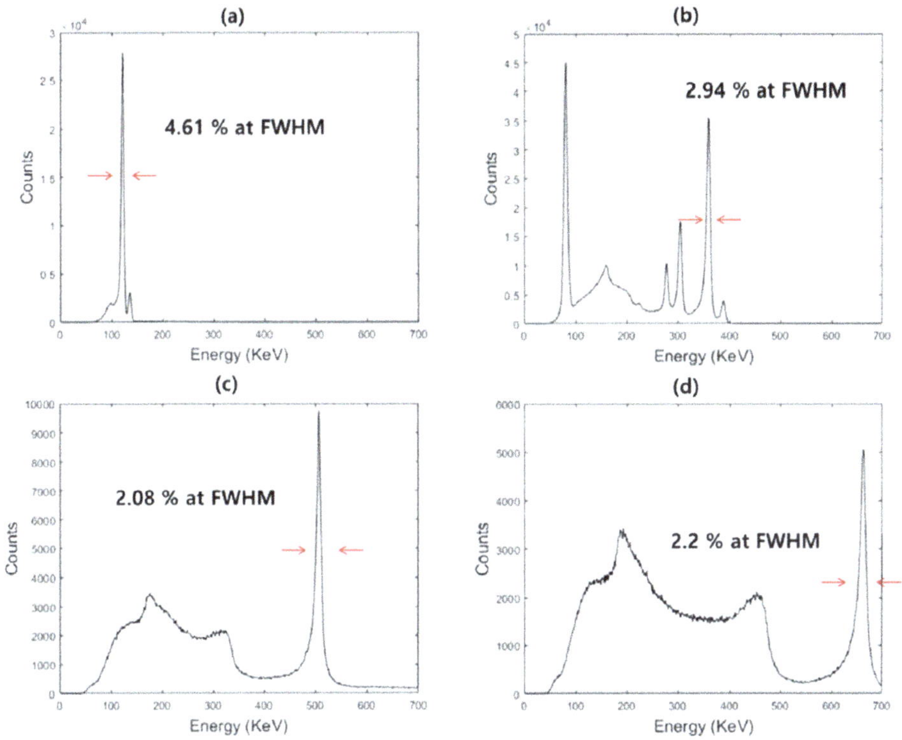

Figure 6.30. Energy spectra obtained with the same detector described in caption to figure 6.29 using different sources: (a) ^{57}Co; (b) ^{133}Ba; (c) ^{22}Na and (d) ^{137}Cs. All data has been corrected pixel-by-pixel to account for charge mobility (see text). (Reprinted from [42] with the permission from the Korean Nuclear Society.)

The energy resolution at 662 keV is 2.2% which is better than any scintillator detector but only marginally so—recently-improved, co-doped $LaBr_3$(Ce) detectors come pretty close (see section 5.2.1). Moreover, scintillators are usually readily available in significantly larger dimensions than CZT. CZT remains, therefore, something of a niche product at the time of writing with applications largely focussing on the potential to produce a detector with high pixellation.

References

[1] Vetter K 2007 Recent developments in the fabrication and operation of germanium detectors *Annu. Rev. Nucl. Part. Sci.* **57** 363–404

[2] Johns P M and Nino J C 2019 Room temperature semiconductor detectors for nuclear security *J. Appl. Phys.* **126** 040902

[3] Kim H *et al* 2020 Thallium bromide gamma-ray spectrometers and pixel arrays *Front. Phys.* **8** 55

[4] Freck D V and Wakefield J 1962 Gamma-ray spectrum obtained with a Lithium-drifted p-i-n junction in germanium *Nature* **193** 669

[5] Claeys C and Simoen E 2007 *Germanium-based Technologies: From Materials to Devices* 1st edn (Amsterdam: Elsevier)

[6] Bruyneel B, Birkenbach B and Reiter P 2016 Pulse shape analysis and position determination in segmented HPGe detectors: The AGATA detector library *Eur. Phys. J.* A **52** 70
[7] Harkness-Brennan L J *et al* 2014 An experimental characterisation of a Broad Energy Germanium detector *Nucl. Instrum. Methods* A **760** 28–39
[8] Ramo S 1939 Currents induced by electron motion *Proc. IRE* **27** 584–5
[9] Shockley W 1938 Currents to conductors induced by a moving point charge *J. Appl. Phys.* **9** 635
[10] Crespi F C L, Camera F, Million B, Sassi M, Wieldand O and Bracco A 2008 A novel technique for the characterization of a HPGe detector response based on pulse shape comparison *Nucl. Instrum. Methods* A **593** 440–7
[11] Agostini M, Bellotti E, Brugnera R, Cattadori C M, D'Andragora A, di Vacri A, Garfagnini A, Laubenstein M, Pandola L and Ur C A 2011 Characterization of a broad energy germanium detector and application to neutrinoless double beta decay search in ^{76}Ge *J. Instrum.* **6** P04005
[12] Warr N *et al* 2013 The MINIBALL spectrometer *Eur. Phys. J.* A **49** 40
[13] Mihailescu L C, Borcea C and Plompen A J M 2007 Data acquisition with a fast digitizer for large volume HPGe detectors *Nucl. Instrum. Methods* A **578** 298–305
[14] Watanabe T, Oi Y, Taki M, Kawasaki K and Yoshida M 1999 Evaluation of peak-to-total ratio for germanium detectors *Appl. Radiat. Isot.* **50** 1057–61
[15] Sangsingkeow P *et al* 2003 Advances in germanium detector technology *Nucl. Instrum. Methods* A **505** 183–6
[16] Castro R M, Vanin V R, Pascholati P R, Maidana N L, Dias M S and Koskinas M F 2004 Developing ^{152}Eu into a standard for detector efficiency calibration *Appl. Radiat. Isot.* **60** 283–7
[17] Sima O and Arnold D 2012 Precise measurement and calculation of coincidence summing corrections for point and linear sources *Appl. Radiat. Isot.* **70** 2107–11
[18] De Felice P, Angelini P, Fazio A and Biagnini R 2000 Fast procedures for coincidence-summing correction in gamma-ray spectrometry *Appl. Radiat. Isot.* **52** 745–52
[19] Hardy J C, Iacob V E, Sanchez-Vega M, Effinger R T, Lipnik P, Mayes V E, Willis D K and Helmer R G 2002 Precise efficiency calibration of an HPGe detector: source measurements and Monte Carlo calculations with sub-percent precision *Appl. Radiat. Isot.* **56** 65–9
[20] Descovich M *et al* 2005 Effects of neutron damage on the performance of large volume segmented germanium detectors *Nucl. Instrum. Methods* A **545** 199–209
[21] Bunting R L and Kraushaar J J 1974 Short-lived radioactivity induced in Ge(Li) gamma-ray detectors by neutrons *Nucl. Instrum. Methods* **118** 565–72
[22] Bhattacharjee T *et al* 2014 Measurement of β-decay end point energy with planar HPGe detector *Nucl. Instrum. Methods* A **767** 19–25
[23] Venhart M *et al* 2017 Application of the Broad Energy Germanium detector: A technique for elucidating β-decay schemes which involve daughter nuclei with very low energy excited states *Nucl. Instrum. Methods* A **849** 112–8
[24] Abt I *et al* 2019 Characterisation of an n-type segmented BEGe detector *Nucl. Instrum. Methods* A **925** 172–83
[25] Schumaker M A and Svensson C E 2007 Compton-suppression and add-back techniques for the highly-segmented TIGRESS HPGe clover detector array *Nucl. Instrum. Methods* A **575** 421
[26] Kshetri R 2012 Modelling of clover detector in addback mode *J. Instrum.* **7** P07008

[27] Eberth J and Simpson J 2008 From Ge(Li) detectors to gamma-ray tracking arrays–50 years of gamma spectroscopy with germanium detectors *Prog. Part. Nucl. Phys.* **60** 283–337
[28] Rizwan U *et al* 2016 Characteristics of GRIFFIN high-purity germanium clover detectors *Nucl. Instrum. Methods* A **820** 126–31
[29] Muralithar S *et al* 2011 Indian national gamma array at IUAC *J. Phys.: Conf. Ser.* **312** 052015
[30] Beausang C W and Simpson J 1996 Large arrays of escape suppressed spectrometers for nuclear structure experiments *J. Phys.* G **22** 527–58
[31] Svensson C E, Amaudruz P, Andreoiu C, Andreyev A, Austin R A E, Ball G C, Bandyopadhyay D, Boston A J, Chakrawarthy R S and Chen A A 2005 TIGRESS: TRIUMF-ISAC gamma-ray escape-suppressed spectrometer *J. Phys.* G **10** S1663
[32] Riley M A, Simpson J and Paul E S 2016 High resolution gamma-ray spectroscopy and the fascinating angular momentum realm of the atomic nucleus *Phys. Scr.* **91** 123002
[33] Zhang N T *et al* 2013 Characteristics of a CsI(Tl) Compton-suppressed clover detector *J. Phys.* G **40** 035101
[34] Baxter A M, Khoo T L, Bleich M E, Carpenter M P, Ahmad I, Janssens R V F, Moore E F, Bearden I G, Beene J R and Lee I Y 1992 Compton-suppression tests on Ge and BGO prototype detectors for GAMMASPHERE *Nucl. Instrum. Methods* A **317** 101–10
[35] Rosiak D *et al* 2019 Escape-suppression shield detector for the MINIBALL spectrometer *Eur. Phys. J.* A **55** 48
[36] Lee I-Y 2013 Gamma-ray energy tracking array: GRETINA *J. Phys. Conf. Ser.* 420 012156
[37] Akkoyun S *et al* 2012 AGATA—advanced gamma tracking array *Nucl. Instrum. Methods* A **668** 26–58
[38] Paschalis S *et al* 2013 The performance of the gamma-ray energy tracking in-beam nuclear array GRETINA *Nucl. Instrum. Methods Phys. Res.* A **709** 44–55
[39] Weisshaar D *et al* 2017 The performance of the gamma-ray tracking array GRETINA for gamma-ray spectroscopy with fast beams of rare isotopes *Nucl. Instrum. Methods Phys. Res.* A **847** 187–98
[40] Korten W *et al* 2020 Physics opportunities with the advanced gamma tracking array: AGATA *Eur. Phys. J.* A **56** 137
[41] Bolotnikov A E *et al* 2015 High-efficiency CzZnTe gamma-ray detectors *IEEE Trans. Nucl. Sci.* **62** 3193–8
[42] Kiam Y, Lee T and Lee W 2019 Radiation measurement and imaging using 3D position sensitive pixelated CZT detector *Nucl. Eng. Technol.* **51** 1417–27
[43] Luke P N 1994 Single-polarity charge sensing in ionization detectors using coplanar electrodes *Appl. Phys. Lett.* **65** 2884
[44] Zhang F and He Z 2007 A prototype three-dimensional position sensitive CdZnTe detector array *IEEE Trans. Nucl. Sci.* **54** 843

Chapter 7

Applications of gamma-ray detection for society, medicine and other areas of science

This chapter provides an overview gamma-ray detection applied to challenges outside the field of experimental nuclear physics; these areas are so broad and wide-ranging that it justifies a chapter of its own. Here, we focus on a few selected applications in the domains of homeland security, nuclear decommissioning, environmental monitoring, oil and gas exploration, medical imaging, and gamma-ray astronomy. These applications draw on many of the techniques and detector technologies described in chapters 5 and 6.

Concepts: societal applications of gamma ray detection; homeland security, 'dirty bomb' detection, gamma-ray imaging, medical imaging, borehole logging, gamma-ray astronomy

Gamma-ray detection is a key tool in experimental nuclear physics and in terms of ionising radiation detection, it probably has the widest applicability to societally-relevant applications. There are several key differences between gamma-ray detection in the laboratory and gamma-ray detection in the field or in a medical setting. First, nearly all wider applications of gamma-ray detection concern a source at rest and not in a moving frame, so Doppler shifts and Doppler broadening are not relevant (except perhaps in an astrophysical setting). In a nuclear physics experiment, the point of gamma-ray emission is generally well defined and determining it is usually not the goal of the experiment. However, in most societal applications the point of emission is not well-defined, indeed the source of gamma-ray emission may be delocalised and spread over a wide area whether hidden in an urban landscape, contaminating a wall, or spread throughout a patient's body. The challenges in the area of applications therefore often focus on *gamma-ray imaging*. Let us describe some of the highlights of the fast-moving field of applications of gamma-ray detection and see how much of the technology used in nuclear physics experiments finds itself reused to societal benefit.

doi:10.1088/978-0-7503-1428-2ch7

7.1 Homeland security

Homeland security has become an increasingly important area in recent years, focussing on safeguarding our urban environment and borders from terrorist threats. One important aspect is the illegal movement and potential dispersal of radioactive material in which gamma-ray detection can play an important role.

7.1.1 Dirty bomb detection

There is significant concern worldwide about the potential for radiological weapons or so-called 'dirty bombs'. The idea is that strong radioactive sources legitimately used in industry or medicine might find their way into the hands of terrorist organisations. This material could then be dispersed in an urban environment by means of a conventional explosive. The threat to life and health from the dispersal of this material might not necessarily be of immediate concern but the public fear of radioactivity could cause mass panic.

The challenge in homeland security clearly has many aspects and the safe stewardship of radioactive materials is obviously the starting point for mitigating the threat of radiological weapons. If radioactive material can be obtained either by defeating security measures in place or by attempting to bring materials in from third countries where precautions may be more lax, then radiation detection will clearly play an important role in monitoring the illicit transport of this material within a state or across national borders.

The D3S ID developed by Kromek is designed to help counter the threat of dirty bombs. It consists of a wearable sensor which can be networked or connected to a mobile phone. Inside, the D3S is a CsI(Tl) scintillator coupled to SiPM readout to record gamma rays and generate a spectrum which can be used to identify the isotope(s) generating the gamma rays. It can therefore both quantify and qualify the nature of the threat from radiological material in the urban environment [1]. This is described in a video presentation by Vincent Tang of DARPA.

7.1.2 Gamma-ray imaging

Gamma-ray imaging in the field has achieved a high degree of sophistication [2] and implementations of it draw broadly from the gamma-ray detector technologies described in chapters 5 and 6. Significantly, it can be effectively merged with visual data and LiDAR ranging to create a composite scene. An example of the technology created in recent years is the High-Efficiency Multimode Imager (HEMI) [2] shown in figure 7.1; it can image the location of different sources comprising different isotopes and their associated gamma-ray emission as shown in figure 7.2. HEMI consists of two planes of 96 1 cm^3 CZT detectors: 32 in the front plane and 64 in the backplane. The front plane has its detector elements arranged in a random pattern providing both the active aperture for the fully populated backplane (for low energies) and a portion of a Compton imaging system, also in combination with the backplane. HEMI exploits the *Compton camera* principle as well as *active coded*

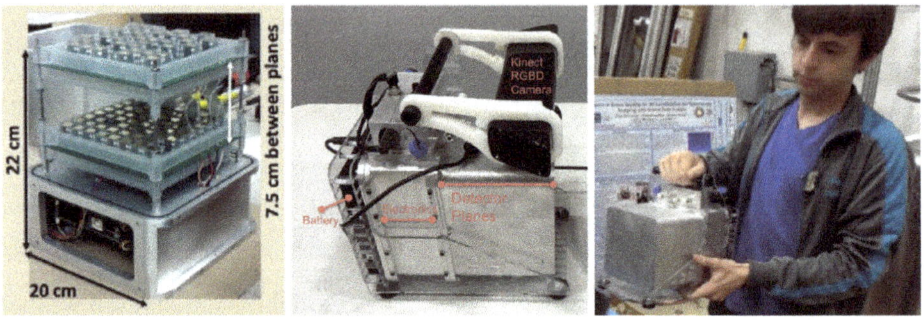

Figure 7.1. The High-Efficiency Multimode Imager (HEMI) consists of two layers of coplanar-grid CdZnTe detectors enabling coded aperture and Compton imaging simultaneously. (Reprinted from [2] with the permission of Elsevier.)

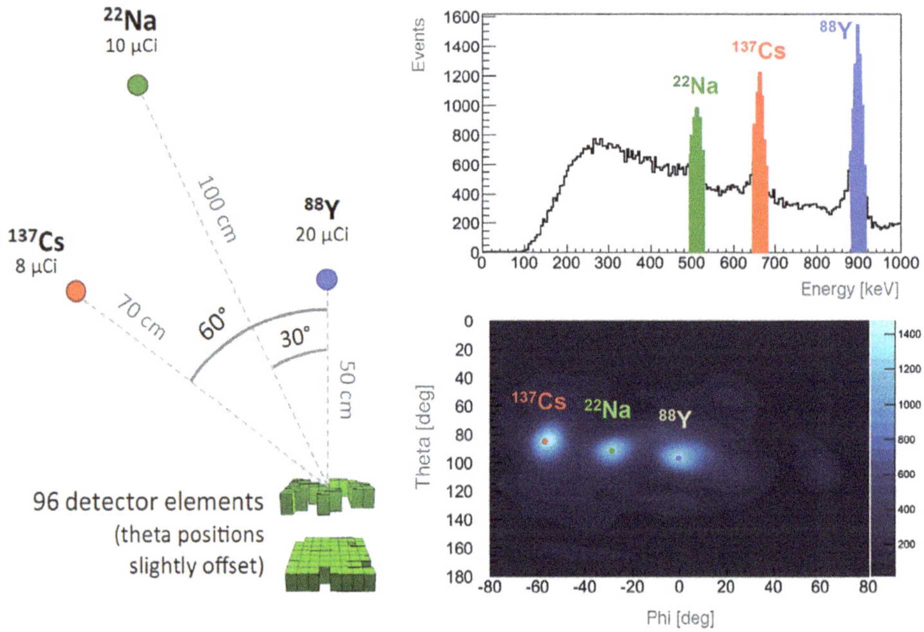

Figure 7.2. Source location reconstruction using the HEMI detector: (left) the location of three standard radioactive sources relative to the detector (top right) the spectrum obtained identifying the photopeaks corresponding to the three different sources (bottom right) reconstruction of the angle of the three sources relative to the detector. (Reprinted from [2] with the permission of Elsevier.)

aperture technology. Since these techniques are particularly important to gamma-ray applications, we describe them in detail below.

Compton camera

The Compton camera is a highly valuable technique for imaging in both security applications and medicine. It is based on the physics of Compton scattering which was outlined in section 2.4.2. The key equation governing Compton scattering was

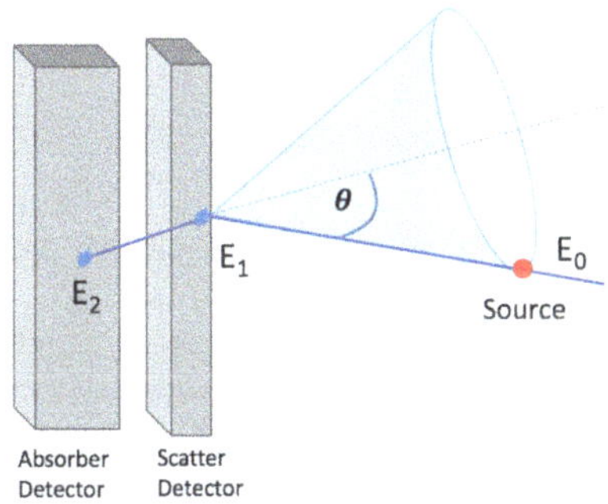

Figure 7.3. Basic principle of a Compton camera where a gamma ray from a source of unknown location scatters in a position-sensitive scatter detector and is absorbed in a position-sensitive absorber detector. The source location can be reconstructed as lying on a cone of half angle, θ (see text). (Reprinted from [4] with the permission of IOP Publishing.)

presented in equation (2.20); it shows that the energies of the initial and scattered photon and the scattering angle are related through conservation of energy and momentum. Consider the arrangement of detectors shown in figure 7.3 and a source of unknown position. Here, the position of the initial Compton scatter is determined in the *scatter detector* from detection of the short-ranged Compton-scattered electron. The energy of this electron is also recorded. In the thicker absorber, the scattered gamma-ray undergoes photoabsorption. The photoelectron is again short ranged and so its energy and interaction position are recorded. These measurements of the respective electron positions and energies can be used in conjunction with equation (2.20) to derive the scattering angle:

$$\cos\theta = 1 - \frac{m_e c^2 E_1}{E_2(E_1 + E_2)}, \tag{7.1}$$

where E_1 is the energy deposited in the scatter detector and E_2 is the energy deposited in the absorber [3]. The solution is that the source must lie on a cone of half-angle, θ, as shown in figure 7.3. While this provides some level of localisation, it is not particularly useful on a single-event basis. However, if many such events are detected originating from the same source then the reconstructed cones can be overlapped to fix the source direction (see figure 7.4). Strategies exist to elaborate this approach to multiple and extended sources. We will return to the Compton camera approach in the context of medical imaging (see section 7.5.2).

Coded aperture imaging

Coded aperture imaging is a sophisticated technique which can produce an image of radiation source(s), particularly gamma-ray sources. At root, it is similar to the concept of a pinhole camera that many of us may have played with as a means of making simple photographs. Figure 7.5 shows a possible implementation of a coded mask system for imaging gamma-ray sources. It employs a highly-pixellated planar detector, in this example, a CZT semiconductor-based detector. Placed on top of the detector is a coded-aperture mask. This mask has a complex design according to

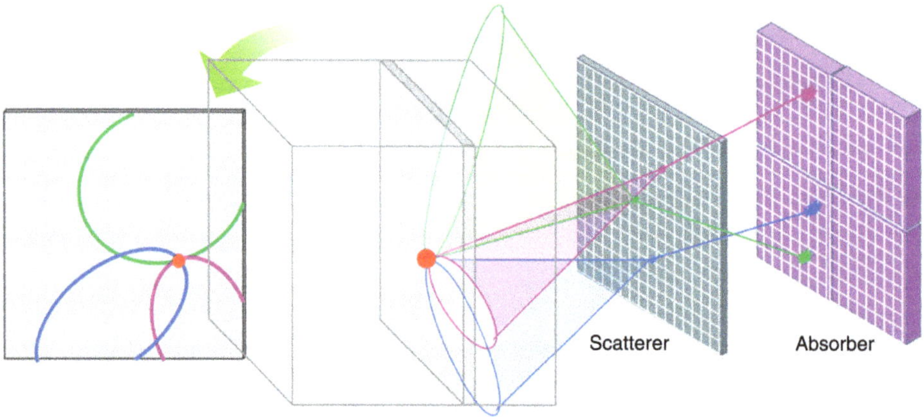

Figure 7.4. Schematic of the Compton camera approach where multiple reconstructed cones can be overlapped to determine the source location. (Reprinted from [5] with the permission of IOP Publishing.)

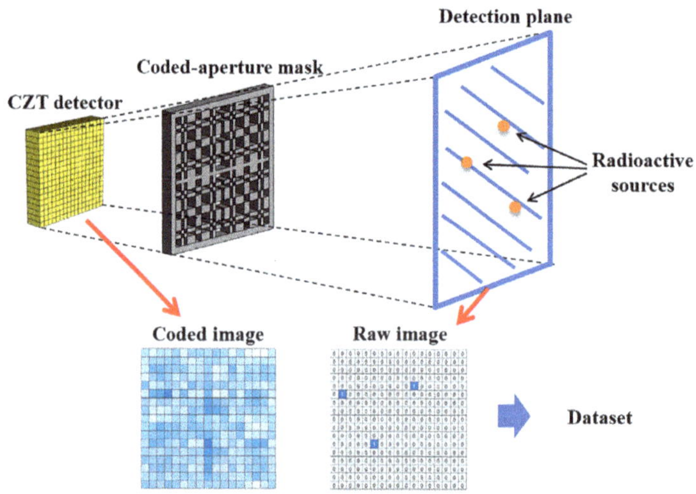

Figure 7.5. The principles of coded-aperture imaging. A highly-pixellated planar gamma-ray detector has a coded-aperture mask placed in front of it. The mask shadows the detector for certain orientations of incoming gamma-rays. The resulting coded image can be converted into a true image through the mathematical process of deconvolution using a definition of the mask geometry. (Reprinted from [7] with the permission of Elsevier.)

various special patterns, e.g. the *modified uniformly redundant array (MURA)* for which mathematical definitions exist [6, 7]. The material the mask is made of is generally dense metal. It casts a shadow onto the detector for certain directions of gamma rays incident from the source(s). This leads to a coded image. Applying a mathematical deconvolution to the coded image using the definition of the mask design can allow a reconstruction of the source location in a horizontal plane (note, that will not produce depth information as that would require *a priori* knowledge of the source strengths). Variations of this technique exist such as *active coded aperture imaging* where the mask is formed from small active detector elements which has certain advantages in efficiency and sensitivity.

7.2 Nuclear decommissioning

There are many interesting applications of gamma-ray detectors in nuclear decommissioning and radioactive waste management [8]. An example is the imaging of radioactive waste drums [9]. Here, fission products such as ^{137}Cs are mixed in with the spent fuel, i.e. uranium. Compton-scattering of the 662 keV gamma ray emitted in the decay of ^{137}Cs generates a strong flux of gamma rays around 100 keV which can fluoresce uranium atoms leading to emission of characteristic x-rays. Figure 7.6 shows the benefit of the high energy resolution of CZT and its good efficiency for low energy x-rays/gamma rays in detecting the uranium x-rays. Using a collimator arrangement, it would be possible to determine the mass of uranium in the waste drum and map out, to some extent, its physical distribution.

7.3 Environmental monitoring

Gamma ray detection is widely used in environmental monitoring. A specific example is worldwide monitoring to show compliance with the Comprehensive

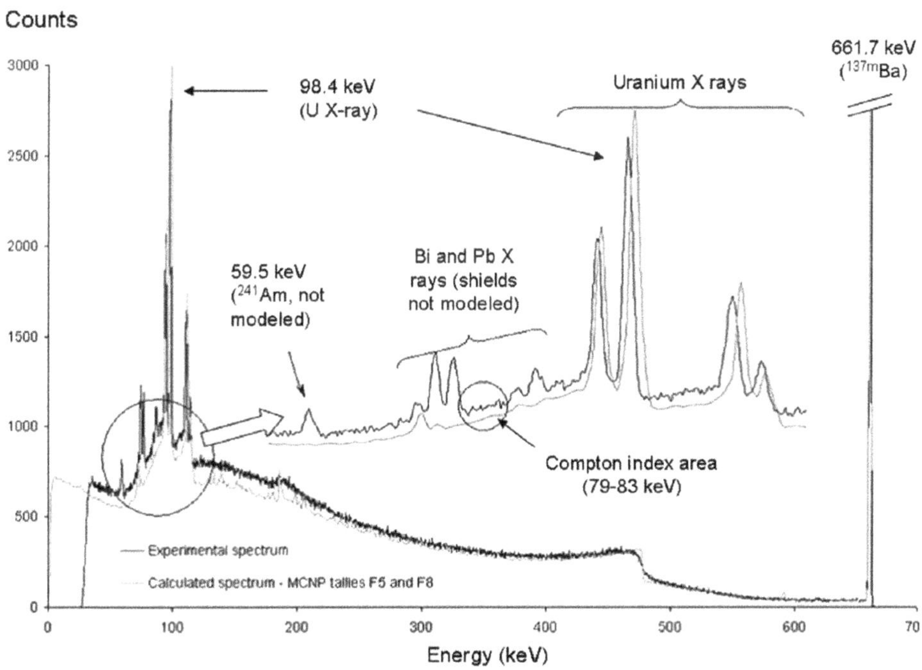

Figure 7.6. Gamma-ray/x-ray spectrum from a nuclear waste drum using a CZT detector with a 1.75 cm × 1.75 cm collimator. The dark line is the measured data while the fainter grey line is the result of a Monte Carlo simulation. Fluorescent x-ray emission from uranium is observed which can be used to map out the geometry of material inside the drum. (Reprinted from [9] with the permission of Elsevier under Creative Commons License.)

Nuclear Test Ban Treaty (CTBT) [10, 11]. Radioactive isotopes of xenon, e.g. ^{133}Xe ($T_{\frac{1}{2}}$ = 5.25 days) and ^{131m}Xe ($T_{\frac{1}{2}}$ = 11.8 days) are produced in the fission of uranium or plutonium and can escape an underground nuclear explosion. Such large quantities of these isotopes are produced that they can even be detected at very large distances from the initial explosion. Air samples are collected and radioxenon is extracted and purified. The isotopes of interest decay by emitting beta particles, gamma rays and conversion electrons. Suitable detection systems therefore employ gamma-ray detection such as NaI(Tl) scintillators and high-purity germanium detectors. Frequently, measurements are made in coincidence with a beta/electron detector e.g. a plastic-scintillator detector [12]. Figure 7.7 provides an example of electron-gamma ray coincidences which can be used to isolate radioxenon isotopes such as ^{133}Xe and ^{131m}Xe.

7.4 Oil and gas, mineral exploration

7.4.1 Sub-sea CT imaging

Tracerco have developed the Discovery system for sub-sea CT imaging of wall deterioration and blockages in undersea pipelines (see related YouTube video). The technique relies on shining a strong ^{137}Cs gamma ray source through the pipe and detecting the gamma rays on the opposite side in an array of CsI(Tl) detectors.

7.4.2 Borehole logging

Gamma-ray detection is used in a number of ways in mineral exploration as part of the technique of borehole logging where a probe with many different instruments is

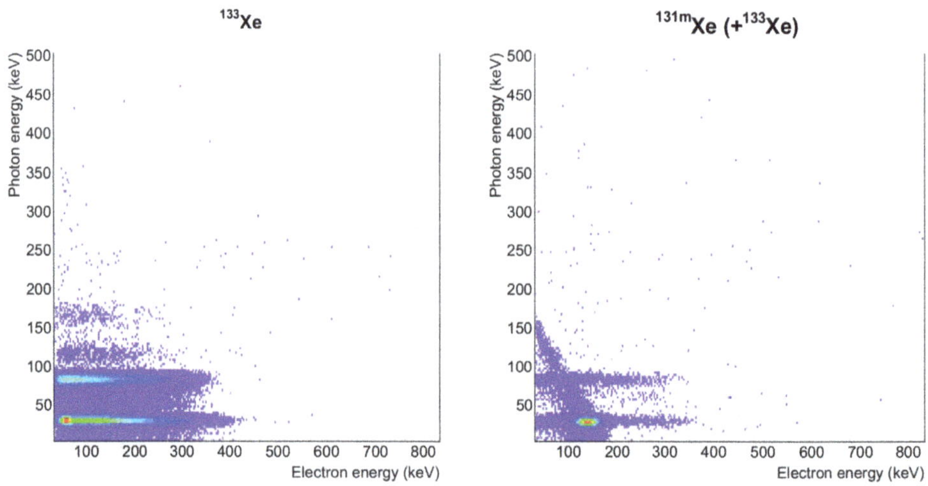

Figure 7.7. Eletron-gamma ray coincidence spectra for samples of radioxenon: (left) a coincidence spectrum for ^{133}Xe which shows features including a beta decay continuum with an endpoint energy of 346 keV in coincidence with 81 keV gamma rays of 32 keV x-rays of the Cs daughter. (Right) a coincidence spectrum for the isomeric state ^{131m}Xe (with some ^{133}Xe present) which shows coincidences between the 129 keV conversion electron and 30 keV x-rays from Xe. (Reprinted from [10] with the permission of Elsevier under Creative Commons License.)

inserted into the ground through a borehole. The technique can be used to measure natural radioactivity, e.g. gamma rays originating from ^{40}K, and the decay chains of uranium and thorium, which is an aid to determining rock strata. It is also commonly used for density measurements where gamma rays originating from a strong ^{137}Cs source are detected following Compton scattering by the surrounding rock/mineral strata[1]. Gamma-ray detection in boreholes is conventionally carried out using NaI(Tl) detectors (see section 5.1.2) coupled to photomultiplier tubes.

7.5 Medical imaging

Gamma-ray detection is widely used in medical imaging. There are two principal methodologies involving gamma detection: positron-emission tomography (PET) and single photon computed tomography (SPECT) which are described below.

7.5.1 PET

PET (positron emission tomography) [13] is a technique where β^+-decaying isotopes, e.g. ^{18}F, are introduced as tracers into the body and selectively taken up in diseased organs. As discussed in section 1.4.2, β^+ decay involves positron emission. Positrons can either annihilate in flight or come to rest and form positronium before subsequent annihilation. Annihilation involves the emission of two 511 keV gamma rays in a near back-to-back geometry. Detecting the two gamma rays determines a line-of-response (LoR). Overlapping these LoRs and applying appropriate mathematical reconstruction techniques leads to an image of the location of the radioactive tracer. The principles behind PET imaging are presented in figure 7.8. A conventional PET scanner comprises sticks of LYSO scintillator coupled to a position sensitive photomultiplier tube. LYSO (see section 5.1.2) is chosen for its high effective Z and fast decay time.

The annihilation gamma rays have an energy 511 keV which means that the main way they initially interact in matter is by Compton scattering. There is therefore some probability they will scatter in the patient, meaning that they fail to be detected or that they are detected but not along the initial LoR. This is the biggest source of undesired background in PET images. It can be mitigated to some extent by using scintillator detectors, e.g. LYSO with high coincident timing resolution but is mostly corrected for by building up a model for scatter correction by imaging the body using some other modality such as CT (x-ray computed tomography). Figure 7.9 illustrates the benefits of overlaying a CT image with the PET image; the former being anatomical and the latter a functional image.

In addition to combined PET/CT, there have been strong advances in PET/MRI[2] imaging [16]. The challenge here is that conventional PMTs will not operate in the high magnetic field (typically ~3 T) of MRI magnets. Early attempts to combine the two techniques into one system used optical fibres to transport scintillation light out

[1] Recall that the probability of Compton scattering depends on the density of the scattering medium—section 2.4.2

[2] Magnetic resonance imaging—a technique which maps out the location of hydrogen in the body giving fine anatomical detail.

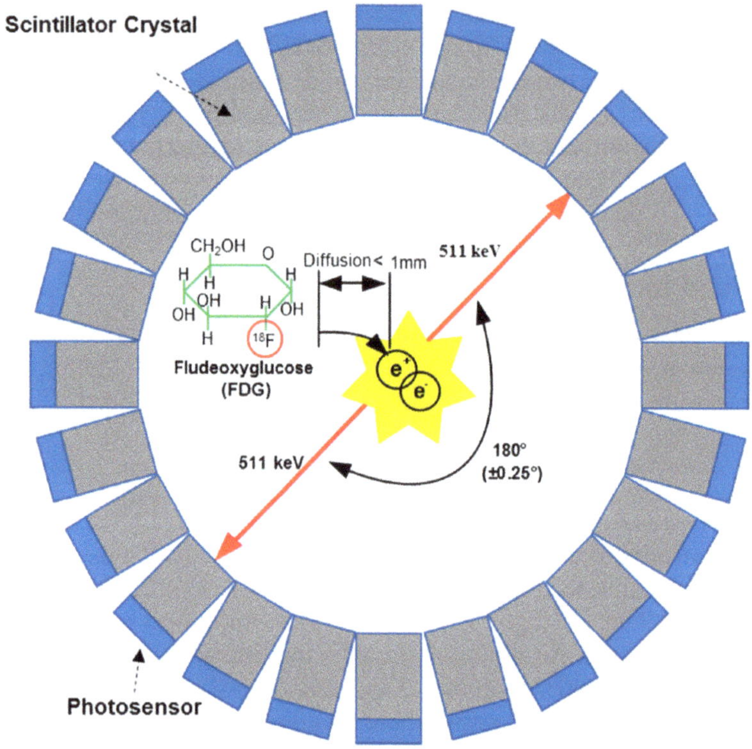

Figure 7.8. The concept behind PET imaging. A radiotracer, ^{18}F-FDG, is introduced into the body. The ^{18}F decays by emitting a positron. The positron travels around 1 mm in the body before coming to rest and annihilating with an electron to produce two 511 keV gamma rays in near back-to-back geometry. The gamma rays are detected with a ring of inorganic scintillator detectors. The geometry of detection defines a line of response (LoR). (Reprinted from [14] with permission from MDPI under Creative Commons License.)

of the field of the magnet to photomultiplier tubes [17]. The development of silicon photomultipliers (see section 5.3.3) has revolutionised this combined medical imaging technique as SiPMs are immune to the effects of high magnetic field. The large PET market drives innovation in SiPM technology and benefits of this are starting to appear in applications in experimental nuclear physics.

Recent innovations in PET imaging include time-of-flight PET (TOF-PET) where the timing resolution of the imaging system is improved to localise the point of emission of the annihilation event through measuring the time difference in photon arrival time [14, 18] (see figure 7.10). The signal-to-noise ratio (SNR) gain in imaging for a TOF-PET system over a non-TOF system is given by [19]:

$$G = \frac{\mathrm{SNR_{TOF}}}{\mathrm{SNR_{non-TOF}}} = \sqrt{\frac{2D}{c \cdot \mathrm{CTR}}}, \tag{7.2}$$

where D is the patient diameter and CTR is the coincidence time resolution of the TOF-PET system. For a diameter, D = 40 cm, and a CTR resolution of 10 ps, the gain, G is 16.4. There are many efforts worldwide to try to drive CTR down into

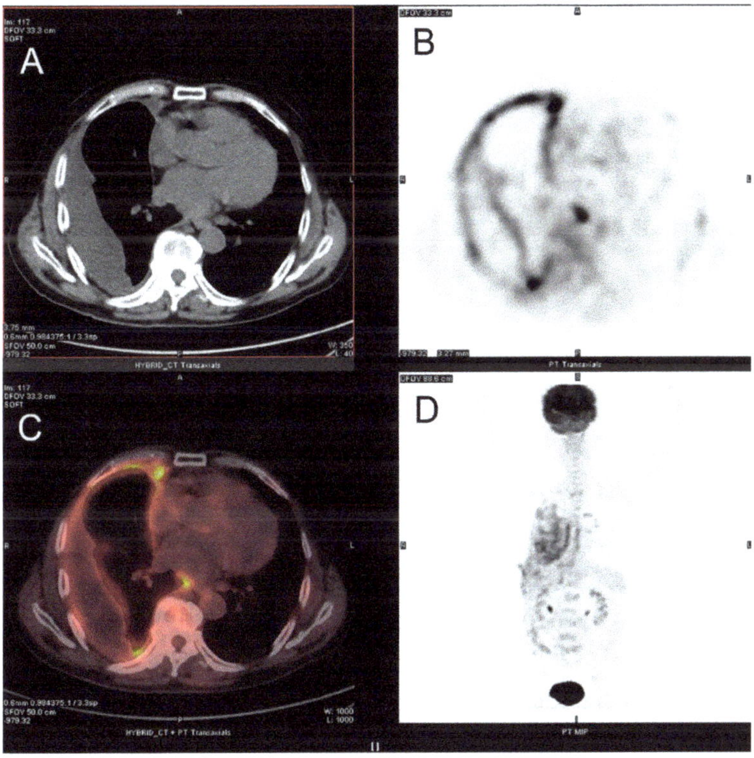

Figure 7.9. Combing PET and axial CT (x-ray computed tomography) imaging: (A) axial CT image of lungs; (B) PET image of lungs; and (C) combined PET-CT image. (Reprinted from [15] with permission of PLOS under Creative Commons License.)

the 10 ps region; the challenging being that this time scale is below the time fluctuations in emission of scintillation photons. Attention is focussing on detection of Cherenkov light (see section 2.3.4) from the interaction of photoelectrons/Compton-scattered electrons in a scintillator crystal which has no time-dependence related to the electronic structure of the scintillator material [20]. However, only around 20 Cherenkov photons are typically generated when an annihilation photon (511 keV) interacts with a standard inorganic scintillator such as LSO used in PET [19].

There is also exciting development work on plastic scintillator-based PET imaging, e.g. the J-PET system which promises to strongly reduce the costs of PET scanners as plastic scintillator is much cheaper than inorganic scintillators such as LYSO.

PET imaging is not confined to medical imaging. Indeed, positron emission particle tracking (PEPT) has been shown to be a powerful application of PET to the industrial domain [21]. Figure 7.11 provides an example of a PEPT measurement where a tracer-containing particle is introduced to a rotating drum containing small glass beads and imaged with conventional PET scanning heads, allowing the position of the tracer particle to be determined dynamically and by extension, its instantaneous velocity vector.

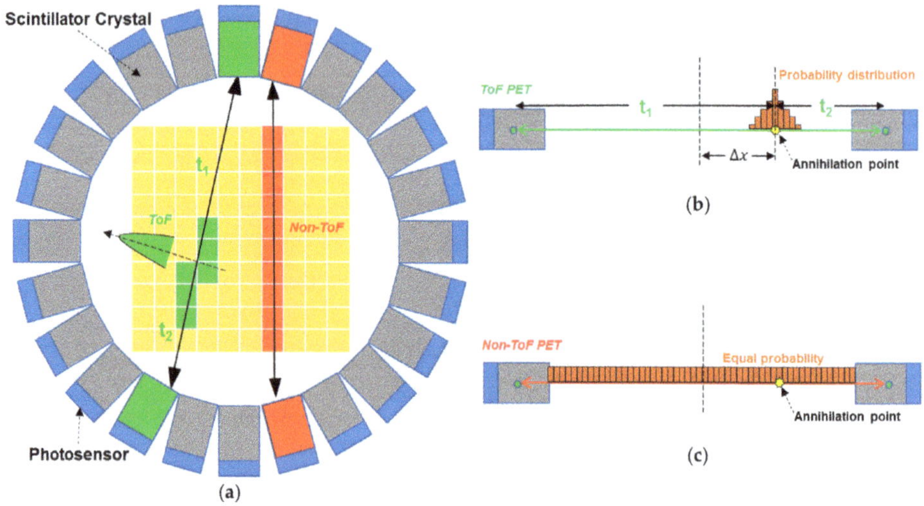

Figure 7.10. Concept of time-of-flight PET (ToF PET): (a) Illustration of a detector ring detecting pairs of gamma photons from annihilation events with (green) and without (red) the ToF technique; (b) the probability distribution of the annihilation positron along the line of response (LoR) in ToF PET; (c) the equal probability of the annihilation position along the LoR in non-ToF PET. (Reprinted from [14] with permission from MDPI under Creative Commons License.)

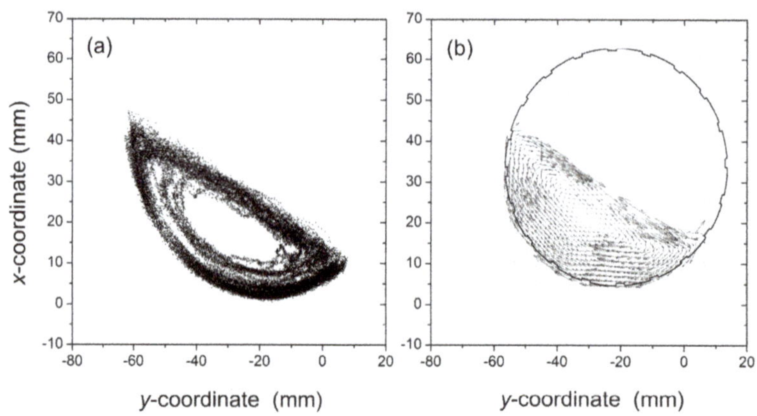

Figure 7.11. An example of a PEPT measurement of (a) a few thousand locations (z-axis projection) of a tracer moving within a small rotating drum filled with glass beads of diameter 5 mm; and (b) the average velocity field determined for the tumbling tracer. (Reprinted from [21] with permission from World Scientific under Creative Commons License.)

7.5.2 SPECT

SPECT (single photon computed tomography) is where a radiotracer, typically, $^{99}Tc^{m}$, a nuclear isomer (see section 1.7.2) with $t_{\frac{1}{2}} \approx 6$ h, is injected into the body and taken up selectively by a diseased organ. In the decay of $^{99}Tc^{m}$, a single gamma ray with an energy of 140 keV is emitted; a gamma ray of this energy interacts with inorganic scintillator detectors predominantly through photoelectric absorption with Compton scattering being minimal. Simply detecting the gamma ray with a relevant gamma-ray detector would not provide any imaging capability. Imaging is conventionally achieved using a collimator system comprising a heavy metal, e.g. a tungsten collimator with many small holes in it. The action of the collimator is to effectively define the incoming direction of the gamma rays which pass through the holes and are detected by a position-sensitive gamma ray detector mounted behind. This leads to the principal limitation of SPECT imaging in that it is highly inefficient since the vast majority of gamma rays are stopped in the collimator as only those with a very specific direction are accepted.

SPECT imaging is generally cheaper and more extensively exploited than PET imaging worldwide because the necessary tracer is cheaper to produce and relatively long-lived. SPECT has also been applied in the nuclear sector where it can be used to image the location of fission products in nuclear fuel, e.g. ^{137}Cs [22, 23].

There are interesting developments to avoid the collimator in SPECT using a Compton camera approach (see figure 7.3) [3, 24]. An example of such a system is ProSPECTus [24] (see figure 7.12). Here, the scatterer is an 8 mm thick Si(Li) double-sided strip detector (discussed in section 8.4.2 as a conversion-electron detector) with 13 strips, 5 mm wide on each side. The absorber detectors comprise a 20 mm thick planar germanium detector with 12 strips on each side and active area

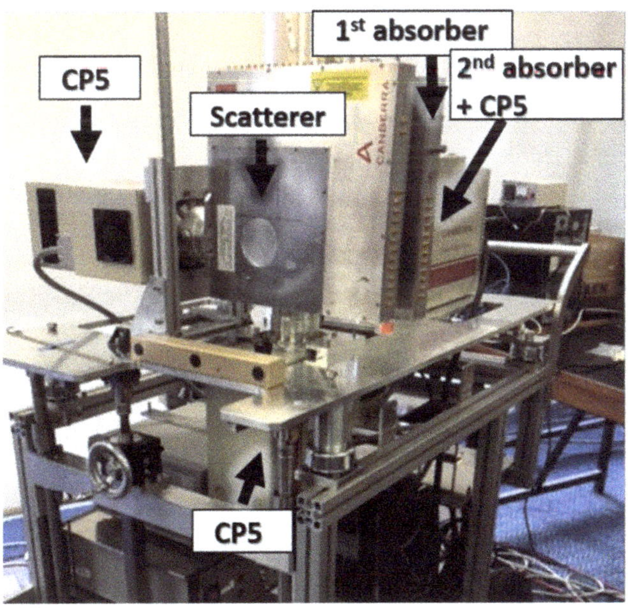

Figure 7.12. Photograph of the ProSPECTus system for SPECT imaging using the Compton camera approach. The scatter detector is an 8 mm thick Si(Li) detector and mounted behind this is a planar germanium DSSD detector and a large volume HPGe detector. (Reprinted from [24] with the permission of IOP Publishing and SISSA Medialab.)

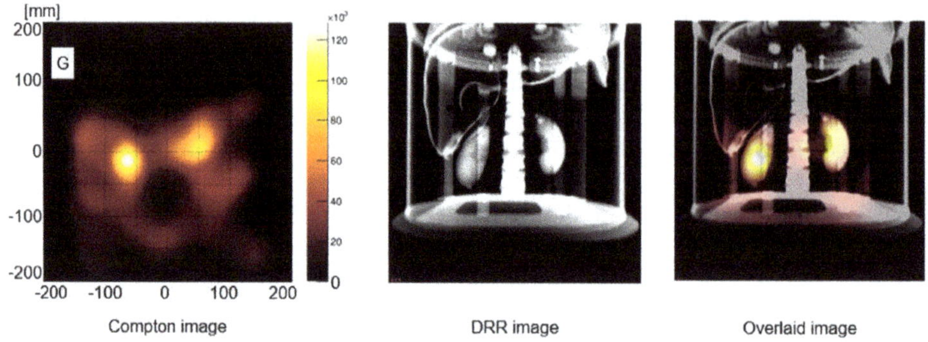

Figure 7.13. SPECT image of a kidney phantom: (left) SPECT image obtained using Compton camera system; (centre) Digital reconstructed radiography (DRR) image (i.e. x-ray CT image); (right) overlay of SPECT image on to DRR image. (Reprinted by permission from Springer Nature [3] under Creative Commons License.)

of $60 \times 60\ \text{mm}^2$ and a large volume HPGe detector. A second system, described in reference [3], comprises a 500 μm thick silicon scatter detector followed by three layers of cadmium telluride (CdTe) absorber detectors. Each detector has dimensions of 32 × 32 mm and is divided into 128 strips on each side. Figure 7.13 shows the power of such a Compton camera SPECT approach in locating activity within a phantom, i.e. a model mock-up using similar density materials, of kidneys within a human torso.

7.5.3 Diagnostics for ion-beam therapy

There is a major initiative worldwide to treat certain cancers with ion beam therapy, principally proton therapy using high-energy proton beams but also heavy-ion therapy, principally ^{12}C beams. The motivation for using ion beams in place of more conventional, and significantly less expensive radiotherapy using x-rays/gamma rays, is that the energy of the ion beam is deposited close to the end of its track in the so-called *Bragg peak* (see section 2.2.4). Gamma-ray detection can be a valuable diagnostic in understanding the physical location of the Bragg peak. This gamma radiation can be promptly emitted, e.g. from inelastic excitation of the beam itself or from inelastic excitation of nuclei along the track of the beam. It can also be delayed with respect to the arrival time of the beam, e.g. through transmutation of ^{12}C in the body into ^{11}C which is a short-lived positron emitter. The short-ranged positrons will annihilate producing, most commonly, two 511 keV gamma rays in near back-to-back geometry. Figure 7.14 and 7.15 show two approaches to measuring the prompt radiation. The first (figure 7.14) uses a series of collimators with gamma-ray detectors behind to provide the required directionality; this has strong synergies with the standard SPECT imaging approach described in section 7.5.2. The second (figure 7.15) is to use a Compton-camera approach to determine the cone of emission on an event-by-event basis. The position sensitivity is reinforced in each case by having a tracking hodoscope which measures the x–y position of the ion beam as it enters the body.

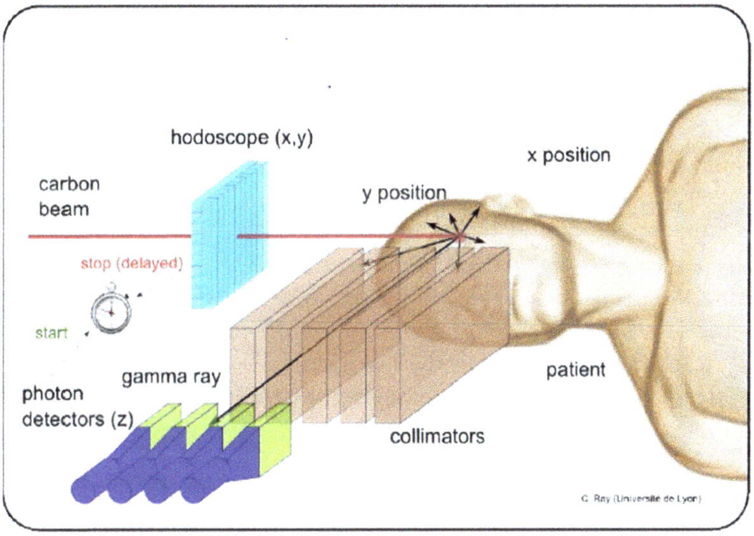

Figure 7.14. Diagnostic of the interaction of a carbon therapy beam within a patient using a collimated gamma-ray detector system. The carbon beam is tracked using a position-sensitive hodoscope. (Reprinted from [9] with the permission of Elsevier under Creative Commons License.)

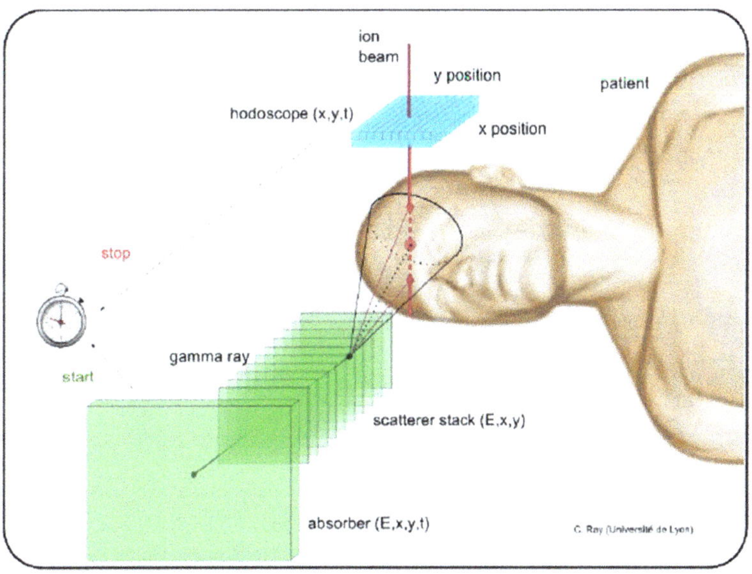

Figure 7.15. Diagnostic of the interaction of a carbon therapy beam within a patient using a Compton camera-based gamma-ray detector system comprising a stack of position-sensitive scatter detectors and an absorber detector. The carbon beam is tracked using a position-sensitive hodoscope. (Reprinted from [9] with the permission of Elsevier under Creative Commons License.)

7.6 Gamma-ray astronomy

Contemporary astronomical imaging encompasses all parts of the electromagnetic spectrum from radio waves to gamma rays. There are several astrophysical mechanisms by which gamma rays can be generated [25]. For example, gamma-ray continua are expected from the bremsstrahlung of high energy electrons. However, there are also situations where discrete line gamma rays are emitted where the same detectors and the techniques discussed in relation to experimental nuclear physics can be applied in their detection [25]. Remarkably, the laws of physics being the same across the distant universe and nuclear structure properties remaining the same, these gamma rays can be readily associated with corresponding gamma ray transitions observed in the laboratory.

Figure 7.16 shows the gamma-ray imaging system inside the INTEGRAL space-based observatory [26, 27] which consists of an array of germanium detectors packed

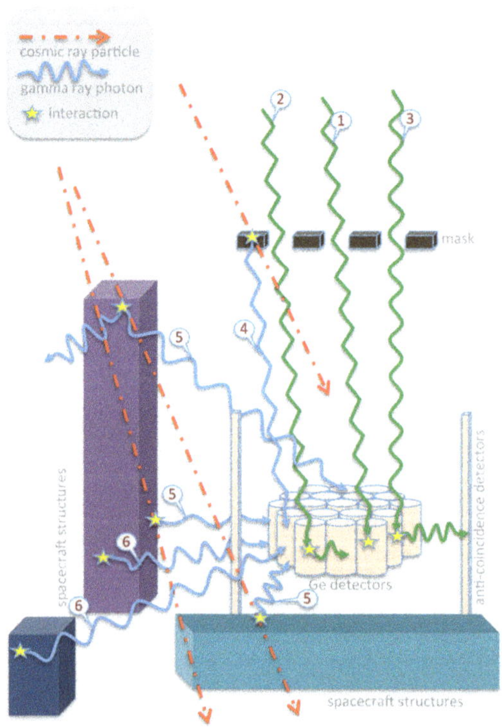

Figure 7.16. Illustration of the gamma-ray imaging system in the INTEGRAL spaced-based observatory, comprising an array of 19 germanium detectors. Cosmic gamma rays of interest enter through the coded mask on the front of the telescope (green wavy lines). They may then either interact fully in one detector, scatter between germanium detectors or be vetoed when they scatter into a wall of anti-coincidence detectors. There are also many other sources of gamma rays in the system (blue wavy lines) representing prompt gamma rays from cosmic-ray interactions in the surrounding materials or delayed gamma-rays from activation. (Credit: [27], reproduced with permission © ESO.)

together side-by-side. The system employs Compton suppression and anti-coincidence shielding in a similar manner to that found in a germanium-detector array in an experimental nuclear physics laboratory (cf section 6.5.1).

Figure 7.17 shows a blow-up of the germanium spectra obtained by INTEGRAL around the 511 keV annihilation peak and the 1.809 MeV peak associated with the

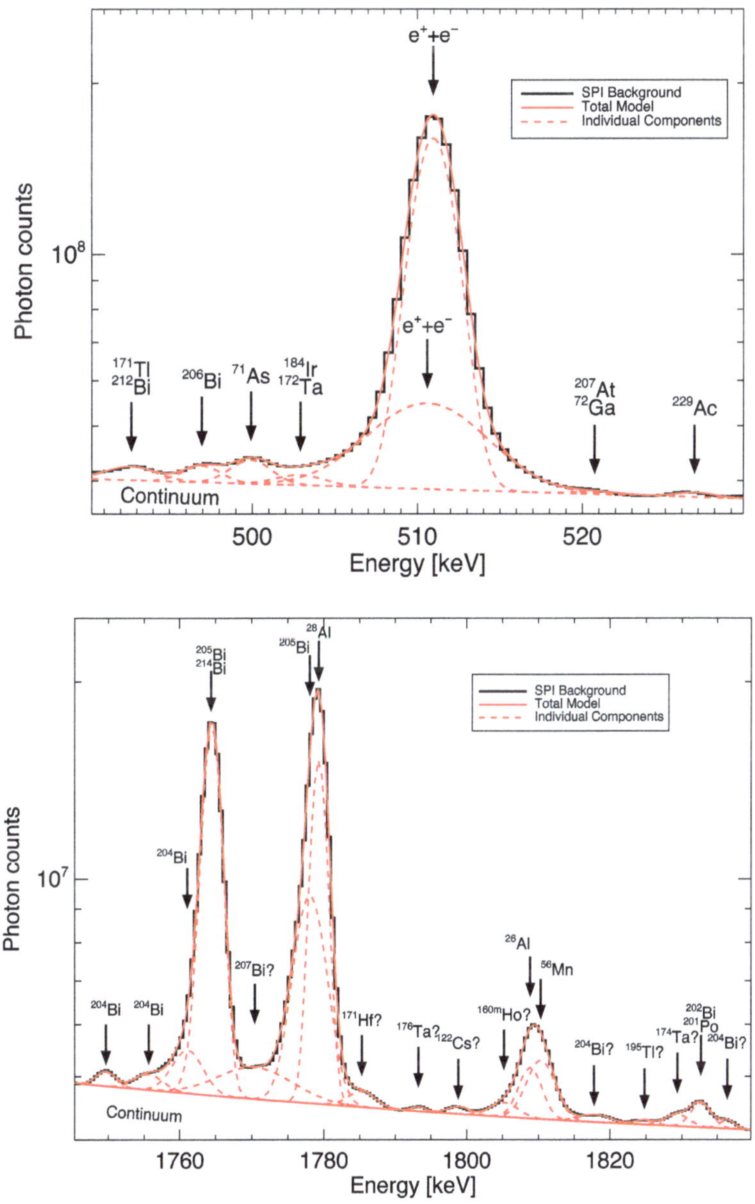

Figure 7.17. Gamma ray spectra obtained with INTEGRAL: (top) the region around the 511 keV annihilation peak and (bottom) the region around the 1.809 MeV peak associated with the decay of ^{26}Al. Peaks associated with various backgrounds are marked with the relevant isotope. (Credit: [27], reproduced with permission © ESO.)

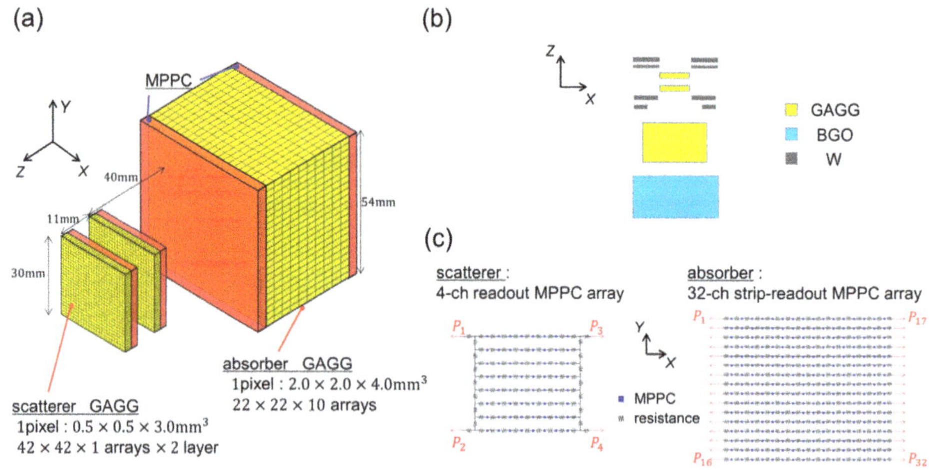

Figure 7.18. Design of a Compton camera system for gamma-ray astronomy: (a) schematic view of the scatterer and absorber. Scatterers are surrounded by collars of tungsten passive shield to prevent direct injection to the absorber. (b) A configuration of the bismuth germanate (BGO) active shield and tungsten passive shield. (c) Schematic view of the MPPC arrays. (Reprinted by permission from Springer Nature [28] under Creative Commons License.)

beta decay of ^{26}Al. The spectra are complex in origin since there are many potential sources of background gamma ray peaks from prompt or delayed activation of the spacecraft components by cosmic rays. The energy resolution of the germanium detectors is clearly very important to this application.

7.6.1 Compton camera for gamma-ray astronomy

In gamma-ray astronomy, it is important to try to localise gamma-ray emission to its source with good angular resolution. In many ways, this is an application similar to some of the applications discussed above under homeland security and medical imaging. It is perhaps unsurprising then that the Compton camera is again suggested as the ideal tool for localising the point of emission of ~1 MeV gamma rays [28]. Figure 7.18 shows the design of a compact Compton camera which could operate in a satellite-borne mission. It makes heavy use of the latest in both scintillator and silicon photomultiplier technology. The system comprises a scatterer and an absorber. Each comprises highly-pixellated arrays of GAGG:Ce scintillator [29] coupled to Hamamatsu MPPC (i.e. SiPM) arrays. The angular resolution of this Compton camera system is 3.4° ± 0.1° at 1.7 MeV [28].

References

[1] Zhao J, Zhang Z and Sullivan C J 2019 Identifying anomalous nuclear radioactive sources using Poisson kriging and mobile sensor networks *PLoS One* **14** e0216131

[2] Vetter K, Barnowski R, Haefner A, Joshi T H Y, Pavlovsky R and Quitter B J 2018 Gamma-ray imaging for nuclear security and safety: Towards 3-D gamma-ray vision *Nucl. Instrum. Methods* A **878** 159–68

[3] Sakai M, Kubota Y, Parajuli R K, Kikuchi M, Arakawa K and Nakano T 2019 Compton imaging with ^{99m}Tc for human imaging *Sci. Rep.* **9** 12906
[4] Turecek D, Jakubek J, Trojanova E and Sefc L 2018 Compton camera based on Timepix3 technology *J. Instrum.* **13** C11022
[5] Kim S M, Sea H, Park J H, Kim C H, Lee C S, Lee S-J, Lee D S and Lee J S 2013 Resolution recovery reconstruction for a Compton camera *Phys. Med. Biol.* **58** 2823–40
[6] Cieslak M J, Gamage K A A and Glover R 2016 Coded-aperture imaging systems: Past, present and future development—a review *Radiat. Meas.* **92** 59–71
[7] Zhang R *et al* 2019 Reconstruction method for gamma-ray coded-aperture imaging based on convoluntional neural network *Nucl. Instrum. Methods* A **934** 41–51
[8] Aspe F, Idoeta R, Auge G and Herranz M 2020 Classification and categorization of the constrained environments in nuclear/radiological installations under decommissioning and dismantling processes *Prog. Nucl. Energy* **124** 103347
[9] Giot M *et al* 2017 Nuclear instrumentation and measurement: a review based on the ANIMMA conferences *EPJ Nucl. Sci. Technol.* **3** 33
[10] Ringbom A *et al* 2014 Radioxenon detections in the CTBT international monitoring system likely related to the announced nuclear test in North Korea on February 12, 2013 *J. Environ. Radioact.* **128** 47–63
[11] Galan M, Kalinowski M, Gheddou A and Yamba K 2018 New improved radioxenon decay data and its implications in nuclear explosion monitoring *J. Environ. Radioact.* **192** 628–34
[12] Reeder P L, Bowyer T W and Perkins R W 1998 Beta-gamma counting system for Xe fission products *J. Radioanal. Nucl. Chem.* **235** 89–94
[13] Vaquero J J and Kinahan P 2015 Positron emission tomography: current challenges and opportunities for technological advances in clinical and preclinical imaging systems *Annu. Rev. Biomed. Eng.* **17** 385–414
[14] Jiang W, Chalich Y and Deen M J 2019 Sensors for positron emission tomography applications *Sensors* **19** 5019
[15] Sun Y, Yu H, Ma J and Lu P 2016 The role of ^{18}F-FDG PET/CT integrated imaging in distinguishing benign pleural effusion *PLoS One* **11** e0161764
[16] Bolus N E *et al* 2009 PET/MRI: the blended-modality choice of the future? *J. Nucl. Med. Technol.* **37** 63–71
[17] Shao Y, Cherry S R, Farahani K, Meadors K, Siegel S, Silverman R W and Marsden P K 1997 Simultaneous PET and MR imaging *Phys. Med. Biol.* **42** 1965–70
[18] Conti M and Benedriem B 2019 The new opportunities for high time resolution clinical TOF PET *Clin. Transl. Imaging* **7** 139–47
[19] Dujardin C, Auffray E, Bourret-Courshesne E, Dorenbos P, Lecoq P, Nikl M, Vasil'ev A N, Yoshikawa A and Zhu R-Y 2018 Needs, trends and advances in inorganic scintillators *IEEE Trans. Nucl. Sci.* **65** 1977
[20] Korpar S, Dolenec R, Krizan P, Pestotnik R and Stanovnik A 2012 Study of TOF PET using Cherenkov light *Phys. Proc.* **37** 1531–6
[21] Buffler A, Cole K, Leadbeater T W and van Heerder M R 2018 Positron emission particle tracking: A powerful technique for flow studies *Int. J. Mod. Phys.: Conf. Ser.* **48** 1860113
[22] Svärd S J, Holcomber S and Grape S 2015 Applicability of a set of tomographic reconstruction algorithms for quantitative SPECT on irradiated nuclear fuel assemblies *Nucl. Instrum. Methods* A **783** 128–41

[23] O'D Parker H M and Joyce M J 2015 The use of ionising radiation to image nuclear fuel: a review *Prog. Nucl. Energy* **85** 297–318

[24] Gutierrez A, Baker C, Boston H, Chung S, Judson D S, Kacperek A, Le Crom B, Moss R, Royle G and Speller R 2018 Progress towards a semiconductor Compton camera for prompt gamma imaging during proton beam therapy for range and dose verification *J. Instrum.* **13** C01036

[25] Pinkau K 2009 History of gamma-ray telescopes and astronomy *Exp. Astron.* **25** 157–71

[26] Winkler C *et al* 2003 The INTEGRAL mission *Astron. Astrophys.* **411** L1

[27] Diehl R, Siegert T, Greiner J, Krause M, Kretchmer K, Lang M, Pleintinger M, Strong A W, Weinberger C and Zhang X 2018 INTEGRAL/SPI γ-ray line spectroscopy: response and background characteristics *Astron. Astrophys.* **611** A12

[28] Hosokoshi H *et al* 2019 Development and performance verification of a 3-D position-sensitive Compton camera for imaging MeV gamma rays *Sci. Rep.* **9** 18551

[29] Yoneyama M *et al* 2018 Evaluation of GAGG:Ce scintillators for future space missions *J. Instrum.* **13** P02023

Chapter 8

Charged particle detection

This chapter provides an overview of the diversity of detectors employed in charged particle detection. The chapter is in two parts. The first part addresses challenges in detection of light ions (e.g. α particles) and heavy ions including fission fragments. We distinguish between counting applications and spectroscopy applications. For the latter, we largely focus on silicon detectors as the workhorse of charged particle spectroscopy and describe some of the key features and design criteria for such detectors. The second part of the chapter focusses on detection of electrons/positrons from beta decay and conversion electrons. Again, silicon detectors of various designs play a key role in these applications. Examples of societal applications of alpha, beta and heavy ion detection will be discussed in various domains from nuclear decommissioning to space science.

Concepts: Counting applications: CsI(Tl), plastic scintillator; Spectroscopy applications: silicon detectors; silicon detector telescopes; pulse shape analysis in silicon; position-sensitive detectors; double-sided silicon strip detectors (DSSDs); operation and calibration of silicon detectors; fission fragment detectors; societal applications of alpha and heavy-ion detection; beta and conversion electron detection and spectroscopy; internal pair spectrometers; societal applications of beta particle detection

8.1 Alpha and heavy ion detection

In section 4.4.1, an outline was given of the different technologies for detection of alpha particles (see figure 4.16) and here this approach is generalised to other light ions such as protons, as well as heavy ions. We distinguish applications on the basis of whether spectroscopy is required or not, in other words, whether we want to essentially *count* the number of charged particles at the expense of energy resolution, or whether we want to measure with some precision (i.e. spectroscopically) the energy of charged particles. The two applications can roughly be broken down into a preference for scintillator detectors, e.g. CsI(Tl) for particle identification and

doi:10.1088/978-0-7503-1428-2ch8

counting, and silicon detectors for measuring particle energies with good resolution. As ever in real life applications, we find exceptions to the rule, e.g. silicon detectors employed for heavy ion detection in Coulomb excitation (see section 3.4.5) where the high intrinsic energy resolution of the detectors is of limited relevance but rather it is the high angular/spatial resolution of segmented silicon detectors that is important. In addition, charged particle detectors are often only components of more complex apparatus, e.g. magnetic spectrometers where charged particles are momentum analysed and so their kinetic energy can be obtained from their position recorded on the focal plane of the spectrometer. Here, the detector does not need to be energy sensitive but, rather, position sensitivity and/or timing resolution is key. We will see similar applications in the area of fast neutron detection in chapter 9, where neutron energies can be measured through time-of-flight using a simple detector with excellent timing resolution.

8.1.1 Counting charged particles

Scintillator detectors are frequently used in counting charged particles emitted following nuclear reactions. Scintillators which are not hygroscopic are the most useful in this application since they can be operated ‘bare’ within a vacuum chamber, minimising energy loss of particles entering them. The energy resolution of scintillators is significantly poorer than silicon detectors (to be discussed in section 8.2 below) but they can still provide limited spectroscopy. A good choice of inorganic scintillator for such applications is CsI(Tl) (see section 5.1.2) since it has two components to its scintillation light: a fast component with a mean decay time of about 400–1000 ns and a slow component with mean decay time around 7 μs. The fast component exhibits a dependence on particle type in that it varies in amplitude and decay time depending on whether the interacting charged particles are protons, alpha particles, etc. This is a characteristic of their respective range and $\frac{dE}{dx}$. Pulse-shape analysis can accordingly be used to distinguish the species of interacting particle. We briefly discuss three detector arrays which make use of CsI(Tl) (and plastic scintillator) detectors as a means of effectively counting evaporated charged particles from fusion-evaporation reactions (see section 3.3.1) to identify the reaction channel in fusion-evaporation reactions:

Microball

The Microball [1] constructed in the 1990s uses an array of CsI(Tl) detectors arranged in the geometry shown in figure 8.1; the detectors are arranged in a near-spherical geometry with the individual detectors inclined towards the target. The geometry diverges at forward angles to introduce more granularity, reflecting the fact that more particles are emitted in the forward direction in most reactions of interest. Scintillation light is recorded with photodiodes (see section 5.3.2). The CsI (Tl) scintillation light signal is split into its fast and slow components using analogue electronics. The ratio of the fast and slow component can be used to distinguish different particle types, e.g. protons, deuterons, alpha particles, etc (see figure 8.2). In the subsequent data analysis, the number and type of emitted charged particles can be counted on an event-by-event basis. Operated in conjunction with an array of

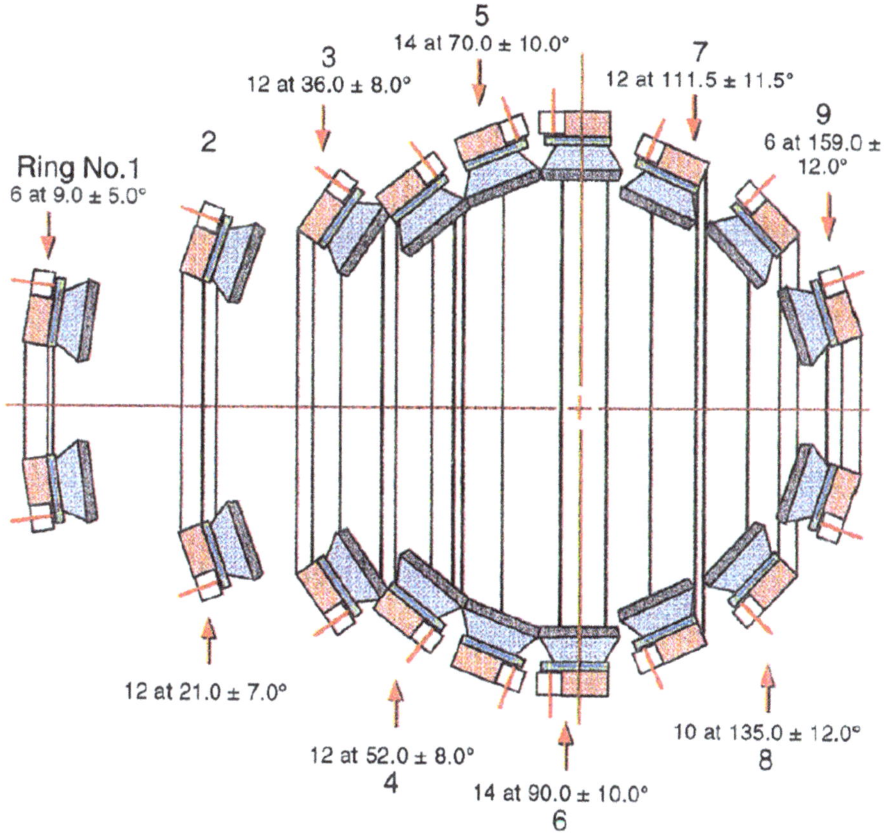

Figure 8.1. Schematic of a vertical section through Microball. The rings of detectors are indicated with their azimuthal angle and the half angle which they subtend. The position of the reaction target is marked with the red cross near the centre. (Reprinted from [1] with the permission of Elsevier.)

high-purity germanium detectors, e.g. Gammasphere (see figure 6.17), this particle count can be used to select different fusion-evaporation channels (see figure 8.3). A key point in the analysis of such data is the occurrence of 'leakthrough', meaning that if, for example, two proton events are selected then this event selection will also likely contain a significant number of events where three or four protons were emitted but the finite efficiency and solid angle coverage of the Microball means that frequently one or more protons were missed.

UoYtube

The UoYtube [2] was constructed for use as a charged-particle veto detector in conjunction with the JUROGAM II germanium detector array at the University of Jyväskylä Finland. The tube consists of 96 CsI(Tl) detectors arranged around six sides of a box with hexagonal cross section (see figure 8.4). The individual detectors comprise a 20 × 20 × 2 mm square tile of CsI(Tl) coupled to an S3590-08 PIN diode from Hamamatsu. Through a digital data acquisition system, pulse shapes can be

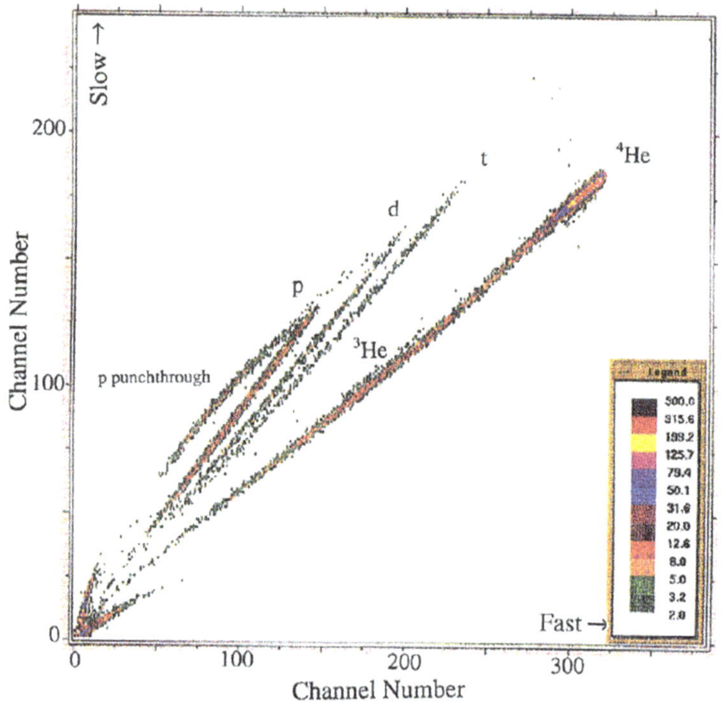

Figure 8.2. Typical charged-particle identification plot for Microball based on plotting the amplitude of the slow component of the CsI(Tl) light output against the fast component. Different loci can be seen corresponding to protons (p), deuterons (d), tritons (t), ^{3}He, and ^{4}He. (Reprinted from [1] with the permission of Elsevier.)

explored. Figure 8.5 shows the different pulse shapes obtained for protons, alpha particles, gamma rays and piled-up events. Comparing the fast and slow components of the pulses measured, different particle types such as protons and alpha particles can be distinguished (see figure 8.6).

JYtube

In some applications with fusion-evaporation reactions, it is desirable simply to veto any event where a charged particle is detected, e.g. when the interest is in the study of exotic proton-rich nuclei which are populated through in fusion-evaporation reactions by neutron emission only. In this case, the pulse shape discrimination afforded by CsI(Tl) is unnecessary. CsI(Tl) is also intrinsically rather slow due to the slow decay component of the light emitted, meaning that arrays of CsI (Tl) suffer pile-up at high rates. This suggests the possibility of replacing CsI(Tl) with plastic scintillator since the signals from plastic scintillator (see section 9.1.2) are very fast compared with CsI(Tl) and can therefore support high detector rates. The JYtube (see figure 8.7)—a veto detector used at the University of Jyväskylä in Finland, uses plastic scintillator elements coupled to 6 mm × 6 mm SensL J-series silicon photomultipiers (the properties of plastic scintillator are described in

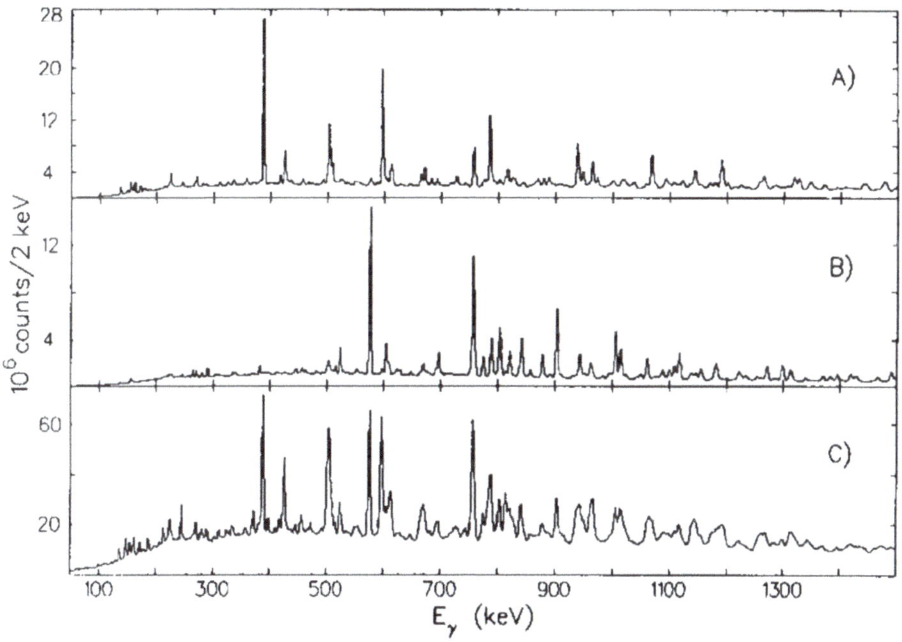

Figure 8.3. Typical gamma ray spectra obtained with the Gammasphere germanium detector array for different charged-particle evaporation channels using Microball. The data is for the fusion-evaporation reaction involving a 130 MeV ^{28}Si beam on a ^{58}Ni target: (a) gamma-ray spectrum selecting the α2p evaporation channel; (b) gamma-ray spectrum selecting the 4p evaporation channel; (c) gamma-ray spectrum without any charged-particle selection. (Reprinted from [1] with the permission of Elsevier.)

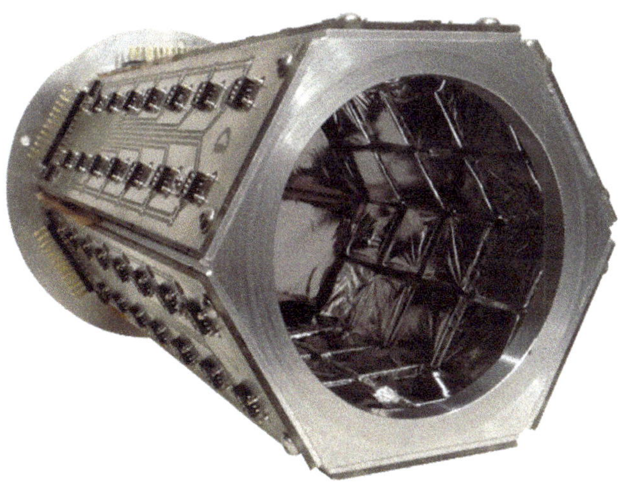

Figure 8.4. The UoYtube comprising 96 CsI(Tl) detectors arranged around six sides of a box with hexagonal cross section. Each detector comprises a 20 × 20 × 2 mm square tile of CsI(Tl) coupled to an S3590-08 PIN diode. (Reprinted from [2] with the permission of IOP Publishing.)

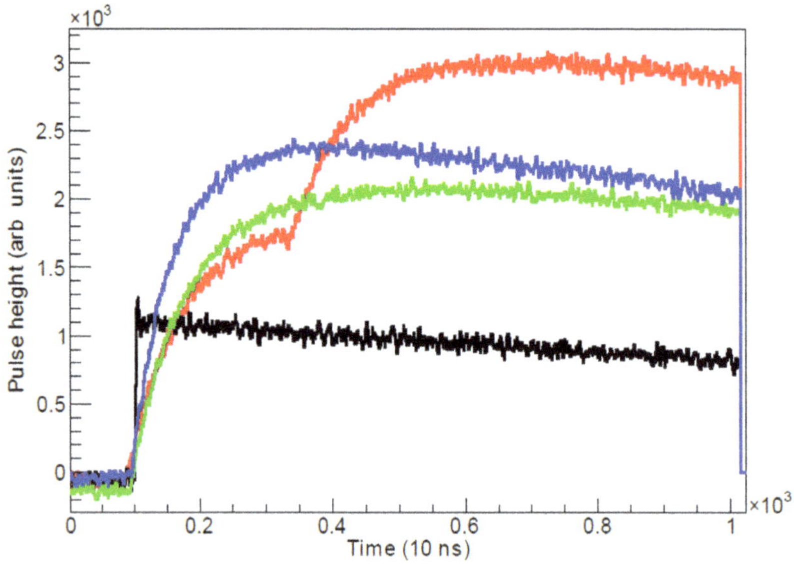

Figure 8.5. Pulse shapes for individual UoYtube detectors for differing particle types: black—a gamma ray; green—a proton; blue—an alpha particle; red—two charged particles in short succession.

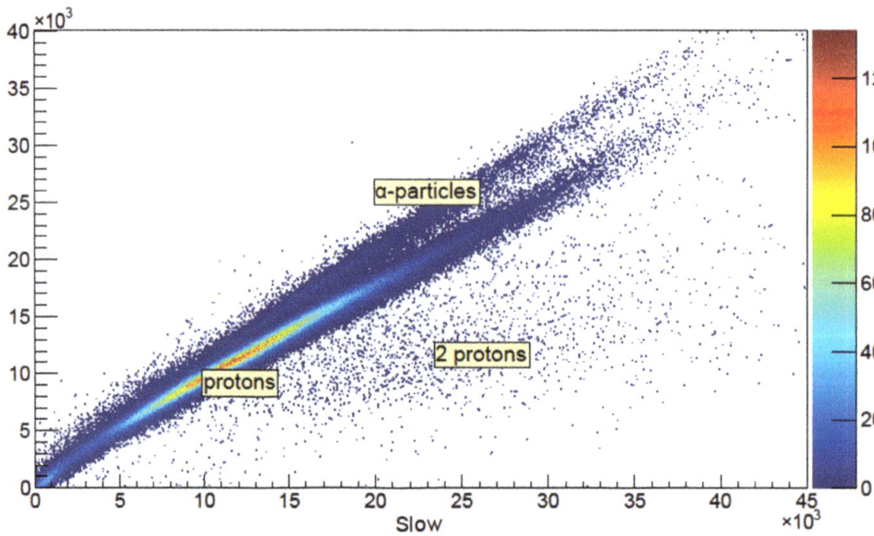

Figure 8.6. The fast signal from the UoYtube (0 → 3 μs) plotted against the slow component 4 → 9 μs). Events due to protons and alpha particles can be distinguished in the figure.

section 9.1.2). The modern technology of SiPMs (see section 5.3.3) is more suitable in this application than photodiodes as the SiPMs have a large gain leading to signals which do not need preamplification and can be directly digitised; this dispenses with the preamplification stage required when using photodiodes.

Figure 8.7. The JYtube comprising a similar hexagonal arrangement to the original UoYtube. Each detector comprises a 20 × 20 × 2 mm square tile of plastic scintillator (Eljen EJ-204) coupled to a prism-shaped perspex light guide and joined to a 6 mm square SensL C-series SiPM. The tube comprises two hexagonal halves which join around the reaction target position. Two hexagonal-shaped endcaps (not shown) complete the encirclement of the target. (Photo credit: the author.)

8.2 Spectroscopy of charged particles: silicon detectors

Silicon detectors are the workhorse of charged particle spectroscopy. In contrast to germanium detectors which can have a significant volume in order to be efficient for gamma-ray detection, silicon detectors are thin, planar, detectors designed to be operated within a vacuum chamber. They provide typical energy resolutions of the order of 30 keV for 5 MeV alpha particles depending on their size and thickness. Elsewhere in this e-book, we have reviewed the basic properties of semiconductors in section 4.3.3 and already described at length, with respect to germanium detectors, many other relevant aspects of semiconductor detectors such as depletion depth (see section 6.2.2). Accordingly, here, we focus on the differences between silicon and germanium, chief of which is silicon's bandgap at room temperature of 1.14 eV compared to 0.67 eV for germanium. The larger bandgap of silicon means that thermal excitation of charge carriers is strongly suppressed compared to germanium meaning that silicon detectors can be operated effectively at room temperature and

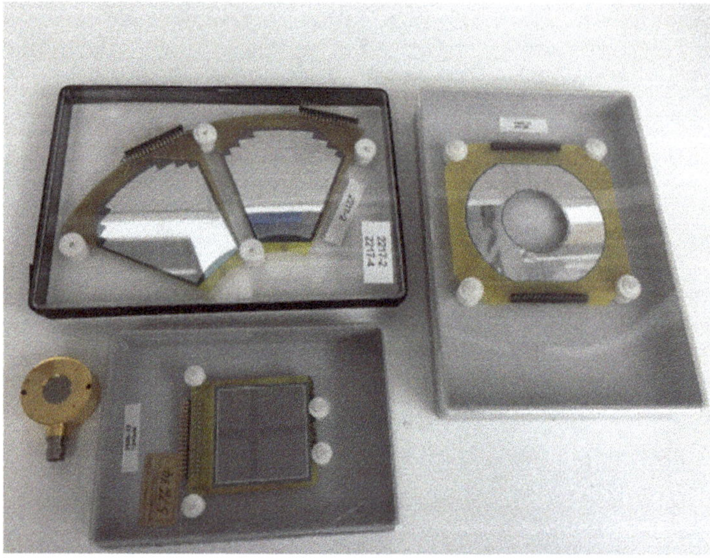

Figure 8.8. Selection of silicon detectors including DSSDs. (Photo credit: the author.)

do not necessarily need to be cooled[1] in the same manner as germanium detectors where it is essential to operate at liquid nitrogen temperatures. This dispenses with the complicated infrastructure needed for the cooling allowing silicon detectors to be mounted in a compact geometry.

8.2.1 Fabrication and design of silicon detectors

The common starting point for silicon detector manufacture is a high-quality silicon wafer, typically 6″ diameter. These wafers are described as detector-grade since they have high-resistivity much exceeding the properties of standard silicon wafers used for other commercial purposes. This high resistivity is important since a key parameter for silicon detectors is the leakage current. Leakage current translates into noise in the detector output and largely originates from thermally generated electron–hole pairs in the material. It scales with the active area of the device meaning that the smallest silicon detectors have the best energy resolution.

The standard 6″ silicon wafer is a size that allows a wide variety of different detector shapes and sizes to be produced (see figure 8.8). The wafers as produced will be either *p-type* or *n-type*—as with germanium detectors, this choice can dictate other properties of the material such as its resistance to radiation damage. To turn the wafer into a diode structure requires the introduction of a doping layer, e.g. p^+ doping for an n-type wafer near one of the surfaces. This is achieved in industrial clean-room facilities through ion implantation [3][2]. The wafers are then metallised to

[1] Although cooling silicon detectors may improve their performance by reducing leakage current and noise-related effects.

[2] Earlier silicon processing used ion diffusion but ion implantation gives superior results and reliable, better quality detectors at scale.

provide electrical contacts, diced out of the wafer and mounted on plastic boards with readout tracks and bonding wires are attached. In some cases, replacement of the plastic with ceramic can provide a cleaner system more suitable for ultrahigh vacuum. A grounded guard ring around the perimeter of the detector is usually employed to ensure a uniform electric field distribution [4, 5]. This loses some element of the detector area which may be important when detectors want to be tiled. Accordingly, efforts are under way to produce edgeless detectors with an 'active-edge' [6].

Silicon detectors can be manufactured in thicknesses from 20 μm [7] up to 1 mm or more. The thinnest detectors have a thickness comparable to the range of a typical alpha particle from a sealed source (see figure 8.10) but such detectors are considerably fragile and need caution in their handling. The thicker detectors are more mechanically robust and with thicknesses of around 1 mm, they can also be used for detection of protons, which have longer range than alpha particles of the same energy, and for conversion electrons as well as low energy x-rays and gamma rays. The latter application to conversion electrons will be discussed more extensively in the second part of this chapter (in section 8.4.2).

Even without an applied bias voltage, a silicon detector will have a thin depletion region arising from thermal diffusion of doping ions leading to a small 'in-built' junction potential. The application of a reverse bias will deplete the detector bulk and the depth of the depletion layer may be estimated using equation (6.1) which was discussed earlier in the context of planar germanium detectors. (Strictly speaking a small correction to the reverse bias voltage should be made by adding the small contribution, ~0.5 V, from the 'in-built' junction potential.) Silicon detectors are usually operated fully depleted. Indeed, the bias voltage often considerably exceeds the voltage for full depletion known as *over-depletion* in order to improve charge collection, reduce capacitance and, hence, optimise energy resolution. The dependence of the capacitance of the fully-depleted detector on the bias potential, V_0, is given by

$$C \propto \frac{1}{\sqrt{V_0}}. \tag{8.1}$$

Choice of a silicon detector for a particular application will depend strongly on the species of charged particles to be detected and their energy. The motivation behind the choice of detector geometry will be discussed more extensively for a range of different applications below.

Some basic silicon detectors suitable for e.g. alpha particle detection are passivated implanted planar silicon (PIPS) detectors; see, for example, the range offered by Mirion. These are conventionally available in a range of thicknesses from 100 to 500 microns, and with a wide range of active areas from 25 mm^2 to 5000 mm^2. In general, the smaller the sensor, the better the performance due to lower leakage

[2] Earlier silicon processing used ion diffusion but ion implantation gives superior results and reliable, better quality detectors at scale.

currents. The smallest detectors can offer an alpha particle energy resolution, i.e. for a ^{241}Am source, of 12 keV, while the largest detectors have energy resolutions of order 35 keV.

8.2.2 Identifying charged particles with silicon detectors

Depending on the application, it may be relevant to use silicon detectors not only to make a spectroscopic energy measurement but also to determine the species of the detected charge particle. Historically, this was mostly achieved by constructing *detector telescopes* with layers of two or more detectors. In recent years, the intrinsic properties of the charge production and collection process in silicon have been harnessed to identify particle species using *pulse shape analysis*, paralleling many of the developments in pulse shape analysis with large volume germanium detectors (see section 6.5.2). These advances, as with germanium detectors, have been driven by the advent of digital data acquisition. Let us describes these two approaches to identification of charged particle species.

Silicon detector telescopes

The concept behind a *detector telescope* arises from considerations of how energy loss $\left(-\frac{\mathrm{d}E}{\mathrm{d}x}\right)$ and range of light charged particles in silicon differs (see discussion in section 2.2.1). Figure 8.9 shows the strong dependence on Z, where there is a large difference in $-\frac{\mathrm{d}E}{\mathrm{d}x}$ when comparing, for example, protons and alpha particles for a given energy. In addition, there is a smaller dependence on isotope where, for example, protons, deuterons (^{2}H) and tritons (^{3}H) exhibit differing energy loss curves. These differences in energy loss are naturally reflected in differing ranges for charged particles in silicon; figure 8.10 presents the range in silicon for light ions as a function of their incident energy. It can be seen that a 1 mm thick silicon detector will fully stop alpha particles up to 30 MeV but a 10 MeV proton will 'punch through' meaning that it will exit the detector having not deposited all its energy. This differing range for particles might be seen as an inconvenience but it can be exploited to create a *detector telescope* which identifies the particle species. A typical telescope arrangement for charged-particle identification comprises two layers of silicon detectors: one thin ($\sim$100 μm) and one thick ($\sim$1 mm), where alpha particles will be fully stopped in the front layer and protons will lose energy in the first layer but deposit most of their energy in the second layer. While the total energy deposited in the two detectors may be the same, discrimination between light charged particle types can be readily obtained by comparing the energy deposited in the front and back detectors (indeed, often, as in the example just discussed, it may be trivial as a particle fully stops in the first layer). Examples of silicon detector arrays that employ the telescope approach are presented in section 8.2.4.

Pulse shape analysis

Pulse shapes from silicon detectors are found to vary as a function of the energy and species (Z, A) of incident ions. This relates to differences in the depth of penetration,

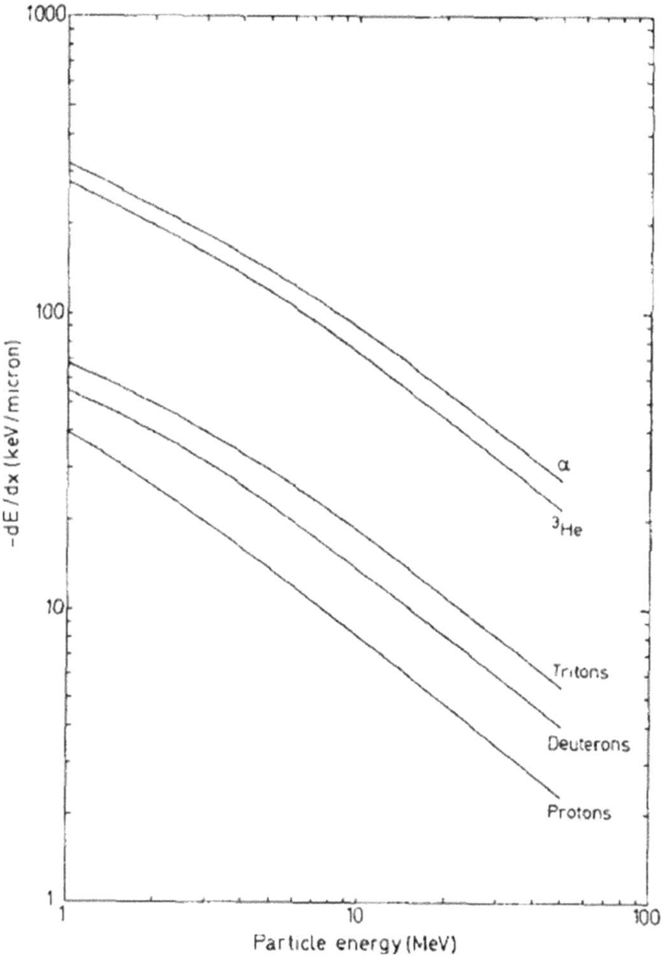

Figure 8.9. Energy loss, $-\frac{dE}{dx}$, of light charged particles in silicon as a function of particle energy. Energy loss is presented in units of (keV μm^{-1}). (Reprinted from [8] with the permission of Elsevier.)

i.e. particle range and the ionisation density profile, i.e. $\frac{dE}{dx}$ along the particle track. The effect of this is subtle in that the ionisation density may be high enough to create plasma conditions in the semiconductor which creates a time delay in the electron–hole separation due to a shielding from the bulk electric field in the semiconductor [9]. Moreover, the electric field in the semiconductor may be distorted by the highly ionised track which has a high conductivity with respect to the bulk material. Various strategies can be applied in terms of the pulse shape analysis. For example, the rise time of the pulse from the charged-sensitive preamplifier can be used to determine the particle type for light ions—note the parallel with the employment of rise time in germanium detectors (see section 6.5.2) for obtaining depth and radius of the point of interaction—but, here, the information obtained is different. Figure 8.11 provides an example of such an analysis, plotting the rise time against the energy of

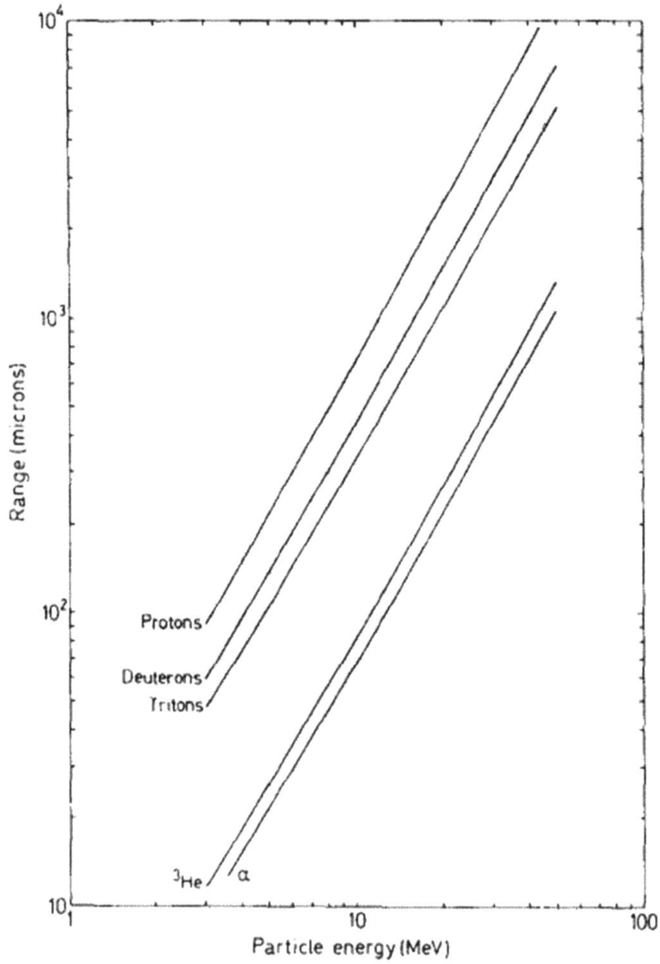

Figure 8.10. Calculated ranges of light charged particles in silicon as a function of particle energy. (Reprinted from [8] with the permission of Elsevier.)

particles to discriminate isotopes up to $A = 12$ [10]. In this example, rise time is defined as the time for the signal to rise from 20% to 80% of the full height. A parallel approach is to investigate the maximum of the current signal from the detector, achieved by differentiating the output of the charge sensitive preamplifier prior to digitisation. Figure 8.12 shows an implementation of this approach with a test experiment using the ISO-FAZIA system (see reference [11] for full details). Separation of charged particles by Z is achieved up to $Z = 36$ and isotopic separation is also possible for particles with $Z < 20$.

8.2.3 Obtaining position sensitivity within silicon detectors

In chapter 3, we outlined some topical examples of studies in experimental nuclear physics such as Coulomb excitation (section 3.4.5) and single-particle transfer

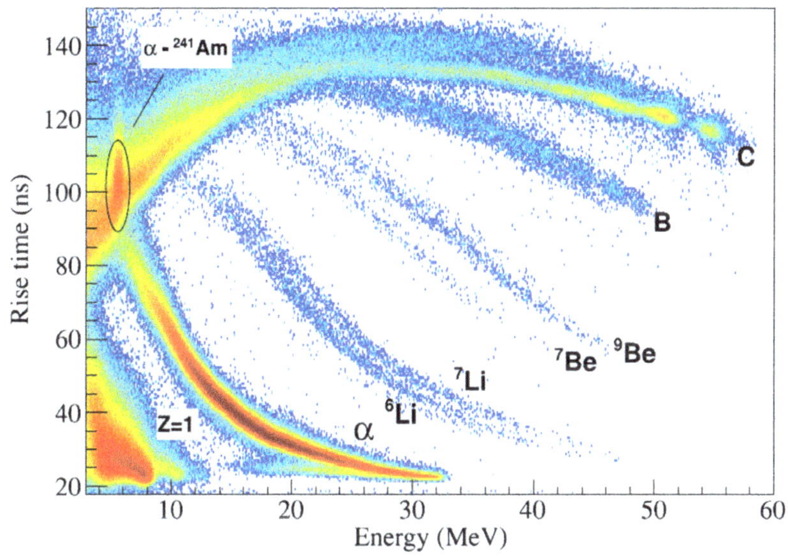

Figure 8.11. Rise time versus detected particle energy for different reaction products from the $^{12}C+^{27}Al$ reaction at 63 MeV. Bands corresponding to different particle types are marked. (Reprinted from [10] with the permission of Elsevier.)

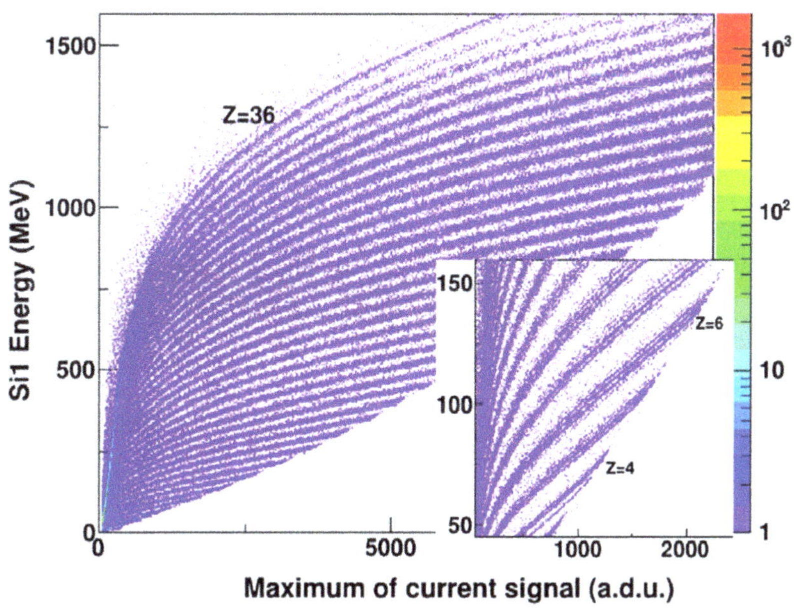

Figure 8.12. Energy recorded in the silicon detector against the maximum of the current signal, obtained from pulse-shape analysis. The experiment used a ^{80}Kr beam at 35 MeV/u incident on a ^{40}Ca target, hence, fragments up to $Z = 36$, corresponding to Krypton, are observed. Isotopic separation is observed up to $Z = 20$; the inset shows the isotopic separation at low Z. (Reprinted from [11] with the permission of Elsevier.)

reactions (section 3.4.6) where it is important to detect both light and heavy ions with high position (or angular) resolution. Some classes of silicon detector are able to provide this functionality: the older-generation of *position-sensitive detectors* which use resistive strip technology, and more modern *double-side silicon strip detectors (DSSDs)*. We will outline the functionality of these two different classes of detectors below along with examples of where they are used.

Position sensitive detectors

A position sensitive detector (PSD) is an older detector design intended to provide information on the position of interaction [12–14]. The operating principle behind some different designs of PSD is shown in figure 8.13. The most common design for a two-dimensional PSD is to have a resistive electrode and read out signals from contacts on the four corners. The position sensitivity is derived from the ratio of the signals on the four corners by means of *resistive charge division*. The cathode on the reverse can be standard and then be used to read out a signal with good energy resolution. The chief drawback of such detectors is that the position and energy calibration is significantly more complex than with double-sided silicon strip detectors, which are discussed in section 8.2.4 below; this is because there is no intrinsic pixellation. The standard technique is to apply a mask with small regularly spaced holes in it in order to characterise the position sensitivity using a source. Position distortion occurs due to the ballistic deficit effect [15]; reconstruction of the distorted signal is discussed in [16].

Figure 8.14 shows an example of a linear array of position-sensitive silicon detectors used as part of the HELIOS spectrometer. The array is used to detect light

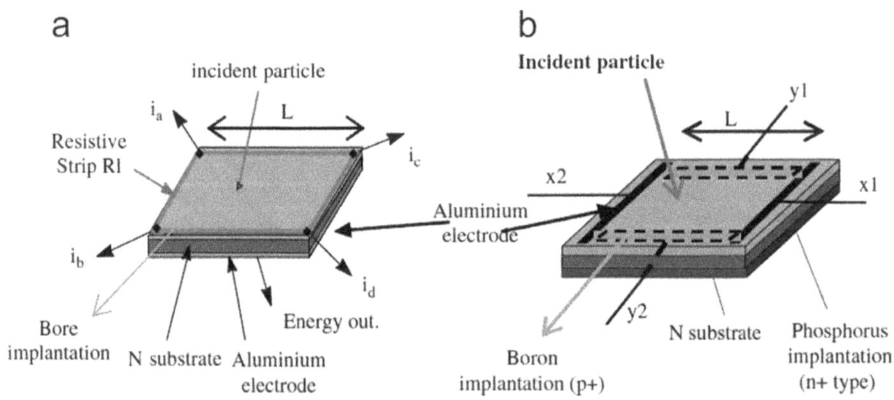

Figure 8.13. Two designs of position sensitive detectors (PSDs): (a) pin-cushion PSD where the front side is a resistive sheet; the signals are taken at the four metallised contacts located on the corners and are used to determine the position. The back side is a metallic contact film used to improve the energy resolution. Without the resistive strip, Rl, the spatial resolution is high but the spatial linearity is very poor. With a resistive strip the response is quite linear while the spatial resolution is degraded; (b) a duo-lateral PSD where both sides are resistive with contacts located at the four edges on both sides (and orthogonally); this kind of PSD offers both a quite good spatial resolution and a good spatial linearity; but the drawback of this detector is the poor energy resolution. (Reprinted from [13] with the permission of Elsevier.)

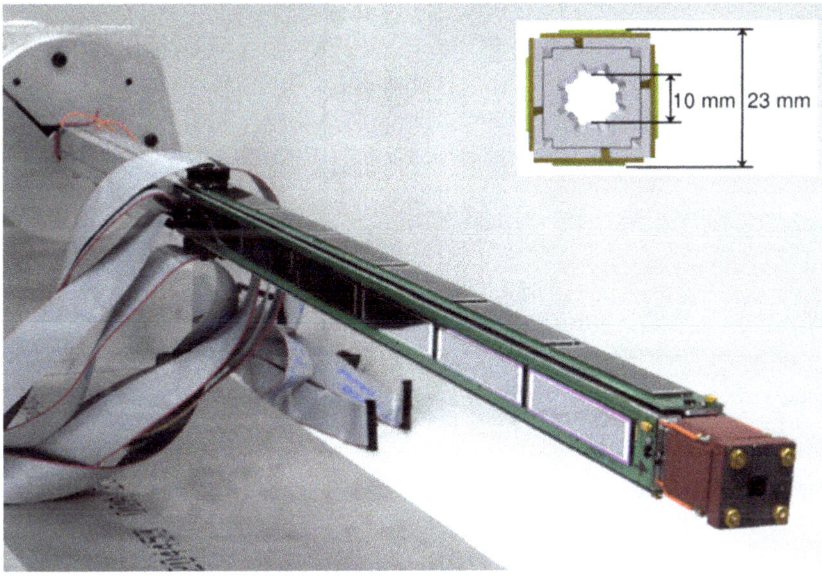

Figure 8.14. Silicon detector array for the HELIOS spectrometer at Argonne National Laboratory. Each detector is position sensitive and has an active area of 9 × 50.5 mm. The inset is a cross-section of the array. (Reprinted from [17] with the permission of Elsevier.)

charged particles following single-particle transfer reactions within the bore of a solenoidal magnet. The particles follow helical trajectories before striking the linear silicon array; this application was discussed in section 3.4.6.

8.2.4 Double-sided silicon strip detectors

A significant advance in development of silicon detectors from the point of view of nuclear physics was the advent of double-sided silicon strip detectors (DSSDs) [18]. The silicon is divided into strips on each side using passivated strips in between typically comprising SiO_2 of about 0.1 mm width. This can be applied either to a rectangular detector achieving x, y strips or to an annular detector where it is divided on one side into rings and the other side into sectors. The dimensions of these strips/rings/sectors can be tuned according to the application of interest. There are several manufacturers of such detectors; the widest range of detector geometries is offered by Micron Semiconductor Ltd.

The division of the silicon detector into strips provides effective pixelisation of the detector which is exploited in the subsequent data analysis, where following appropriate calibration of the strips, events detected in strips on the two sides can be associated by requiring that they correspond in energy within a certain tolerance set according to the experimental conditions and the relevant energy resolution. Not all events can be reconstructed in such a fashion because there is non-negligible probability that an incident particle hits the passivated region between strips. When this happens, two effects can occur: (1) electron trapping so incomplete charge collection results and (2) charge sharing where the charge is collected on two

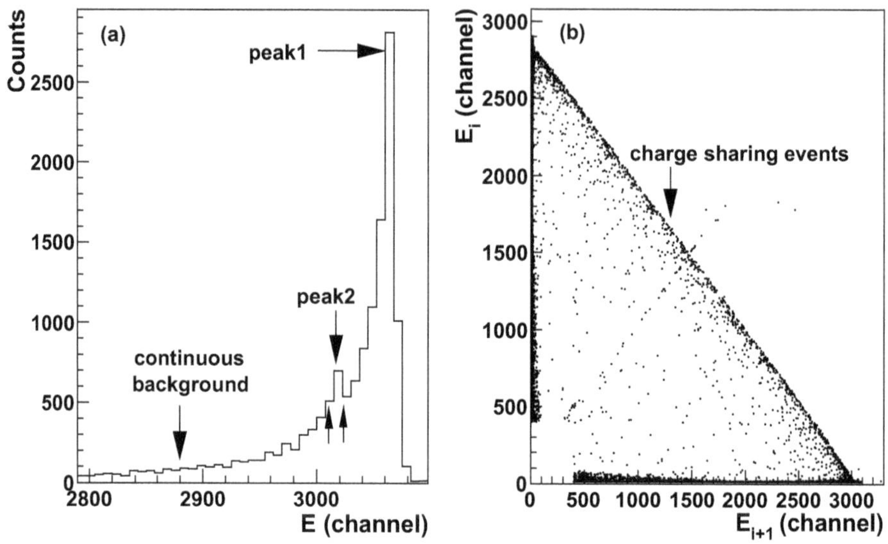

Figure 8.15. Illustration of the effects introduced by inter-strip passivation on a DSSD illuminated by a ^{241}Am source (left) incomplete charge collection within a single strip leading to a low energy tail on an alpha peak (right) charge sharing shown by plotting the apparent energy recorded in two adjacent strips. (Reprinted from [19] with the permission of IOP Publishing.)

adjacent strips [19]. These two effects are shown in figure 8.15. These design-related features need careful handling in the subsequent data analysis and some events are inevitably discarded reducing efficiency.

Implantation detectors

DSSD detectors are commonly used for implantation at the focal plane of a recoil spectrometer; such a spectrometer separates fusion residues from scattered beam using a combination of electric and magnetic fields; this application was briefly introduced in section 3.3.1. The benefit of a silicon detector in this application is that it can record both the implantation of the recoiling nucleus and its subsequent decay, if the decay is via emission of a particle with a short range, e.g. an alpha particle. The intrinsic segmentation of the DSSD into effective pixels allows implantation and subsequent decay to be correlated so long as the decay process is fast relative to the implantation rate per pixel. This technique is known as recoil-decay tagging (RDT). Typically in RDT studies, the heavy ion implanted has a relatively low energy. Given that the energy loss for such a heavy ion is high in silicon compared to light ions such as protons and alpha particles, the heavy ion will implant close to the front surface of the silicon detector. This short range is much less than the thickness of any practical silicon detector. The thickness of the detector used can vary and this is dictated by the characteristics of the subsequent decay of interest:

- Where the decay concerns emission of an alpha particle or proton then a thin detector will be optimal as the thinner the detector, the better the achievable energy resolution. Typical thicknesses employed are in the range 150–700 μm. In some cases involving heavy nuclei, the alpha emission will reach excited states in the daughter nucleus which then decay by emission of conversion electrons. Since both the alpha particle and electrons are emitted from the same implanted atom within the silicon detector, the energy deposited by the alpha particle and the conversion electron may effectively sum which is a feature that needs to be taken account of in the analysis [20].

- Where the decay concerns emission of a beta particle, energy resolution is of no importance because beta particles follow a broad distribution of energies—the Fermi–Kurie distribution (as described in section 1.4). Here, the identification of the beta particle emission is important as a means of correlating events. However, beta particles deposit much less energy in a given range of silicon than alpha particles. In this case, DSSDs with thicknesses up to 1 mm are more suitable so that energy loss by the beta particle is higher in order to identify beta particles above the noise.

An example of an implantation system is the GREAT spectrometer [21] at the University of Jyväskylä which is designed for implantation and decay studies. Two rectangular DSSDs are mounted side by side and cover the focal plane of the RITU recoil separator (see figure 8.16). The photograph also shows a planar germanium detector (see section 6.4.1) which is installed behind the DSSDs and can record low-energy gamma rays or x-rays, as well as beta particles.

Figure 8.16. Focal plane of the GREAT spectrometer at University of Jyväskylä. The planar germanium detector is lowered to show the two DSSDs. (Photo credit: the author.)

Scintillator alternatives to silicon implantation detectors

As an aside, it is worth mentioning innovations using scintillators as replacements to DSSDs (and other applications where silicon detectors are used). Scintillators lack the high energy resolution of silicon detectors but they have a number of advantages such as being radiation hard, and dependent on the scintillator they can have very fast time response allowing fast counting. A good example of a replacement scintillator is YAP:Ce [22, 23] which is a material readily produced in thin sheets—a common application for it is as an electron or x-ray screen. YAP:Ce has a fast decay time of 28 ns. This makes it possible to resolve two successive events within 100 ns of each other. This has initially been applied to a specialist application of searching for two fast successive alpha decays following implantation of a recoiling heavy ion [24]. Figure 8.17 provides a schematic of a novel implantation detector based on YAP:Ce; the system can achieve sub-mm position resolution in x and y which is comparable to that of a silicon-based DSSD [24].

Annular detectors

Annular detectors are an elegant design for nuclear physics experiments with accelerated beams [25]. With the beam directed down the centre of the hole in the annular detector, the rings will subtend the same angle with respect to the beam direction. An example of such an annular detector is the S3 pattern from Micron Semiconductor Ltd which is shown in figure 8.18. The S3 detector is electrically segmented into 24 rings on the junctions side and 32 sectors (wedges) on the ohmic side. Data is usually collected independently for the two sides of the detector and can be reassociated in the offline analysis to determine more precisely the point of interaction from requiring that the energy for an event recorded on the two sides matches. An exception to this is in the case of *charge sharing* where the interaction took place at the point of the electrical segmentation and charge is recorded in the two adjoining annular rings or wedges.

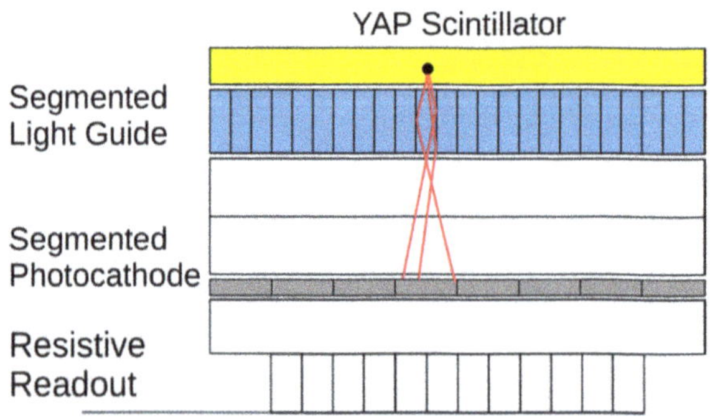

Figure 8.17. YAP:Ce scintillator detector for implantation studies. The YAP:Ce crystal is a monolithic slab with 50.8 mm diameter and a thickness of 1 mm. The scintillation light enters a segmented, quartz, light guide with 24 × 24 pixels, and falls on to a Hamamatsu position-sensitive photomultiplier tube (PSPMT) with 64 segments. (Reprinted with permission from [24]. Copyright (2019) by the American Physical Society.)

Figure 8.18. Micron S3 detector mounted on a ceramic mount. (Photo credit: the author.)

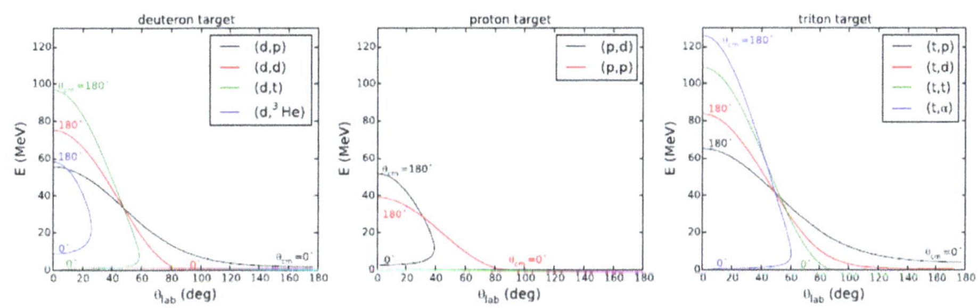

Figure 8.19. Kinematics of various transfer reactions for a ^{54}Ca beam at 10 MeV/u impinging on a light ion target: deuteron (left), proton (centre) and triton (right). The kinematics are shown as a function of lab angle, each locus has the centre-of-mass angle labelled at the two ends: $\theta_{cm} = 180°$ and $\theta_{cm} = 0°$. (Reprinted from [27] with the permission of IOP Publishing.)

Examples of silicon detector arrays for transfer reaction studies

Silicon detector arrays are ideal for nuclear reaction studies such as single-particle transfer reactions (see section 3.4.6) where they are often combined with an array of high-purity germanium detectors. This technique is increasingly being applied to studies with radioactive beams in inverse kinematics, meaning a high-Z beam on a low-Z target, e.g. deuterated plastic target. Reviews of this topic and the relevant methodologies are given in references [26] and [27]. Annular silicon detectors are very suitable for this application based on their geometry which provides good angular resolution. Measurement of particles as a function of angle with respect to the beam direction is critical to this application. Figure 8.19 illustrates the kinematics of typical light-ion transfer reactions in inverse kinematics; most light ion ejectiles are detected at small angles (up to ~60°) where $\theta = 0$ is the beam direction. A large portion of this angular range can be readily covered with an annular silicon detector (or detector telescope) at forward angles. Moreover, such annular detectors have a hole in the middle, allowing the beam to pass through the silicon detector.

Annular silicon detectors can be used to identify which excited states are produced in the transfer reaction and the probability of excitation as a function of angle from the measured intensity of light charged particles, e.g. protons, deuterons, alpha particles detected in the various annular rings. In practice, the achievable energy resolution for the charged particles is limited by the properties of the accelerated radioactive beams employed and the parameters on the targets used since these often need to be thick to maximise the number of reactions observed with low intensity radioactive beams. Such arrays of annular silicon detectors are therefore commonly coupled with high-purity germanium detectors (see chapter 6) or other scintillator-based gamma-ray detectors with good intrinsic energy resolution, e.g. $LaBr_3(Ce)$ detectors (see section 5.2.1). The germanium detectors record γ rays emitted from the excited states populated in the reaction. Selecting coincidences between these characteristic gamma rays and the light charged particles can be used to separate the different reaction channels and effectively recover the energy resolution for charged particles. There are several such systems in operation worldwide; some examples include TIARA [28], SHARC [29] and CAKE [30]. Here, we focus in more detail on a specific example, the STELLA array [31].

STELLA

STELLA [31] is a dedicated station built to measure the $^{12}C+^{12}C$ fusion reaction which is one of the key processes governing the evolution of massive stars [32]. STELLA is designed to measure evaporated charged particles, i.e. protons and alpha particles in coincidence with gamma rays from excited states. A 3D CAD drawing of STELLA is presented in figure 8.20. The evaporated charged particles are measured with three annular silicon strip detectors: two of Micron S3 design and one of Micron S1 design (see figure 8.21). One S1 and S3 form a telescope to discriminate protons from alpha particles. STELLA has been coupled to the FATIMA $LaBr_3(Ce)$ array described in section 5.4.2. This has offered yet another means of discriminating between light ions such as protons and alpha particles using the excellent, fast rise time properties of the $LaBr_3(Ce)$ detectors as a time reference. Figure 8.22 shows the energy of light charged particles detected from $^{12}C+^{12}C$ fusion as a function of the time difference between the detection of a coincident gamma ray with FATIMA and a charged particle in the annular silicon DSSD. Two loci are observed—one corresponding to protons and one to alpha particles; the separation achieved is because of the differing rise time in the silicon detector of the proton and alpha particle-induced events. It is worth emphasising that it is the sub-ns timing resolution of the $LaBr_3(Ce)$ detectors that makes this possible. Large germanium detectors have too poor a timing resolution to allow this approach to be applied.

8.2.5 Specialist applications of silicon detectors—storage rings

Nuclear physicists are increasingly exploring the potential of storage rings for nuclear reactions studies [33]. Here, highly-charged, high energy ion beams circulate in a ring

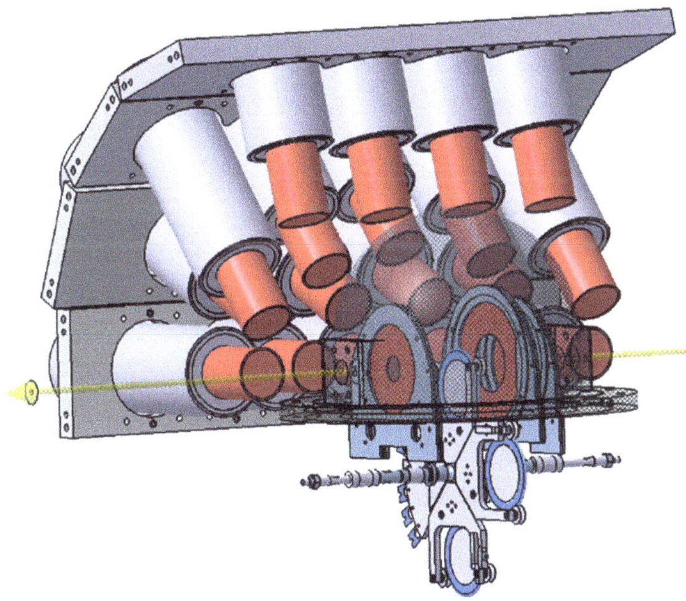

Figure 8.20. 3D-rendered CAD drawing of STELLA. The volume of the scattering chamber is shown semi-transparent in order to allow its interior to be viewed, comprising three annular silicon detectors, and a target holding/rotating mechanism. The large circular carbon target foils are shown with blue frames on their perimeter. The direction of the ^{12}C beam (from right to left in the diagram) is shown with the yellow arrow. Only half of the $LaBr_3(Ce)$ gamma-ray array is depicted for clarity of presentation.

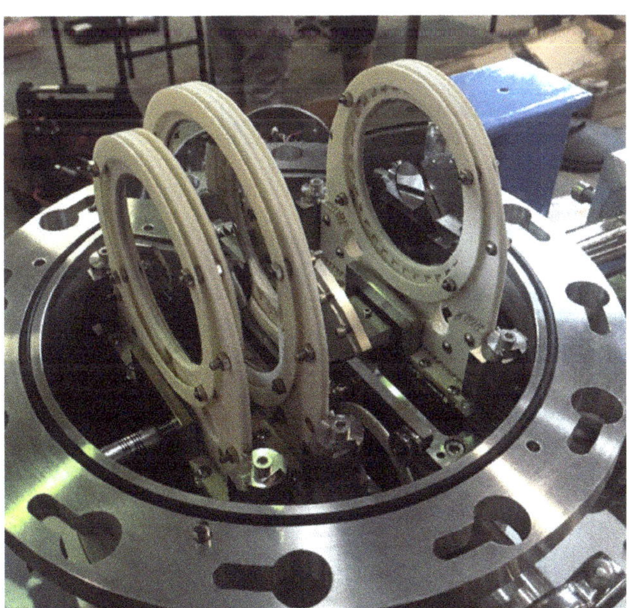

Figure 8.21. Annular silicon detectors: 2 off Micron S3 and one Micron S1 (the larger area detector in the centre) mounted inside the reaction chamber of STELLA. (Photo credit: the author.)

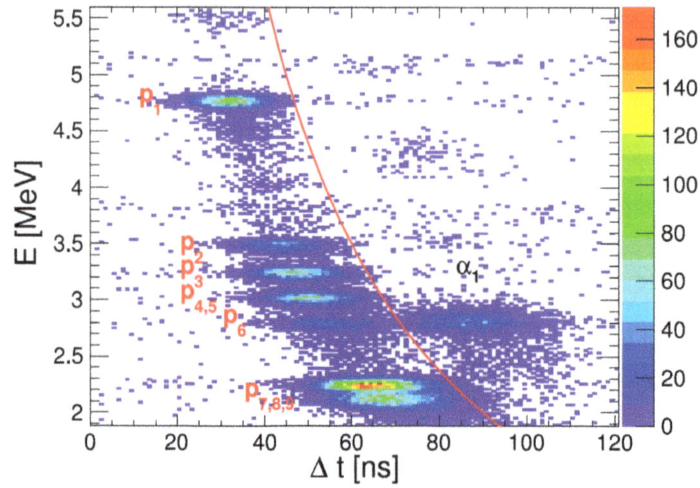

Figure 8.22. Distinguishing evaporated protons and alpha particles using time difference between the silicon signal and $LaBr_3(Ce)$ array in STELLA. (Reprinted from [31] with the permission of Elsevier.)

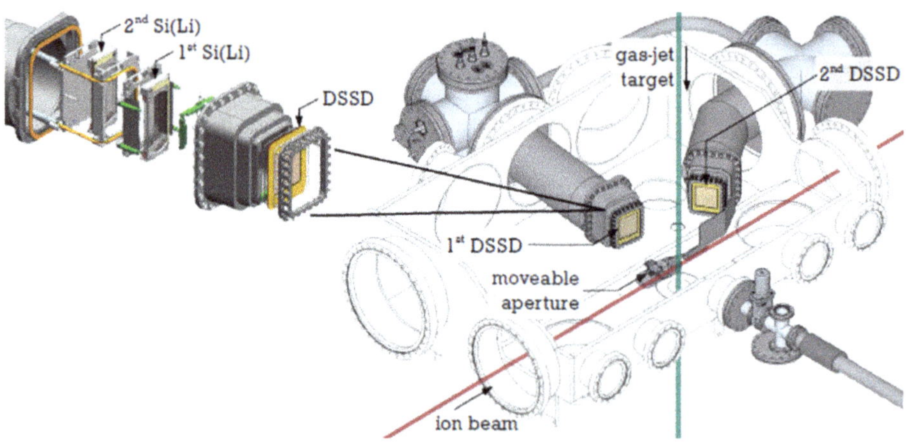

Figure 8.23. Design of DSSD telescopes to detect reactions in a storage ring. (Reprinted from [34] with the permission of IOP Publishing.)

and can initiate scattering or transfer reactions with, e.g. light nuclei in a gas-jet target. Storage rings operate in ultrahigh vacuum (UHV) conditions, i.e. with a pressure below 10^{-10} mbar. A conventional silicon detector would be unsuitable for such conditions as they are generally manufactured on a plastic frame. The plastic would outgas spoiling the UHV conditions and the detector could not be baked. Specialist DSSD detectors have been developed for the ESR storage ring at GSI [34]. These detectors are on a specialist ceramic mount so they can be baked. They form an effective window between the UHV of the storage ring and an auxiliary (lower quality) vacuum in a pocket behind them where further detectors such as thick Si(Li) detectors can be mounted to form a detector telescope (see figure 8.23).

8.2.6 Operation and calibration of silicon detectors

Having reviewed the basic physics of silicon detectors and some of their many variants and applications, it is appropriate to say something more specific about how to operate them and achieve the best performance. Firstly, a suitable, low-noise, regulated bias supply will be needed which is capable of handling the necessary voltage and currents. The bias voltage required depends on the silicon detector in use but is of order 10–500 V. Ideally, the supply should have the capability of reading out leakage current to high precision since this is one of the key operating parameters and a major shift in leakage current is one of the first indications that there is a problem. It's important to note, that if the silicon detector is not in a temperature-stabilised environment then a diurnal cycle will be commonly seen in the leakage current due to the change in ambient temperature over the course of the day and night. This is normal and should be monitored during operation. The bias supply should also have the capacity to ramp up the bias voltage slowly and ramp off the voltage should there be a power outage although this is not such a big concern as it is for germanium detectors which operate on much higher voltage. A suitable low-noise preamplifier will also be required. Preamplifers are also available in high density units with 16–64 channels of input which is advantageous for DSSDs commonly employed in experimental nuclear physics. Preamplifier output can then go to a suitable shaping amplifier and then to an MCA, or directly to a digitiser unit. There are several options on the market for relevant NIM (or other standard) electronics. In nuclear physics, a commonly-found electronics platform for silicon detectors is that supplied by Mesytec who offer, for example, a four-channel bias supply with high precision current readout, the MHV-4, and high-density preamplifiers, e.g. the MPR-16.

Basic setup of a silicon detector

Testing of a silicon detector will commonly take place in a vacuum chamber using an alpha source; alpha particles lose energy in air and so a good vacuum is required to minimise such energy loss. A silicon detector is conventionally supplied with details on its depletion voltage and an *I–V* curve which plots the leakage current as a function of applied bias. It is a common source of misconception and so worth explaining clearly here that the situation with regard to biasing the detector is rather different to that for a germanium detector. With a germanium detector, a recommended bias voltage is provided and operating above this voltage will be risky. Silicon detectors can often be safely operated significantly above the nominal depletion voltage (sometimes two or three times higher than the nominal value) and their performance can continue to increase as charge collection becomes more rapid in the increased electric field. The key is to monitor the leakage current and observe that it is following the expected diode response. A further difference to germanium detectors is that most silicon detectors will operate at some level even with no bias applied due to the 'built-in' bias potential and the thin depletion region setup by it, albeit with very poor energy resolution.

A first step would be to verify the supplied *I–V* curve while monitoring the output signal on an oscilloscope, increasing the bias voltage in small steps to allow the system to settle. This should verify that the detector is performing according to specification. Energy calibration of a silicon detector is conventionally carried out with a triple-alpha source (see section 3.1.1) to ensure linearity. In the same procedure, the energy resolution can be determined from fitting the peak widths. Given that each of the three main alpha peaks has a small satellite peak below it, being able to resolve these satellites visually in a histogram immediately implies the energy resolution is good. There are some subtleties to energy calibration of silicon detectors that must be borne in mind. Firstly, silicon detectors frequently have a dead layer (SiO_2) and metallisation, e.g. aluminium on their surfaces. An alpha particle from a source loses energy continuously along its track and so will already have lost some energy before it enters the active volume of the silicon detector. This needs to be corrected for otherwise there would be a discrepancy in the suggested energy of alpha particles emitted from a heavy ion implanted into a DSSD, for example. Such energy losses can be calculated and removed from the calibration using, e.g. SRIM (see section 2.2.8). A second important effect is that the pulse height in a silicon detector generated by a charged ion of a given energy differs depending on the ion species [35]. It has been suggested that this relates to the energy required to generate an electron–hole pair in silicon which depends on the charge of the ion (see figure 8.24). The correction is not particularly large—of order 1% between protons and alpha particles. However, such a correction needs to be applied if, for example, a silicon detector is calibrated, as is usual, with an alpha particle source and then latterly used to measure protons.

Detector damage

Silicon detectors suffer from radiation damage after being exposed to a high fluence of charged particles, leading to them often being considered a consumable item in experimental nuclear physics. Radiation damage of a silicon detector will lead to decreased pulse height and degradation of energy resolution which continues to the point where the detector is no longer useful for spectroscopic measurements. Damage effects are most severe in a flux of heavy ions [36] and are signalled by a strong (and irreversible) increase in leakage current. There are other ways in which leakage current can increase which are unrelated to damage. For example, if the detector has been insufficiently grounded then it can become charged up in an in-beam experiment by the spray of delta electrons from the target which can appear to greatly increase the leakage current. This leakage current will disappear if the detector is turned off and allowed to recover.

Radiation damage in silicon is associated with the generation of *Frenkel defects* in the crystal which act as charge-trapping centres leading to electron–hole recombination. For DSSDs, damage to the interstrip region can lower its resistance leading to incomplete charge collection and cross-talk between strips [36]. One of the issues with radiation damage is that it can be highly localised, particularly if the flux of particles is high in a specific angular range, for example. Nevertheless, even if the damage is local, it can spoil the properties of the detector as a whole. There are

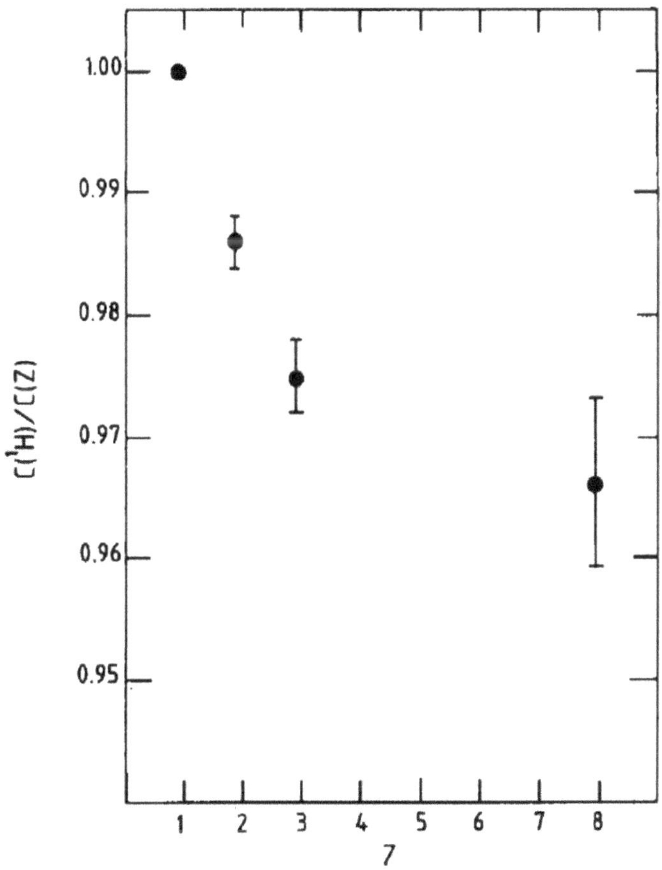

Figure 8.24. Apparent electron–hole pair creation energy ratios for ^{4}He, ^{7}Li and ^{16}O relative to ^{1}H as a function of atomic number, *Z*. (Reprinted from [35] with the permission of Elsevier.)

strategies which can mitigate radiation damage; an example applied to an annular DSSD is discussed in reference [37]. The left-hand side of figure 8.25 shows the impact of radiation damage on an annular DSSD. In each case, the pulse height is plotted for alpha source events in a given sector (or ring) as a function of ring (or sector) on the other side of the DSSD. Since the damage is localised in the inner rings (high ring number), the pulse height is significantly degraded for events registering in the inner rings. The approach taken to remedy this was to consider the DSSD as a large number of pixels based on each ring/sector combination and apply an algorithm to correct the pulse height on a pixel by pixel basis. This largely recovers the resolution of the DSSD as shown on the right-hand side of figure 8.25.

8.2.7 Alternatives to silicon

Diamond and silicon carbide detectors

Diamond detectors have been used to a minor extent in nuclear physics [38]. They are more widely employed in high-energy particle physics, principally as beam

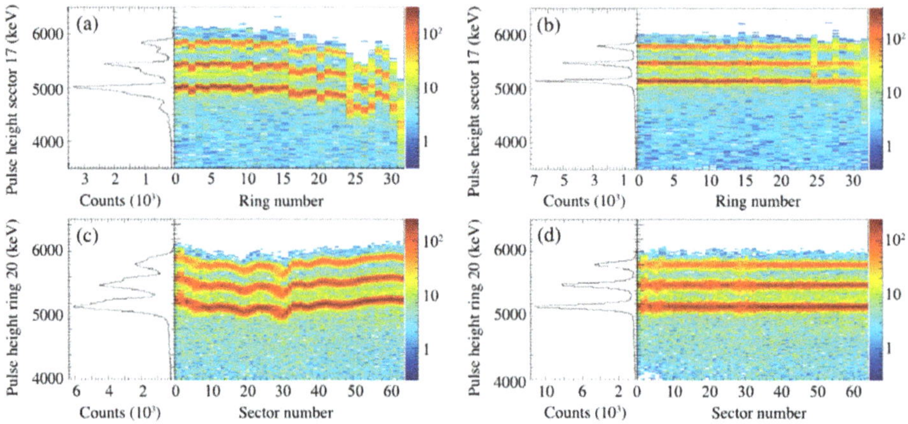

Figure 8.25. Pulse heights from triple-alpha source recorded by radiation damaged DSSD: (a) events across the set of 32 rings correlated with an event recorded in sector 17. (b) The same events corrected on a pixel-basis (see text for details). (c) Events recorded across the 64 sectors correlated with an event recorded in ring 20. (d) The same events corrected on a pixel-basis (see text for details). (Reprinted from [37] with the permission of Elsevier.)

monitors. At very high energies, e.g. for >0.1 GeV protons, the radiation hardness of diamond detectors can be ten times higher. However, diamond detectors can suffer radiation damage more easily than silicon detectors at low energies—the more typical domain of nuclear physics [39, 40]. Silicon carbide detectors have also been discussed as a radiation-hard alternative to conventional silicon [41].

Optical time-projection chamber

In the examples discussed above, we have seen the routes to spectroscopic measurements of alpha particles and heavy ions and how position sensitivity can be used to infer the angle of emission of particles in nuclear reaction studies. Here, we discuss a fascinating example of where it is possible to 'image', albeit indirectly, the tracks of alpha particles (and other heavy ions) in a gas volume. This can be important for study of break-up reactions where, for example, a high energy photon beam interacts with ^{16}O and breaks it up into ^{12}C + ^{4}He [42]. The three-dimensional imaging is achieved with an Optical Readout Time Projection Chamber (O-TPC) (see figure 8.26). Within the gas volume, ionisation occurs and electrons drift, through a Frisch grid, into an amplification volume. The resulting electron avalanches and ionisation generate optical photons which can be imaged using CCD cameras leading to a two-dimensional projection. The third dimension is obtained from the projected time-structure recorded by photomultiplier tubes (PMTs). Figure 8.27 shows an example of the image reconstruction of ^{12}C dissociated into three alpha particles using the O-TPC.

8.3 Applications relevant to fission

Studies of fission are motivated by a focus on our fundamental understanding of nuclear fission as well as applications to conventional nuclear fission energy

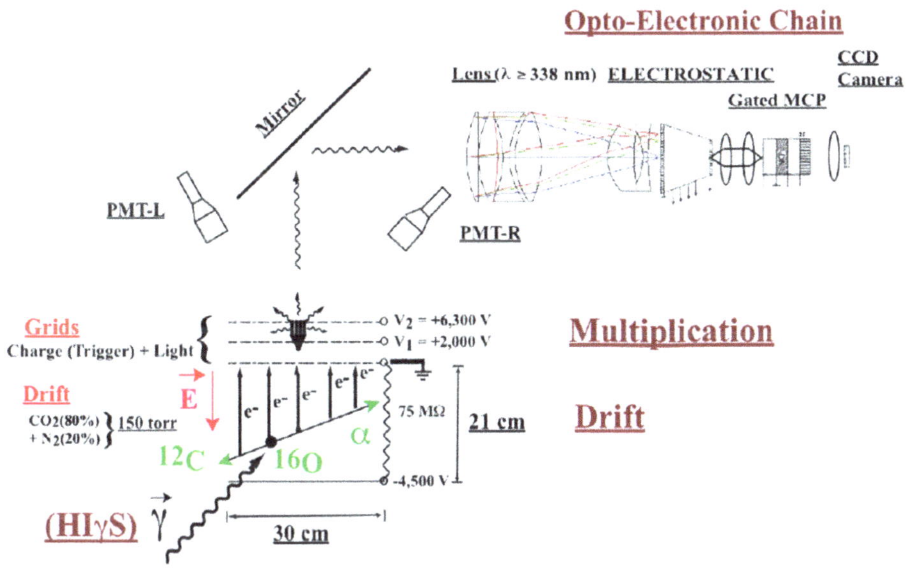

Figure 8.26. A schematic diagram of the O-TPC with an ^{16}O target nucleus dissociated by a gamma-ray beam from the HIγS facility. The gas mixture and pressure as well as the operating voltages are indicated. The drift electrons that provide the third (time projection) upward dimension are shown schematically. (Reprinted from [42] with the permission of IOP Publishing.)

generation and to fission of exotic short-lived nuclei relevant to nucleosynthesis. The underlying physics of the fission process was reviewed in section 1.5. From this perspective, it is of high interest to understand the cross-sections for neutron-induced fission which can be achieved by simply counting the number of fission events. A secondary objective would be to determine the fragment species, i.e. Z and A so that the fission mechanism can be understood in detail. Let us first look at basic system for fission counting before moving to more elaborate apparatus which allow the fission fragment mass distribution to be obtained and finally systems which can also identify the Z of fission fragments.

8.3.1 Counting fission events

In order to obtain cross sections for neutron-induced fission, a simple counter is needed for fission events with high efficiency. At the nToF facility in CERN (see section 3.5.1), the MicroMegas detector [43] is used. Figure 8.28 shows a schematic of the MicroMegas detector. The sample, in this case, ^{235}U is mounted directly inside the chamber. Neutrons interact with the ^{235}U inducing fission. The geometry ensures a very high probability that one of the fission fragments will enter the gas volume; such a heavy charged ion will create significant ionisation of the fill gas. The electrons will drift slowly towards the anode, and ions towards the cathode. The key feature of the detector design is a micro-mesh situated close to the anode which separates the drift region from a narrow amplification region with high electric field where electron multiplication will occur. The micromesh enables a high gain ($\sim 10^4$)

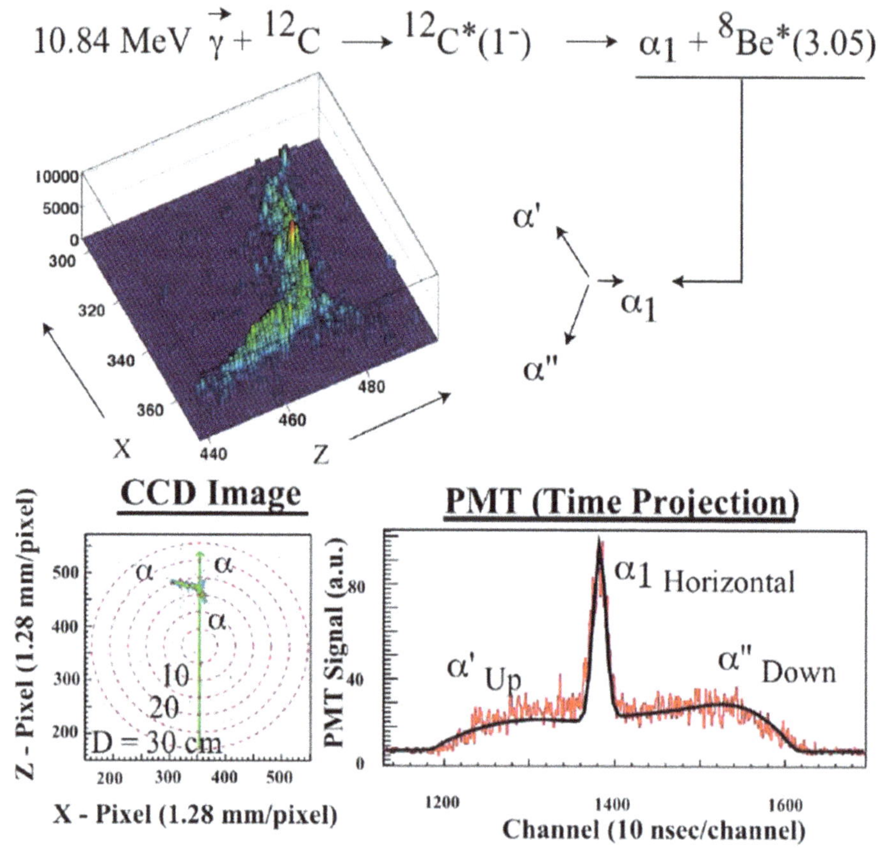

Figure 8.27. Three alpha-particle event from the dissociation of ^{12}C measured at 100 Torr. The projected 2D track and time of this event are shown in the lower part of the figure together with the fitted line shape of the light detected from the emerging three alpha particles. The geometry of the reconstructed event is shown schematically with α_1 emitted horizontally and the ^{8}Be decay products, α' and α'', emitted upward and downward, respectively. (Reprinted from [42] with the permission of IOP Publishing.)

and fast timing. Such a detector can also be used to register light charged particles from, e.g. (n,α) and (n,p) reactions.

8.3.2 Identifying the fission fragment mass distribution

There is a wide variety of apparatus used for fission studies, tuned to the needs of the specific experimental study since a wide range of experimental techniques are employed where accelerated ion beams or fast/thermal neutrons are used in producing a fissioning system [45]. The parameters of interest in such studies are principally the kinetic energies and velocities of the fission fragments and their angle with respect to the beam direction. An ideal detector would therefore have excellent angular resolution, energy resolution and timing resolution. Timing resolution is important because it allows time-of-flight of fragments to be measured and hence the velocity of fission fragments to be reconstructed. All available detectors naturally

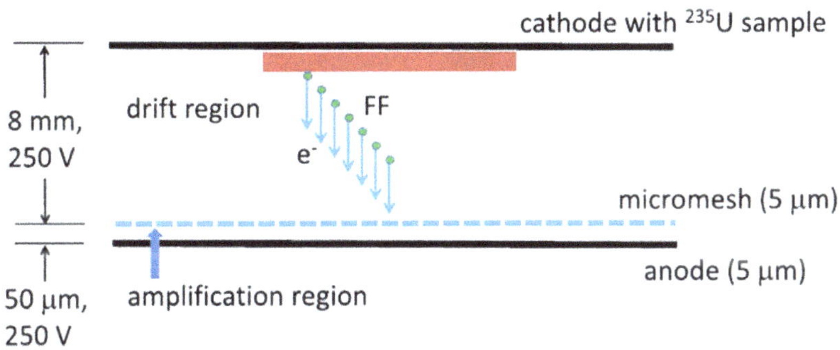

Figure 8.28. Schematic setup of a MicroMegas detector used at nToF. These detectors are defined by two gas volumes: the drift region, typically a few mm in thickness, and the amplification region, a few tens of microns thick. These regions are separated by a micro-mesh layer of a few microns. Electrons produced by the charged particles emitted from the sample are drifting to the mesh and are triggering an avalanche signal in the high electric field of $\geqslant 10$ kV cm^{-1} of the amplification region. (Reprinted from [44] with the permission of Elsevier.)

have some limitations against these idealised requirements. For example, the timing resolution of silicon detectors is not particularly good, which provides a limitation to time-of-flight measurements. They also suffer from pulse-height defect when detecting heavy ions, which depends on the mass and charge of the ions. Moreover, as discussed above, silicon detectors are not especially radiation hard and so their performance can noticeably degrade. These considerations mean that gas-ionisation based detectors are often preferable for fission studies given their high rate capability, intrinsically fast timing and relative radiation hardness. Many of these considerations are applicable to detection of heavy ions in other experimental settings not involving fission.

Fission fragment detection with silicon detectors: beta-delayed fission

An application where fission fragment detection with silicon detectors is appropriate is in the study of beta-delayed fission where low fission event rates are expected and so radiation damage is not of high concern [46]. Figure 8.29 illustrates such an implementation where silicon detectors are used for fission fragment detection. A low-energy radioactive beam of ^{180}Tl, at the very low rate of 150 atoms s^{-1}, is implanted into thin carbon foils on a 'windmill' system which rotates stepwise. ^{180}Tl beta decays to ^{180}Hg. A small fraction of this decay branch goes to highly excited states which spontaneously fission. The fission fragments are detected in two annular silicon detectors mounted either side of the implantation foil. MINIBALL germanium cluster detectors are mounted behind one of the silicon detectors to measure coincident gamma rays. The system allows the measurement of the individual and total fission fragment kinetic energies which affords insight into the mass asymmetry in the fissioning system.

Fission fragment detection with MWPCs

Figure 8.30 shows a setup for measuring fission fragments following a multi-nucleon transfer (MNT) reaction where a number of nucleons are transferred onto a heavy

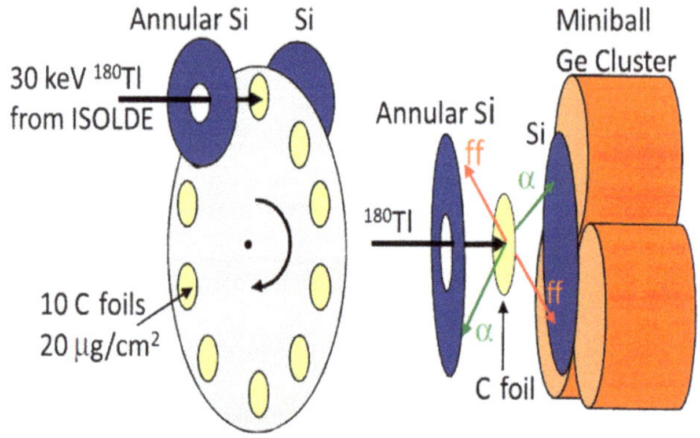

Figure 8.29. Detector set ups for study of β-delayed fission. (Left) a 'windmill' setup where a radioactive beam of ^{180}Tl is implanted at low energy into thin carbon foils on a windmill which can be rotated. (Right) a blow-up of the left. (Reprinted with permission from [46]. Copyright (2010) by the American Physical Society.)

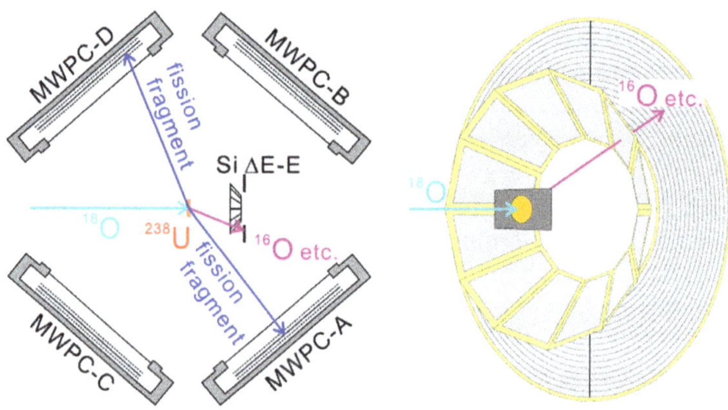

Figure 8.30. A typical detection setup for in-beam MNT-induced fission experiments at JAEA. In the case shown, the MNT reaction ^{18}O+^{238}U was studied, whereby four MWPCs: MWPC-A–MWPC-D were used to detect the coincident fission fragments, while a silicon ΔE–E telescope array was exploited to determine the transfer channel. In the case of the fusion–fission experiments, the ΔE–E telescope array is removed. (Reprinted from [45] with the permission of IOP Publishing.)

system such as ^{238}U to produce a heavier, excited nucleus which then fissions. Information on the number of nucleons transferred can be measured using an array of silicon ΔE–E telescopes (see section 8.2.2) to identify the beam-like residual nucleus and, hence, the fissioning system. The fission fragments are recorded in four multi-wire proportional counters (MWPCs). MWPCs [47] are a well-established variant of the proportional counter discussed in section 4.3.1. The gas-filled proportional counter contains a crossed array of anode wires. When the fission fragment enters the MWPC volume, it causes ionisation and the charges are collected on the neighbouring wires. This provides information on the position of interaction within the gas volume. MWPCs generate a fast timing signal (~1 ns) which supports time-of-flight measurements[3] and high rate capability; they are also intrinsically radiation hard.

8.3.3 Determining both A and Z of fission fragments

A key challenge for fission fragment spectrometry and other areas of experimental nuclear physics is to determine the Z of heavy ions. In section 2.2.1, we discussed the physics of stopping of heavy charged particles in matter, in particular the Bethe–Bloch formula (equation (2.1)) which shows an explicit dependence on the atomic number of the particles. We also discussed the appearance of the characteristic Bragg peak at the end of a heavy ion track as the energy loss increases strongly at low velocities. This property can be exploited using a Bragg curve spectrometer [48]; a schematic of such a spectrometer is shown in figure 8.31. The spectrometer comprises an ionisation chamber with a Frisch grid close to the anode. The heavy ion loses energy and stops within the ion chamber volume. The current signal from the ion chamber represents a time inverse of the stopping process. The ionisation from the end of the track appears first, followed by the rest of the track moving backwards in time. The two key points in the signal are the position of the Bragg peak and the initial energy loss $\left(\frac{dE}{dx}\right)$ as shown on figure 8.31. This technique works best for ions with higher energies, e.g. few MeV/u; separation of Z becomes challenging for $Z > 30$.

This technology is applicable to fission fragment detection but, here, the situation is more complex as the fragments have typical kinetic energies of 1 MeV/u. At such

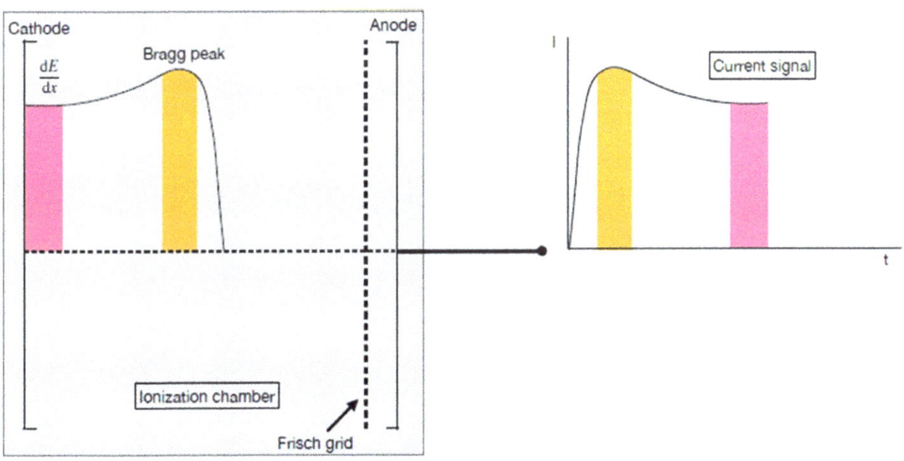

Figure 8.31. Concept of a Bragg curve spectrometer comprising an ionisation chamber with a Frisch grid close to the anode. The current signal shows a time-reversed profile of the track of ionisation.

[3] In this specific example, what is measured is the time difference between the two fission fragments recorded in separate MWPCs.

energies, the ion may not be fully ionised and charge exchange can occur with the fill gas in an ionisation chamber. This variation in the effective charge states of the ions serves to 'wash out' the Bragg peak. Figure 8.32 shows examples of this effect from fragments of the same Z and different mass, and for close-lying isotopes of different Z. The identification of fission fragments by Z becomes, therefore, very challenging but not impossible.

Let us describe a system which employs a Bragg detector and other detectors to determine fission fragment Z and A. The FiFI spectrometer, designed by the University of Manchester, is an example of a $2E2v$ (two energy—2 velocity) spectrometer where there are two detector arms which measure the kinetic energy and velocity of the fission fragments. Since, the fission fragments are travelling at non-relativistic velocities, their kinetic energy, E_k is connected to their velocity by the conventional formula

$$E_k = \frac{Av^2}{2}. \tag{8.2}$$

This provides a route to identifying their atomic mass, A, if their kinetic energy and velocity can be measured with sufficient precision. Indeed, if the masses of the two fragments, A_1 and A_2 could be determined with high enough precision then the number of neutrons emitted in the fission process could also be established event-by-event; this, however, is significantly challenging.

Each of the two arms of the FiFI spectrometer comprise a set of detectors: a START detector, a Bragg detector volume (similar to the example described above) and a STOP detector (see figure 8.33) [49]. The START and STOP detector are used to determine the time-of-flight of fission fragments through the spectrometer arm with high precision. When fission fragments pass through these detectors, electrons

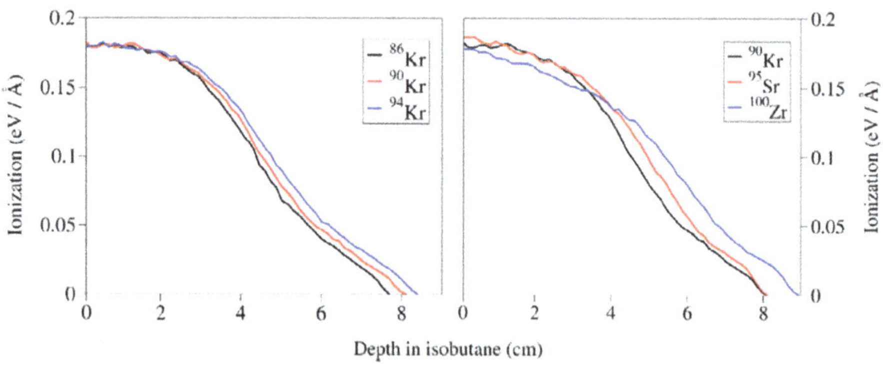

Figure 8.32. Profiles of simulated average energy loss with range in 100 mbar of isobutane by fragments with kinetic energy-to-mass ratios of 1 MeV/u. Panel (a) shows energy loss by ^{86,90,94}Kr fragments illustrating the effect mass has on stopping. Panel (b) shows ^{90}Kr, ^{95}Sr and ^{100}Zr energy loss, demonstrating differences in energy loss with change in atomic number and mass for typical ^{235}U fission fragments. The profiles were simulated and averaged over 20 fragments of each type using SRIM-2013. (Reprinted from [49] with the permission of Elsevier under Creative Commons License.)

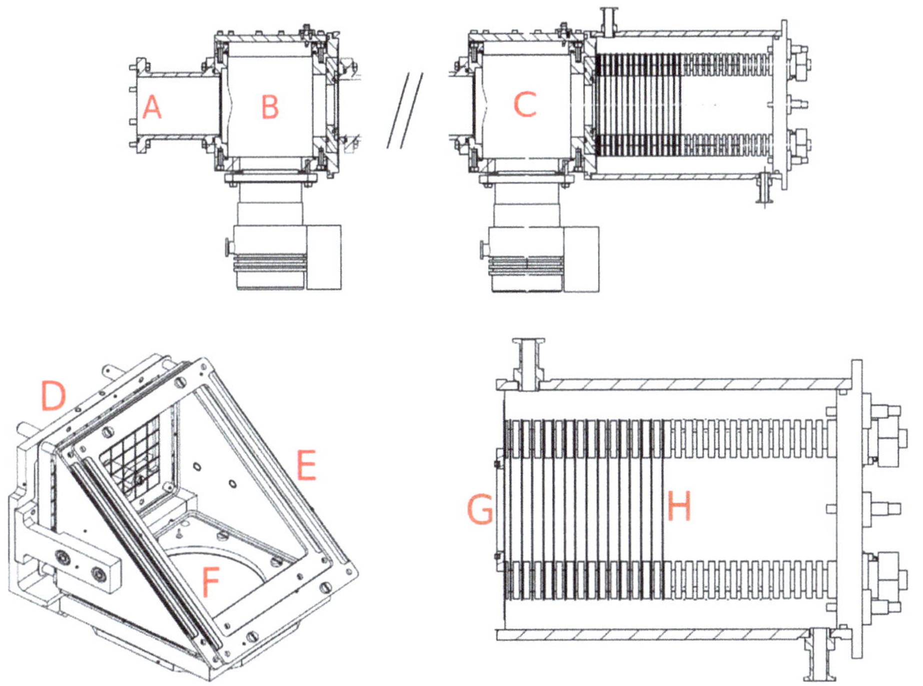

Figure 8.33. A schematic diagram of the fission fragment identification arm. The entry point of a FF beam is marked with letter A. Letters B and C denote volumes where the START and STOP detector assemblies are positioned. The positions of the vacuum pumps are shown beneath those volumes. The 1 m-long ToF section has been truncated for the purposes of the diagram. An expanded view of the timing detectors is shown in the bottom left part of the diagram. Letter D marks the location where the mesh supporting the emission foil is located. Letter E denotes the plane onto which the electrostatic mirror is attached. The MCP is located at letter F. Note that the timing detector in the diagram is shown upside down compared to its actual orientation inside FiFI. An expanded view of the BD is shown in the bottom right corner. The mylar window is at letter G, followed by a gas-filled volume containing the field-shaping rings, the Frisch grid and an anode (position H). ToF foils were removed during data collection for the results presented in this work. (Reprinted from [49] with the permission of Elsevier under Creative Commons License.)

are emitted from a thin foil of aluminium (10 ± 1 μg cm^{-2} thick), called a secondary electron emission (SEE) foil, held at a −1.75 kV potential. Backscattered electrons enter the space between the foil and the electrostatic mirror, which consists of two wire grids separated by 1 cm. A potential of −3.5 kV is applied across the gap between the grids in order to deflect the electrons into the microchannel plate (MCP) assembly. The two MCPs are $\approx$1 mm-thick lead-doped glassplates with $\approx$25 μm-wide tubular channels, which extend from one side of the plate to the other at an angle to the surface. A voltage of −4.8 kV is applied across the MCPs, accelerating the electrons which enter it in order to cause ejection of secondary electrons in the collisions with the walls. The collisions in the MCPs multiply the number of electrons by a factor of $\approx 10^6$. The timing resolution of the time-of-flight component of FIFI is around 1.2 ns.

8.3.4 Societal applications of alpha particle/heavy-ion detection

There are relatively fewer societal applications of charged-particle detection such as alpha-particle detection compared to the broad range of applications of gamma-ray detection discussed in chapter 7. Alpha spectrometry is relevant to nuclear, environmental and biological samples; a recent review of these applications is found in reference [50]. Silicon detectors such as PIPS detectors are frequently used in such applications. Here, we look at two specific applications of heavy charged particle detection relevant to nuclear decommissioning and space science.

Decommissioning of pipework at nuclear installations

The decommissioning of nuclear sites requires reliable techniques for the detection, identification, and quantification of radioactive substances. Discriminating between disposable and contaminated waste has a significant impact on the cost of decommissioning. In particular, there is no simple, low-cost solution available for assessing the presence of radioactive sources that are essentially alpha-only emitters, such as ^{239}Pu, on the inner walls of pipework. The range of alpha particles in media is intrinsically short, and as such, alpha-emitters are undetectable from the outside, as any radiation occurring on the inside of the pipe is absorbed by the wall (even gamma radiation from ^{235}U excited states). The emitting isotope may even be embedded in tiny cracks in the pipe wall, meaning that alpha particles may lose energy as they emerge. Consequently, alpha-particle detectors need to be positioned as closely as possible to the inner walls in order to achieve the greatest detection efficiency and be able to detect very low radiation doses.

A detector system has been developed [51] based on flexible crystalline silicon that can conform to the inner diameter of pipework for the direct detection of alpha-particles. This solution not only offers the potential for 360° inner diameter coverage but can also be adapted to pipes of different diameters. Being as close as possible to the inner pipe walls, maximum sensitivity and uniform efficiency can be achieved, even for low energy alpha-particles, where other approaches may struggle. As a proof-of-principle, it was demonstrated that a sheet of flexible crystalline silicon of 50 μm thickness, as shown in figure 8.34, forms the basis for a high-performance in-pipe alpha-particle detector (see figure 8.35). The sensor can be more generally adapted to any curved surface, such as the exterior of barrels filled with radioactive waste. It can also address the requirements of some nuclear physics experiments that require non-planar detectors.

Cosmic ray telescope

In section 7.6, we described aspects of gamma-ray astronomy and the strong synergies between the techniques of experimental nuclear physics, in particular the application of high-purity germanium detectors and anti-coincidence shields. Here, we describe an example relating to charged-particle detection in the area of Space Science where the techniques and technology employed also have strong connections with nuclear physics. The example is the Low-Energy Telescope (LET) which is one of four sensors that make up the Solar Energetic Particle (SEP) instrument of the

Figure 8.34. At thicknesses of about 50 μm, silicon becomes flexible and can be bent into cylindrical or other curved geometries. (Reprinted from [51] with the permission of AIP Publishing under Creative Commons License.)

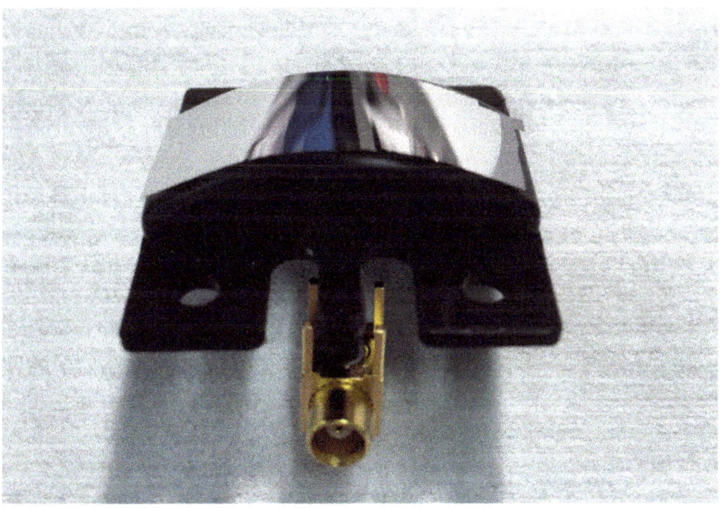

Figure 8.35. A curved silicon detector (50 μm) thick designed for use as a pipeline inspection system for alpha-contamination. (Photo credit: the author.)

IMPACT investigation for NASA's STEREO mission [52] (figure 8.36). The LET is designed to measure the elemental composition, energy spectra, angular distributions, and arrival times of charged particles from the Sun: protons and helium ions over the energy range ~1.5–13 MeV/u and heavier ions up to nickel over the energy

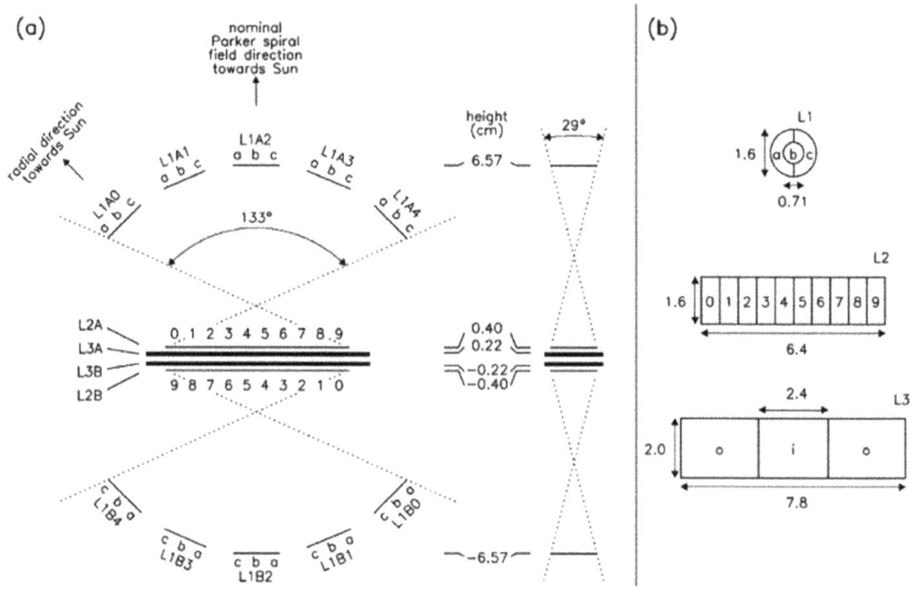

Figure 8.36. Schematic of the Low-Energy Telescope (LET), part of the STEREO-IMPACT mission. (a) Arrangement of silicon detector telescopes within the LET; (b) the shape and dimensions of the three telescope elements, L1, L2 and L3. Dimensions are in mm. The L1 detectors are 24 μm thick, L2 are 50 μm thick and L3 are 1000 μm thick. (Reprinted by permission from Springer: [52].)

range from ~2 to ~30 MeV/u. This is a very similar energy range to that encountered in low energy nuclear physics although the directionality of the particles is clearly different. The detection and identification of ions with LET is achieved with an array of silicon detector telescopes (see section 8.2.2). The detectors making up the telescopes, L1–L3, have differing thicknesses: L1 detectors are 24 μm thick, L2 are 50 μm thick and L3 are 1000 μm thick. The principle here is the same as in the nuclear physics example: an ion with incident energy, E, will transit one (or more) of the thin detectors and deposit an energy ΔE in a thickness, L. In the case where the particle is not normally incident to the detector surface, L can be adjusted to take account of the angle of incidence. The stopping detector records the residual energy, E', where $E = E' + \Delta E$. The particle's corresponding atomic number, Z, can be obtained from [52]

$$Z \approx \left(\frac{k}{L(2+\varepsilon)^{a-1}}\right)^{\frac{1}{(a+1)}} (E^a - E'^a)^{\frac{1}{(a+1)}}, \tag{8.3}$$

where $2 + \varepsilon$ is the mass to charge ratio of the nuclei under consideration and k and a are constants which model the range-energy relationship for heavy ions in silicon (see section 2.2.1). Figure 8.37 shows Monte Carlo simulations of the energy deposited in the different elements of the detector telescopes which shows interesting parallels with some of the nuclear physics examples discussed above. The readout of the LET system illustrates interesting aspects of digital electronics. There are

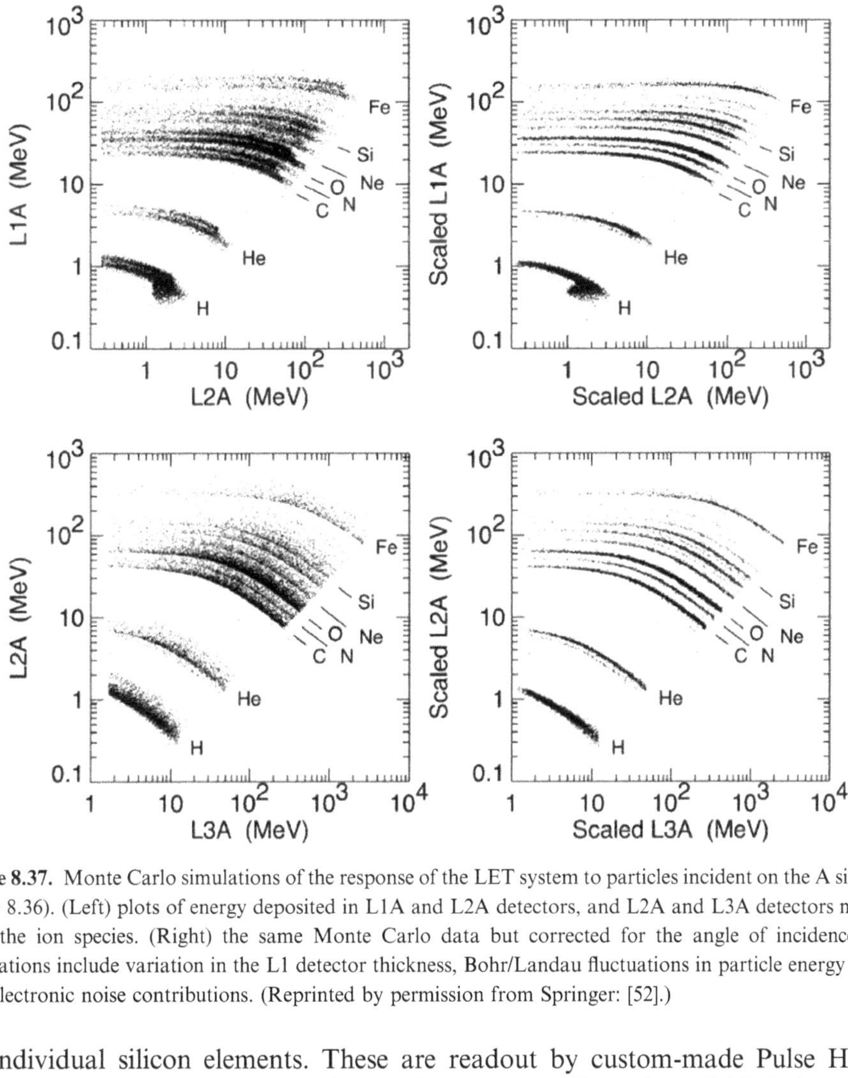

Figure 8.37. Monte Carlo simulations of the response of the LET system to particles incident on the A side (see figure 8.36). (Left) plots of energy deposited in L1A and L2A detectors, and L2A and L3A detectors marked with the ion species. (Right) the same Monte Carlo data but corrected for the angle of incidence. The simulations include variation in the L1 detector thickness, Bohr/Landau fluctuations in particle energy losses, and electronic noise contributions. (Reprinted by permission from Springer: [52].)

54 individual silicon elements. These are readout by custom-made Pulse Height Analysis System Integrated Circuits (PHASICs)—see discussion of ASICs in section 10.3. A Minimum Instruction Set Computer (MISC) automatically translates these measured energies into particle identification for ~12 energy intervals at event rates around 1000 events s^{-1} [52].

8.4 $\beta^{+/-}$ and electron detection

8.4.1 Beta decay spectroscopy

As discussed in section 1.4, β spectra are largely uninteresting from a spectroscopic point of view since beta decay is a three-body decay and the energy distribution of beta particles is continuous up to the end-point. The location of the end-point is of interest, however, since it can be used to determine the mass difference between parent and daughter nuclei. In contemporary studies of exotic nuclei, relatively few ions are produced of a given exotic nucleus and it is desirable to measure the beta decay half life and learn about the details of the decay in terms of which daughter states are populated in the decay observed through their subsequent gamma-ray emission. This often requires a detector which can detect implantation of high energy heavy ions and then detect their beta decay. These successive events can be correlated via their position. Let us discuss three systems that can provide this functionality, observing how they draw on a diverse set of detector technologies.

DSSD-stack implantation detector

The first example is a stack of nine DSSDs (Micron Model W1-1000) used at RIKEN-RIBF in Japan, where each DSSD was segmented into 16 strips in the horizontal and vertical axes as a position-sensitive β-counting system (see figure 8.38). Incoming heavy ions are degraded in energy and then implanted into one of the DSSDs in the stack. The subsequent beta decay can be correlated with the implantation by detecting the beta particle in the same DSSD pixel. The beta particle will not be fully stopped in the DSSD but will lose energy as it passes through the DSSD and where that energy exceeds the threshold set on the detector it can be registered as a beta particle. Depending on the implantation depth of the heavy ion within the DSSD, the efficiency for β particle detection varies from 40% to 80% [53].

YSO-based implantation detector

An alternative concept uses a scintillator-based, pixellated implantation-decay detector constructed from the inorganic scintillator, YSO. Figure 8.39 provides a

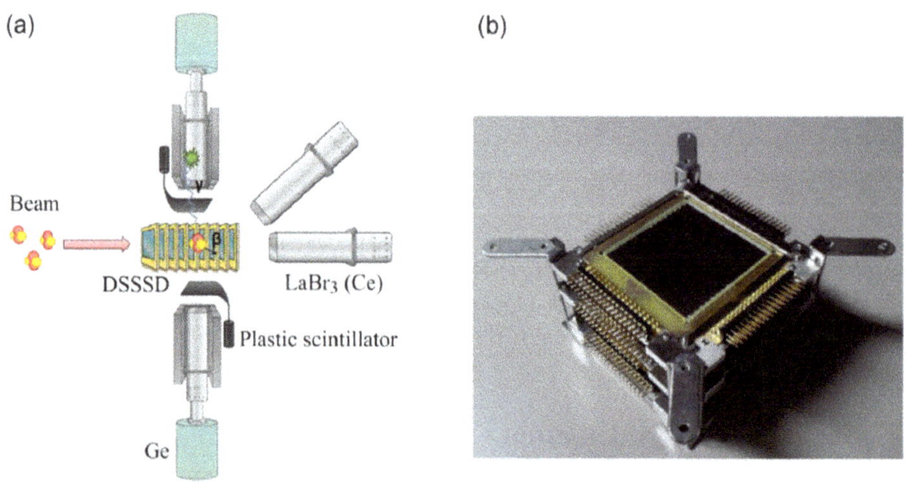

Figure 8.38. Apparatus for correlating beta decay of radioactive nuclei used at RIBF-RIKEN in 2009: (a) incoming high energy radioactive ions are stopped in a stack of double-sided silicon-strip detectors (DSSD) and then subsequently beta decay; the position of implantation and decay can be correlated. γ rays are detected by four Compton-suppressed clover-type Ge detectors arranged around the DSSDs in a close geometry. (b) Photograph of the stack of nine DSSDs. (Reprinted from [53] with permission from OUP under Creative Commons License.)

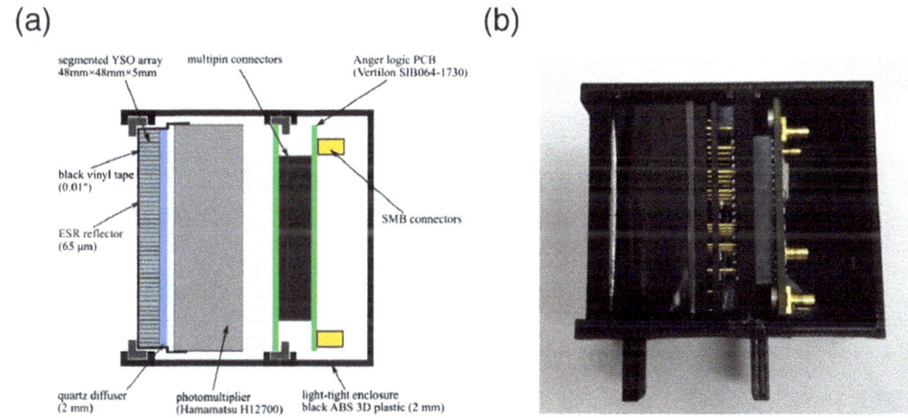

Figure 8.39. (Left) schematic of a YSO-based beta-implant detector. (Right) photograph of the detector. (Reprinted from [54] with the permission of Elsevier.)

schematic of such a device [54]. YSO is chosen for its high effective atomic number ($Z = 35$) and density (4.4 g cm^{-3}) which can stop beta particles in a relatively short range. YSO is radiation hard and so will not be damaged as rapidly as a more conventional silicon implantation detector. Moreover, it can support high counting rates due to its fast decay time (42 ns). In the example shown, the YSO is pixellated into sticks of $1 \times 1 \times 5$ mm.

CAITEN revolving pixellated plastic scintillator system

The third example to be discussed (see figure 8.40) is an innovative design using a rotating highly-pixellated cylindrical array of plastic scintillators[4] called CAITEN which has also been used at RIKEN-RIBF in Japan [53]. There are 40 000 pixels of plastic scintillator each with dimensions of $6 \times 6 \times 20$ mm^3. Energetic radioactive ions are implanted into a region of the detector. The detector is rotating at a rate between 0 and 60 rpm to reduce the density of non-correlated decay events in a given pixel; this is the same philosophy as that behind earlier tape-transport systems where radioactive ions are stopped on a tape and transported to detectors where the decay is subsequently observed. The scintillation light is recorded with a cylindrical array of 24 position-sensitive photomultiplier tubes (PSPMTs). Each PSPMT has a segmented light guide mounted in front of it. The light is transported across an air gap of around 3–4 mm between the scintillator barrel and the light guides.

8.4.2 Conversion electron spectroscopy

Conversion electron spectroscopy (see section 1.7.3) is more interesting from a spectroscopic perspective than beta decay because the peaks are discrete *and* have different components, i.e. K, L etc, whose relative intensity is an indicator of the

[4] See section 9.1.2 for a full description of plastic scintillators.

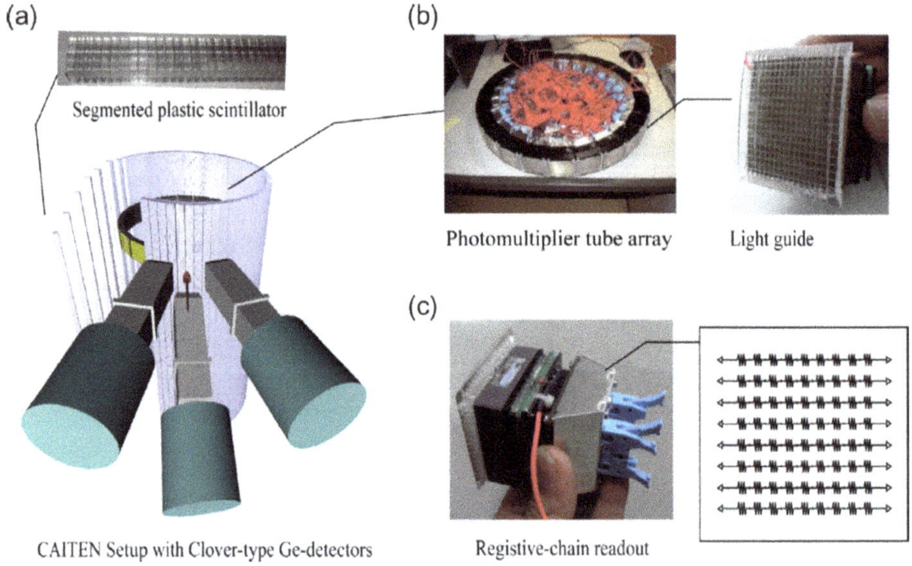

Figure 8.40. CAITEN beta-counting system (see text for more detailed description): (a) segmented plastic scintillators. (b) Position-sensitive photomultiplier tube array. (c) Resistive channel readout of the PSPMTs. (Reprinted from [53] with permission from OUP under Creative Commons License.)

multipolarity of the transition concerned (see section 1.7.3). Lithium-drifted silicon (or Si(Li)) detectors [55] are often used for electron detection since the addition of lithium allows the depletion layer to be much deeper and hence, the detector to have a depth more appropriate to the range of electrons. Figure 8.41 shows a conversion electron spectrum from a ^{207}Bi source acquired with a 6.5 mm thick Si(Li) detector. The achievable energy resolution corresponds to 0.77 keV at 481 keV which is sufficient even to resolve the separate L1, L2, etc components.

While thick Si(Li) detectors are suitable for measurement of conversion electrons following beta decay, detection of conversion electrons is more challenging in in-beam experiments, e.g. fusion-evaporation as described in section 3.3.1, principally due to the fact that an accelerated ion beam striking a reaction target produces a flood of delta electrons which would swamp a silicon detector mounted near the target position. The SACRED spectrometer [57] built at the University of Jyväskylä in the 1990s has design features to avoid these issues. Electrons are transported away from the target using a solenoidal field. They pass through a high voltage barrier which can repel the lower energy delta electrons allowing the higher energy conversion electrons to pass through. The conversion electrons are detected in a 5×5 array of 5 mm square PIN diodes. The array is cooled to around −20 °C to strongly reduce the leakage current in the PIN diodes. A typical conversion electron spectrum obtained with SACRED in an in-beam study is shown in figure 8.42. Clearly, K and L components of different transitions can be resolved but the energy resolution is much lower than that of the Si(Li) spectrum shown in figure 8.41 which

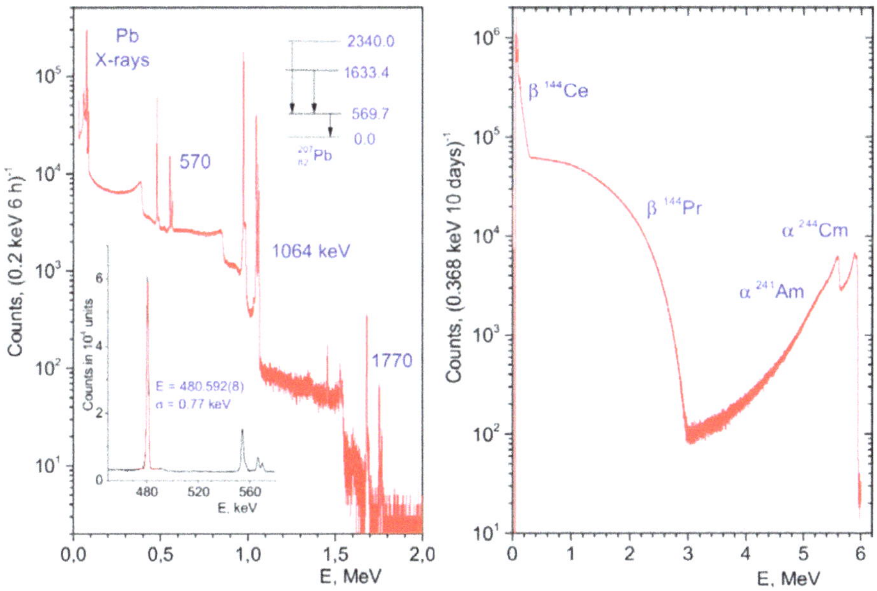

Figure 8.41. (Left) conversion electron spectrum from a ^{207}Bi source using a 6.5 mm thick Si(Li) detector. The inset shows the K, L, ··· peaks corresponding to conversion electrons for the 570 keV transition. (Right) beta decay spectra for the ^{144}Ce–^{144}Pr decay. (Reprinted from [56] with the permission of IOP Publishing.)

is partially due to the intrinsic resolution of the detectors but also attributable to Doppler-broadening of the peaks as this data is from an in-beam experiment (see further discussion of this effect in section 3.3.1).

SPEDE [59] is an example of a next-generation electron spectrometer designed for Coloumb excitation studies, specifically for high-Z nuclei where internal conversion is an important process (see section 3.4.5). It comprises an annular silicon detector, segmented into 24 segments: 3 rings and 8 sectors (see figure 8.43). Variants can be produced with variable thickness up to 1.5 mm depending on the energy range of interest for the conversion electron study; a typical thickness is 500 microns. The preamplifiers are mounted directly on the circuit board around the silicon detector and surface-mount field effect transistors (FETs). The circuit board and silicon detector sit on a copper cooling block allowing the system to be cooled to around 5 °C for improved operation with lower leakage currents. Delta electrons are handled in this case by biasing the target holder up to +5 kV.

Electron-gamma ray coincidence detection

Since many experimental studies populate highly excited nuclear states which emit many coincident gamma rays (and conversion electrons in competition with gamma ray emission), there is high interest in detecting gamma ray-conversion electron coincidences particularly for studies of high-Z nuclei where the probability for conversion electron emission is significantly higher (see section 1.7.3). The SAGE

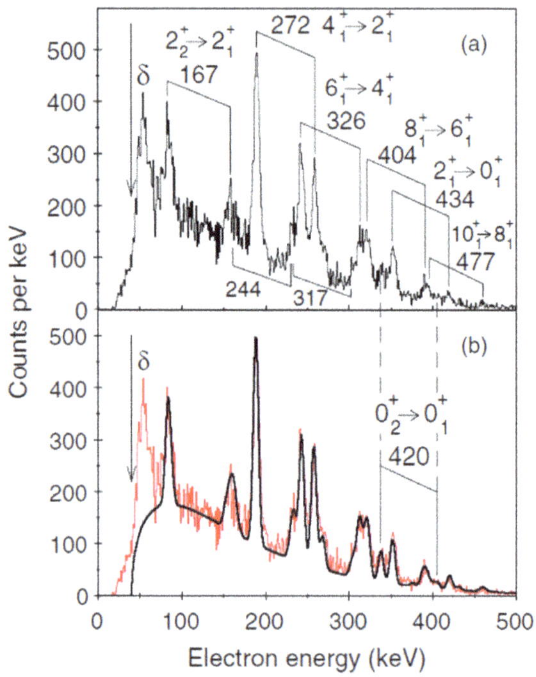

Figure 8.42. Conversion electron spectrum for ^{180}Hg measured with SACRED. (top) The K and L peaks are marked for various transitions. (bottom) A simulation of the data. (Reprinted with permission from [58]. Copyright (2011) by the American Physical Society.)

spectrometer [60, 61] is an evolution of the earlier SACRED design (see above) and uniquely allows coincident detection of gamma rays and conversion electrons in in-beam studies. Figure 8.44 shows the design of how the electron spectrometer component of SAGE interfaces with a germanium detector array while figure 8.45 shows an expanded view of the electron spectrometer itself. A societal example of electron-gamma ray coincident detection for measuring atmospheric Radioxenon was already provided in section 7.3.

8.4.3 Pair spectrometer

Internal pair conversion, as discussed in section 1.7.3, is only a subject of interest for transitions exceeding 1.022 MeV. Given that most transitions in heavier nuclei are below this energy threshold, it makes internal-pair spectrometry a somewhat specialist application focussed on light nuclei. It is of high interest to measure the angular correlation between the emitted electron and positron since this contains information of the multipolarity of the transition. Figure 8.46 provides an illustration of the design of such a pair spectrometer which has six detector modules, each comprising a plastic-scintillator $\Delta E - E$ telescope which can distinguish high energy electrons or positrons from gamma rays [62]. In order to achieve good angular resolution each $\Delta E - E$ telescope has a position-sensitive, multi-wire proportional counter (MWPC) in front of it. The efficiency of such a coincidence detector system is relatively low, of order 1% (see bottom of figure 8.46) which makes such studies very challenging.

An alternate system for pair spectrometry is the Super-e spectrometer at the Australian National University (ANU) [63] (see figure 8.47). Positron–electron pairs are produced at the target position following a nuclear reaction. The positrons and electrons are transported in helical trajectories in a magnetic field to an array of thick, segmented Si(Li) detectors which detect both electrons and positrons. Baffles placed within the spectrometer can be moved to select the 'energy bite' in terms of which electrons/positrons are transported.

8.4.4 Societal applications of beta detection

PET isotope assay

Radioisotopes for positron emission tomography (PET) are produced at cyclotron facilities and subsequently prepared as various molecules in solution such as ^{18}F-FDG. A standard task in the assay of such materials is to establish their activity and their purity. Measurements of activity are conventionally based on detection of annihilation photons (511 keV) following the β^+ decay of the radioisotope by

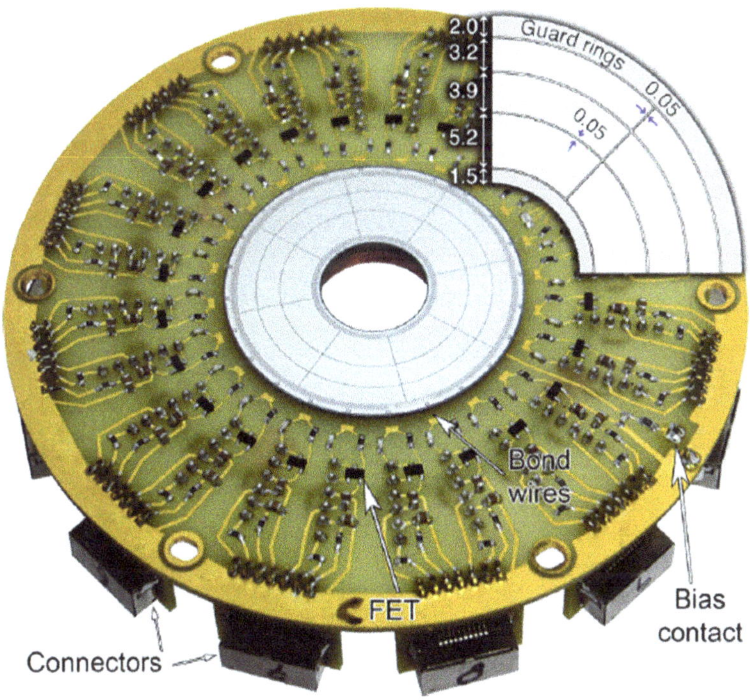

Figure 8.43. The SPEDE spectrometer—an annular silicon DSSD for conversion electron measurements in-beam. (Reprinted by permission from Springer Nature [59].)

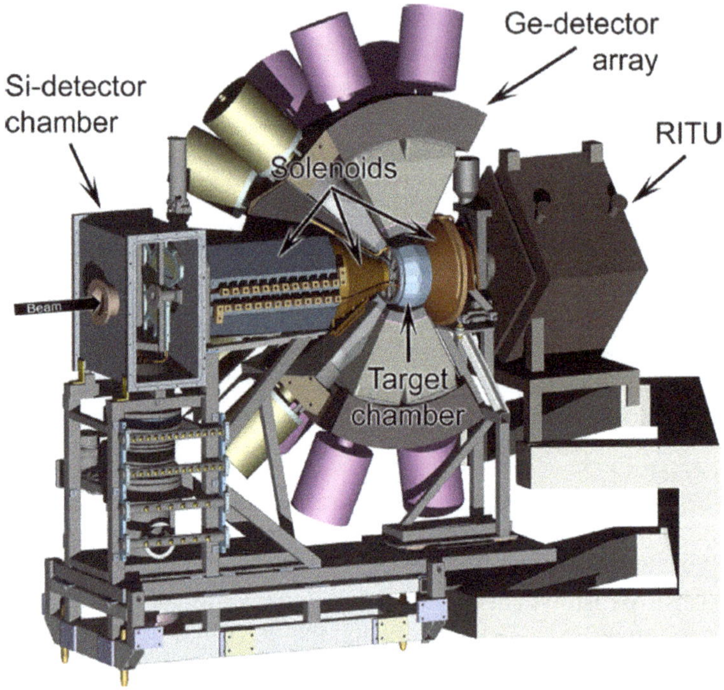

Figure 8.44. The SAGE spectrometer—a unique spectrometer for simultaneous conversion electron/gamma-ray detection. (Reprinted by permission from Springer Nature [60].)

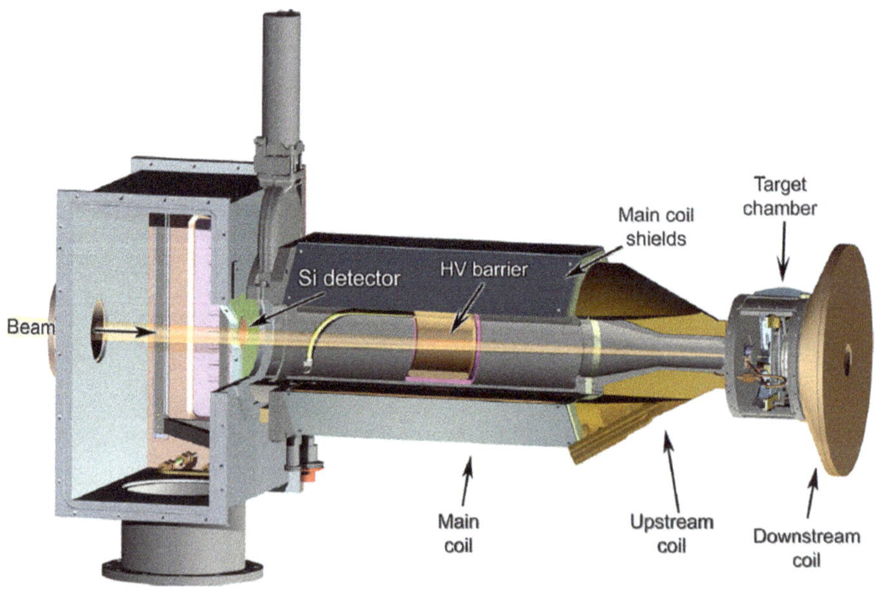

Figure 8.45. The electron spectrometer part of SAGE. (Reprinted by permission from Springer Nature [60].)

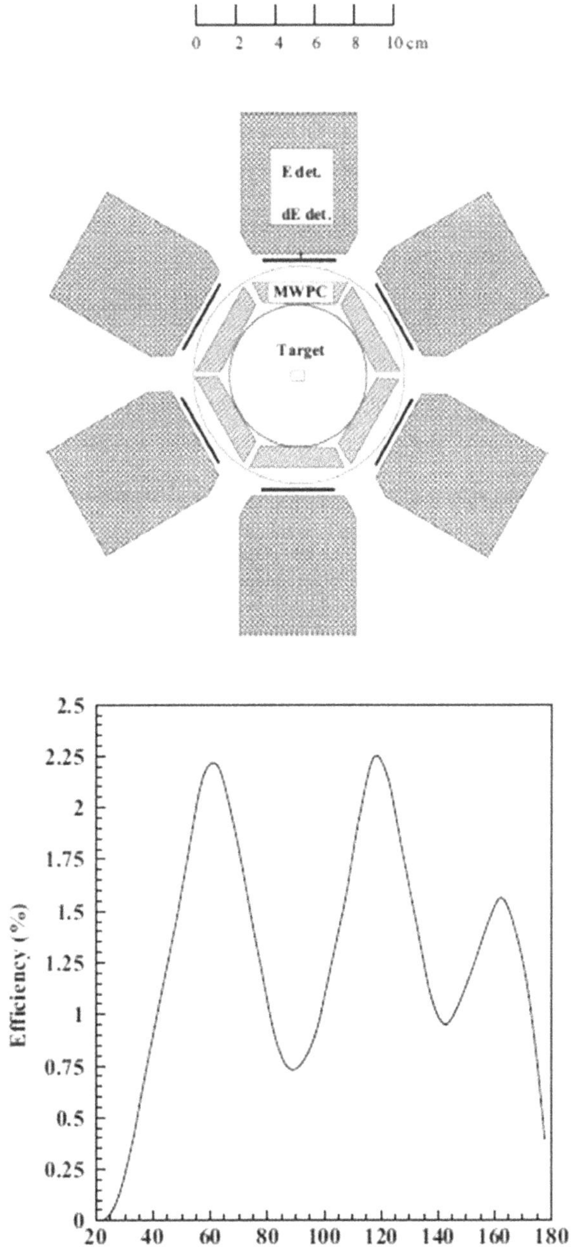

Figure 8.46. (top) Design of a coincidence pair spectrometer with high angular resolution. There are six detector modules, each with an MWPC for position sensitivity backed up by a $\Delta E - E$ telescope of plastic scintillator. (bottom) The efficiency of the pair spectrometer as a function of the opening angle of the emitted positron–electron pair. (Reprinted from [62] with the permission of Elsevier.)

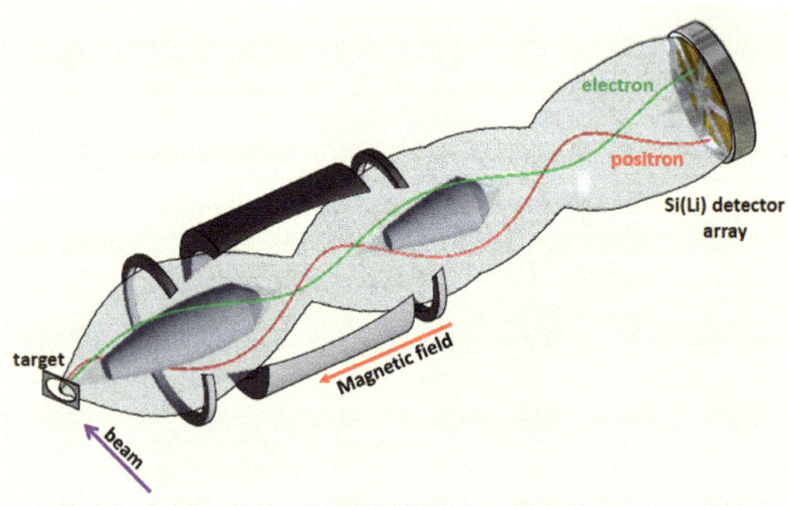

Figure 8.47. Schematic of the Super-e magnetic pair spectrometer comprising a superconducting solenoid, baffles and an array of thick Si(Li) detectors. (Reprinted from [63] with the permission of EDP Sciences under Creative Commons License.)

positron emission. This is carried out using a standard gamma-ray detector, e.g. based on a NaI(Tl) scintillator. There is increasing interest in the PET community in 'dose-on-demand' systems where single doses of a specific radiotracer would be produced on site when required. Allied to this is the desire to miniaturise the apparatus involved in handling the radiotracer using microfluidic chips. Recent advances in photodetector technology such as SiPMs offer a route to a compact detector which can directly detect the positrons emitted from the radiotracer [64, 65]. Figure 8.48 shows a prototype detector system based on plastic scintillator (see section 9.1.2) with a thickness of 1 mm and a 6 mm square Sensl C-series SiPM (see section 5.3.3). At this thickness, the background from 511 keV annihilation photons is negligible as the probability for gamma rays to interact with plastic scintillator is very low (the main mode of interaction for plastic scintillator is Compton scattering). This system has been tested by injecting a plug of radiotracer solution sequentially through a conventional NaI(Tl) annihilation gamma-ray detection system and then into the microfluidic chip with its integrated positron detector. Figure 8.49 shows that both detector systems are capable of recording the radiotracer activity.

Nuclear decommissioning

As discussed in section 3.1.2, ^{90}Sr is a common beta-only emitter used as a radioactive source. This radioisotope is commonly found in nuclear decommissioning sites as it is a long-lived isotope ($t_{\frac{1}{2}}$ = 28.8 years) produced in nuclear fission [66]. Leaks and spills of material can lead to it being found in groundwater at nuclear installations. Conventionally, in order to assess such contamination, samples of

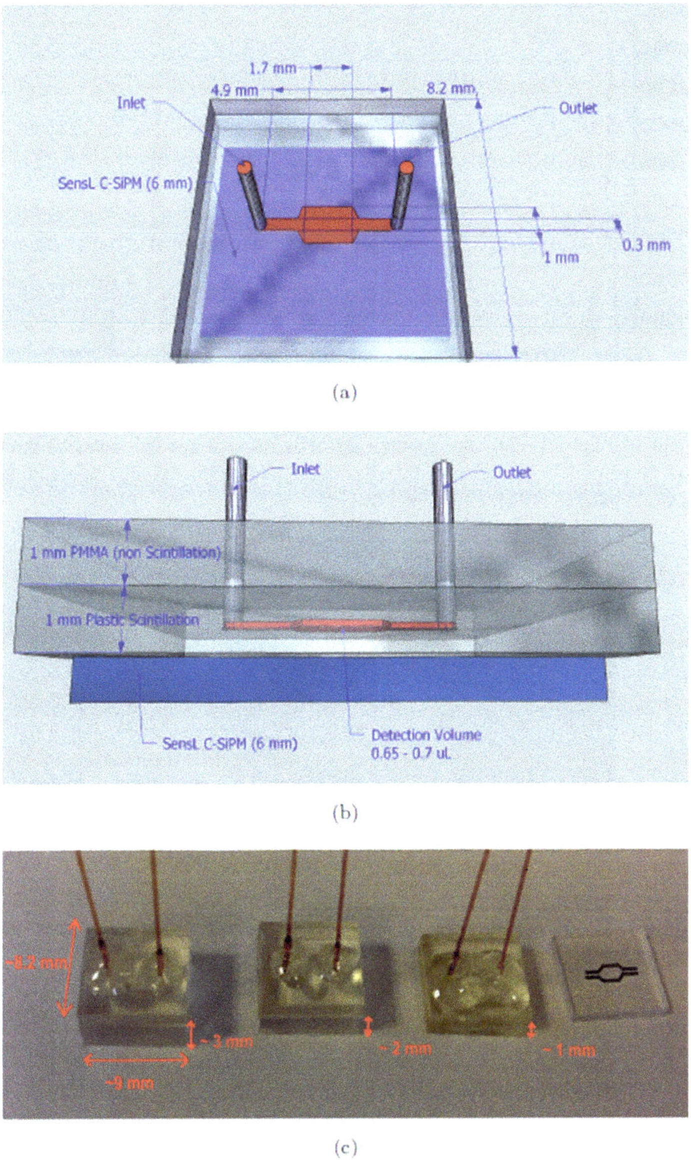

Figure 8.48. A micro-fluidic chip-based detector for PET isotope assay. (a) Top-down design of the chip. (b) Side view showing PMMA element with microfluidic channel, plastic scintillator and SiPM. (c) Photographs of the assembled chip detectors. (Reprinted from [64] with permission from University of York.)

water are obtained, manipulated and processed in a laboratory. It would be highly advantageous to have a direct *in situ* measurement. Figure 8.50 shows a concept of an *in situ* detector using a gallium arsenide (GaAs) PIN diode [66]. GaAs has a large bandgap (1.42 eV) meaning that it can operate at room temperature without suffering appreciably from thermal noise; it is also a radiation-hard material.

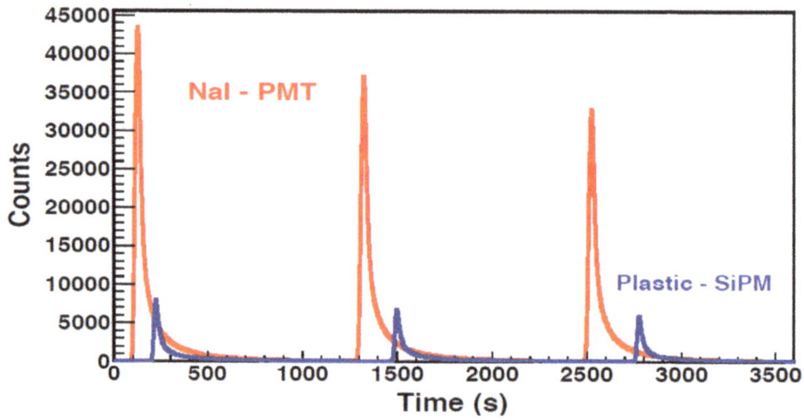

Figure 8.49. A plug of 10 μℓ ^{18}F-PDG in solution (157 MBq ml^{-1}) is injected sequentially through a system comprising a NaI(Tl) gamma ray detector with lead shielding, and a microfluidic chip with integrated plastic scintillator detector coupled to SiPM. (Reprinted from [64] with permission of University of York.)

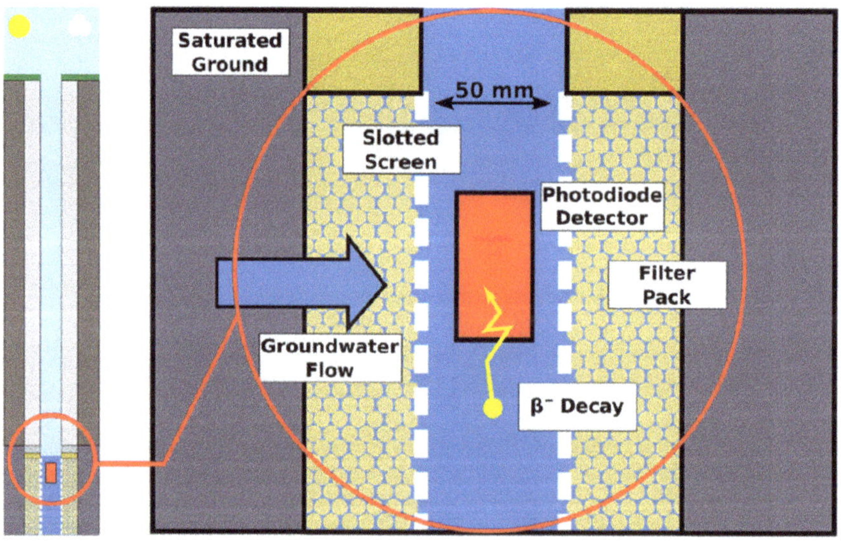

Figure 8.50. Concept for a groundwater beta detector based on an *in situ* GaAs photodiode. (Reprinted from [66] with the permission of Elsevier.)

References

[1] Sarantites D G, Hua P-F, Devlin M, Sobotka L G, Elson J, Hood J T, LaFosse D R, Sarantites J E and Maier M R 1996 'The Microball' Design, instrumentation and response characteristics of a 4π-multidetector exit channel-selection device for spectrocopic and reaction mechanism studies with Gammasphere *Nucl. Instrum. Methods* A **381** 418–32

[2] Henderson J *et al* 2013 Enhancing the sensitivity of recoil-beta tagging *J. Instrum.* **8** 04025

[3] Hyder A 1981 Ion-implanted and bakeable silicon detectors *Nucl. Instrum. Methods* **187** 595–8

[4] Beck G A *et al* 1997 Radiation-tolerant breakdown protection of silicon detectors using multiple floating guard rings *Nucl. Instrum. Methods* A **396** 214–27

[5] Evensen L, Hanneborg A, Avset B and Nese M 1993 Guard ring design for high voltage operation of silicon detectors *Nucl. Instrum. Methods* A **337** 44–52

[6] Koybasi O *et al* 2020 Edgeless silicon sensors fabricated without support wafer *Nucl. Instrum. Methods* A **953** 163176

[7] Valtonen E *et al* 2016 Fabrication of a thin silicon detector with excellent thickness uniformity *Nucl. Instrum. Methods* A **810** 27–31

[8] Skyrme D J 1967 The passage of charged particles through silicon *Nuclear Instrum. Methods* **57** 61–73

[9] Tove P A and Seibt W 1966 Plasma effects in semiconductor detectors *Nucl. Instrum. Methods* **51** 261–9

[10] Mahata K *et al* 2018 Particle identification using digital pulse shape discrimination in a nTD silicon detector with a 1 GHz sampling digitizer *Nucl. Instrum. Methods* A **894** 20–4

[11] Pastore G *et al* 2017 Isotopic identification using Pulse Shape Analysis of current signals from silicon detectors: Recent results from the FAZIA collaboration *Nucl. Instrum. Methods* A **860** 42–50

[12] Laegsgaard E 1979 Position-sensitive semiconductor detectors *Nucl. Instrum. Methods* **162** 93–111

[13] Solal M C 2007 The origin of duo-lateral position-sensitive detector distortions *Nucl. Instrum. Methods* A **572** 1047–55

[14] Banu A *et al* 2008 Performance evaluation of position-sensitive silicon detectors with four-corner readout *Nucl. Instrum. Methods* A **593** 399–406

[15] Qiao R, Li J, Wang J, Dai X and Le Y L 2012 Charge collection and equalization for position-sensitive silicon strip detector *Nucl. Instrum. Methods* A **686** 38–42

[16] Cowin R L and Watson D L 1997 Surface charge flow in two-dimensional position sensitive silicon detectors *Nucl. Instrum. Methods* A **399** 365–81

[17] Lighthall J C *et al* 2010 Commissioning of the HELIOS spectrometer *Nucl. Instrum. Methods* A **622** 97–106

[18] Davinson T 2004 Multistrip silicon detectors for studies with low energy RNBs *Nucl. Phys.* A **746** 188c–94c

[19] Han J-L *et al* 2014 Performance of a double sided silicon strip detector as a transmission detector for heavy ions *Chinese Phys.* C **38** 056202

[20] Theisen C, Lopez-Martens A and Bonnelle C 2008 Internal conversion and summing effects in heavy-nuclei spectroscopy *Nucl. Instrum. Methods* A **589** 230–42

[21] Page R D *et al* 2003 The GREAT spectrometer *Nucl. Instrum. Methods Phys. Res.* B **204** 634–7

[22] Moszynski M, Kapusta M, Wolski D, Klamra W and Cederwall B 1998 Properties of the YAP:Ce scintillator *Nucl. Instrum. Methods* A **404** 157–65

[23] Yamamoto S, Kamada K and Yoshikawa A 2018 Use of YAP(Ce) in the development of high spatial resolution radiation imaging detectors *Radiat. Meas.* **119** 184–91

[24] Xiao Y *et al* 2019 Search for α decay of ^{104}Te with a novel recoil-decay scintillation detector *Phys. Rev.* C **100** 034315

[25] Ostrowski A N *et al* 2002 CD: A double side silicon strip detector for radioactive nuclear beam experiments *Nucl. Instrum. Methods* A **480** 448–55

[26] Catford W N 2014 *What Can We Learn From Transfer, and How is Best to Do It?* (Lecture Notes in Physics vol 879) (Berlin: Springer)
[27] Wimmer K 2018 Nucleon transfer reactions with radioactive beams *J. Phys.* G **45** 033002
[28] Labiche M *et al* 2010 TIARA: A large angle silicon array for direct reaction studies with radioactive beams *Nucl. Instrum. Methods* A **614** 439–48
[29] Diget C A *et al* 2011 SHARC: Silicon highly-segmented array for reactions and coulex used in conjunction with the TIGRESS gamma-ray spectrometer *J. Instrum.* **6** P02005
[30] Adsley P *et al* 2017 CAKE: the coincidence array for K600 experiments *J. Instrum.* **12** T02004
[31] Heine M *et al* 2018 The STELLA apparatus for particle-gamma coincidence fusion measurements with nanosecond timing *Nucl. Instrum. Methods* A **903** 1–7
[32] Fruet G *et al* 2020 Advances in the direct study of carbon burning in massive stars *Phys. Rev. Lett.* **124** 192701
[33] Litvinov Y A *et al* 2013 Nuclear physics experiments with ion storage rings *Nucl. Instrum. Methods Phys. Res.* B **317** 603–16
[34] Kröll T *et al* 2016 Nuclear reactions in the storage ring ESR with EXL *J. Phys.: Conf. Ser.* **724** 012026
[35] Lennard W S, Geissel H, Winterbon K B, Phillips D, Alexander T K and Forster J S 1986 Nonlinear response of Si detectors for low-Z ions *Nucl. Instrum. Methods* A **248** 454–60
[36] Livingston K, Woods P J, Davinson T and Shotter A C 1996 Heavy ion radiation damage in double-sided silicon strip detectors *Nucl. Instrum. Methods* A **370** 445–51
[37] Kaya L *et al* 2017 Characterization and calibration of radiation-damaged double-sided silicon strip detectors *Nucl. Instrum. Methods Phys. Res.* A **855** 109–17
[38] Manfredotti C 2005 CVD diamond detectors for nuclear and dosimetric applications *Diam. Relat. Mater.* **14** 531–40
[39] de Boer W, Bol J, Furgeri A, Mueller S, Sander C, Berdermann E, Pomorski M and Huhtinen M 2007 Radiation hardness of diamond and silicon sensors compared *Phys. Status Solidi* **9** 3004–10
[40] Zambonni I, Pastuovic Z and Jaksic M 2013 Radiation hardness of single crystal CVD diamond detector tested with MeV energy ions *Diam. Relat. Mater.* **31** 65–71
[41] Litrico G *et al* 2018 Silicon carbide detectors for nuclear physics experiments at high luminosity *J. Phys. Conf. Ser.* **1056** 012032
[42] Gai M *et al* 2010 An optical readout TPC (O-TPC) for studies in nuclear astrophysics with gamma-ray beams at HIγS *J. Instrum.* **5** P12004
[43] Giomataris Y, Rebouregard P, Robert J and Charpark G 1996 MICROMEGAS: a high-granularity position-sensitive gaseous detector for high particle-flux environments *Nucl. Instrum. Methods* A **376** 29–35
[44] Colonna N, Gunsing F and Käppeler F 2018 Neutron physics with accelerators *Prog. Part. Nucl. Phys.* **101** 177–203
[45] Andreyev A N, Nishio K and Schmidt K-H 2018 Nuclear fission: a review of experimental advances and pheonomenology *Rep. Prog. Phys.* **81** 016301
[46] Andreyev A N *et al* 2010 New type of asymmetric fission in proton-rich nuclei *Phys. Rev. Lett.* **105** 252502
[47] Charpak G, Bouclier R, Bressani T, Favier J and Zupancic C 1968 The use of multiwire proportional counters to select and localise charged particles *Nucl. Instrum. Methods* **62** 262–8

[48] Gruhn C R *et al* 1982 Bragg curve spectrometry *Nucl. Instrum. Methods Phys. Res.* A **196** 33–40

[49] Sosnin N V *et al* 2020 Fission fragment atomic number measurements using Bragg detectors *Nucl. Instrum. Methods* A **967** 163397

[50] Aggarwal S K 2016 Alpha-particle spectrometry for the determination of alpha emitting isotopes in nuclear, environmental and biological samples: past, present and future *Anal. Methods* **8** 5353

[51] Schuster C S, Smith B R, Sanderson B J, Mullins J T, Atkins J, Joshi P, McNamara L, Krauss T F and Jenkins D G 2017 Flexible silicon-based alpha-particle detector *Appl. Phys. Lett.* **111** 073505

[52] Melwaldt R A *et al* 2008 The low-energy telescope (LET) and SEP central electronics for the STEREO mission *Space Sci. Rev.* **136** 285–362

[53] Nishimura S 2012 Beta-gamma spectroscopy at RIBF *Prog. Theor. Exp. Phys.* **2012** 03C006

[54] Yokoyama R *et al* 2019 Segmented YSO scintillation detectors as a new β-implant detection tool for decay spectroscopy in fragmentation facilities *Nucl. Instrum. Methods* A **937** 93–7

[55] Rahab H, Keffous A, Akam A and Menari H 1999 The preparation of lithium drifted silicon Si(Li) detector *Adv. Perform. Mater.* **6** 53–6

[56] Alekseev I E *et al* 2017 A silicon detector based beta-spectrometer *J. Phys.: Conf. Ser.* **934** 012056

[57] Butler P A, Jones P M, Cann K J, Cocks J F C, Jones G D, Julin R and Trzaska W H 1996 Electron spectroscopy using a multi-detector array *Nucl. Instrum. Methods* A **381** 433–42

[58] Page R D *et al* 2011 In-beam conversion electron spectroscopy of ^{180}Hg *Phys. Rev.* C **84** 034308

[59] Papadakis P *et al* 2018 The SPEDE spectrometer *Eur. Phys. J.* A **54** 42

[60] Pakarinen J *et al* 2014 The SAGE spectrometer *Eur. Phys. J.* A **50** 53

[61] Papadakis P *et al* 2011 The SAGE spectrometer: A tool for combined in-beam gamma-ray and conversion electron spectroscopy *J. Phys.: Conf. Ser.* **312** 052017

[62] Gulyas J *et al* 2016 A pair spectrometer for measuring multipolarities of energetic nuclear transitions *Nucl. Instrum. Methods* A **808** 21–8

[63] Kibédi T, Stuchberry A E, Dracoulis G D and Robertson K A 2012 Towards the pair spectroscopy of the Hoyle state in ^{12}C *EPJ Web Conf.* **35** 06001

[64] Yavuzkanat N 2015 Development of novel scintillation detection techniques for use in nuclear physics and medical applications *PhD Thesis* University of York

[65] Tarn M D *et al* 2018 Beta detection of strontium-90 and potential for direct *in situ* beta detection for nuclear decommissioning applications *Chem. Eur. J.* **24** 13749

[66] Turkington G, Gamage K A A and Graham J 2018 Beta detection of strontium-90 and potential for direct *in situ* beta detection for nuclear decommissioning applications *Nucl. Instrum. Methods* A **911** 55–65

IOP Publishing

Radiation Detection for Nuclear Physics

Methods and industrial applications

David Jenkins

Chapter 9

Neutron detectors

This chapter provides an overview of the most commonly encountered technologies used in both fast and thermal neutron detection. In terms of fast-neutron detection, organic scintillators are the most commonly employed technology. The physical principles behind organic scintillators are reviewed, leading to an explanation of how they can distinguish neutrons from gamma rays on the basis of the fast and slow components of the scintillation light. Different types of organic scintillator are reviewed, focussing on liquid scintillators and plastic scintillators which are largely used for fast neutron counting and neutron time-of-flight measurements, respectively. Attention then turns to thermal neutron detectors, which operate on rather different principles: using nuclear reactions on light nuclei to produce charged particles, which can ionise a gas or produce scintillation light. The classic detector technology for thermal neutron detection is a ^{3}He gas-filled proportional counter. The scarce availability and high price of ^{3}He has focussed attention more recently on alternatives using different neutron-converter materials such as ^{6}Li and ^{10}B. The principles behind such detectors are reviewed.

Concepts: Organic scintillators: liquid scintillators and plastic scintillators; fast neutron discrimination; neutron detector arrays; thermal neutron detection; ^{3}He tubes; ^{3}He alternatives based on ^{6}Li or ^{10}B; societal applications of thermal neutron detection

9.1 Fast neutron detectors

Fast neutrons are produced in fission, by certain specialist radioactive sources and in nuclear reaction studies in the laboratory (see section 3.5). Fast neutrons are not themselves ionising and so fast neutron detection chiefly relies on transferring some of the neutron's kinetic energy to a charged particle which is itself ionising. The largest energy transfer possible, kinematically speaking, is with a proton since the proton and neutron have a near-identical mass. Neutrons scattering off heavy nuclei will only impart a small amount of energy[1]. These considerations explain why most fast neutron detectors are based on the principle of proton (or in some cases, deuteron)-recoil. An ideal detector, therefore, would be a hydrogen-rich scintillator material and this is the class of *organic scintillators*, outlined in section 4.3.2 and revisited in more detail here. Reviews of the functionality of organic scintillators may be found in the monograph of J B Birks [1] and elsewhere, e.g. [2]. Organic scintillators come in a variety of forms including *organic crystals*, *liquid scintillators* (see section 9.1.1) and *plastic scintillators* (see section 9.1.2). The earliest organic scintillators were crystals such as anthracene and stilbene. These have fast decay time of the order of 10 ns. However, they cannot be readily machined and historically, were difficult to obtain in large sizes, although more recent methods have started to surmount this obstacle [3]. Accordingly, their use has been somewhat limited. Moreover, their light emission is anisotropic [3]. Nevertheless, anthracene is the brightest, i.e. highest light output per unit deposited energy of traditional organic scintillators, corresponding to 17 400 photons MeV^{-1}, and so it is common for organic scintillators to have their performance expressed as % anthracene.

Quenching in organic scintillators

An interesting feature of organic scintillators is *quenching*[2]. Where scintillator excitation centres are spaced far apart and interaction between them is minimal, then it can be shown for particles with a low stopping power, e.g. electrons above 100 keV that the yield of scintillation light, L, in an organic scintillator is proportional to the energy deposited i.e.

$$\frac{\mathrm{d}L}{\mathrm{d}r} = S\frac{\mathrm{d}E}{\mathrm{d}r}, \tag{9.1}$$

where S is the scintillation efficiency. This is not the case for heavily ionising particles, e.g. protons, α-particles, heavy ions where the light output is quenched according to Birks' law [4]:

$$\frac{\mathrm{d}L}{\mathrm{d}r} = S\frac{\frac{\mathrm{d}E}{\mathrm{d}r}}{1 + kB\frac{\mathrm{d}E}{\mathrm{d}r}}, \tag{9.2}$$

where kB is Birks' constant comprising k, the energy transfer probability and B, a constant of proportionality associated with the number of damaged molecules [5]. This relationship breaks down for particles with a very high linear energy transfer (LET) because it does not explicitly account for the spatial configuration of the fluorescent and damaged molecules [5]. Figure 9.1 shows the strong deviation from linearity in light yield for high LET ions, implied by Birk's law.

These considerations regarding the quenching of scintillators by transiting ions, i.e. equation (9.2), lead to the useful concept of electron equivalent energy,

[1] Imagine comparing the collisions of two equal mass snooker balls to a collision between such a snooker ball and a medicine ball.

[2] It is not clear how applicable this is to inorganic scintillators.

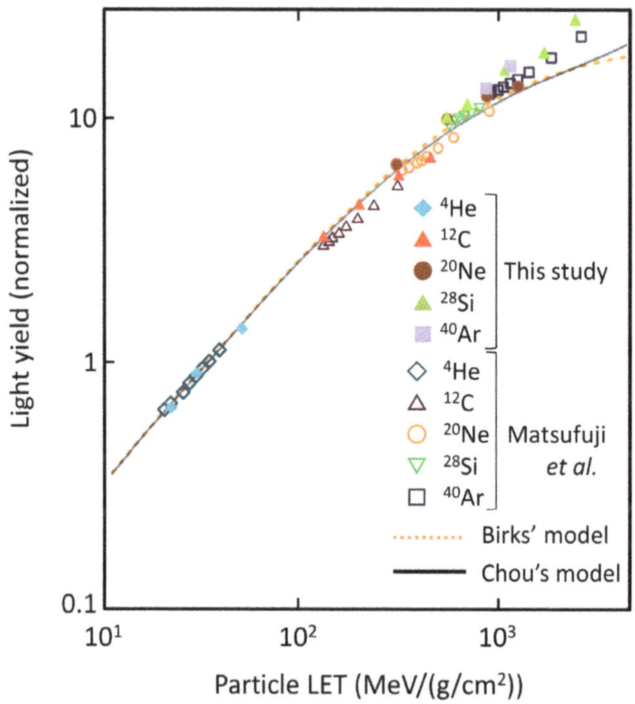

Figure 9.1. Light yield from an NE-102A organic liquid scintillator detector irradiated by heavy ions (150–550 MeV/u) as a function of linear energy transfer (LET). The dashed line is the Birk's model discussed in the text which deviates from the data at the highest values of LET. (Reprinted from [5] with permission from PLOS under Creative Commons License.)

commonly written as keV_{ee} or MeV_{ee} this notation will be seen extensively in some of the figures presented in this chapter.

Potential for neutron-gamma discrimination

The excitation and decay modes of organic scintillators naturally lend them to the task of neutron-gamma ray discrimination. In practice, this translates into the differing manner in which recoil protons and photoelectrons/Compton scattered electrons interact in material (see chapter 2). A gamma ray interacts discontinuously in a material transferring its energy to one or a few photoelectrons or Compton-scattered electrons, whereas an ion such as a proton, alpha particle or heavy ion deposits energy all along its track and the effective $\frac{dE}{dx}$ is large. The excitation densities in an organic scintillator are correspondingly higher for the interaction of recoil protons from a neutron interaction. This leads to a suppression of the singlet, i.e. fluorescence decay mode and an enhancement of the triplet, i.e. phosphorescence decay mode. The net effect of this is to enhance the long-lived component of the scintillation light relative to the short-lived component. Figure 9.2 (bottom) shows the effect of this mechanism on the components of the scintillation light for three common organic crystals: anthracene, stilbene and P-terphenyl. In the data analysis,

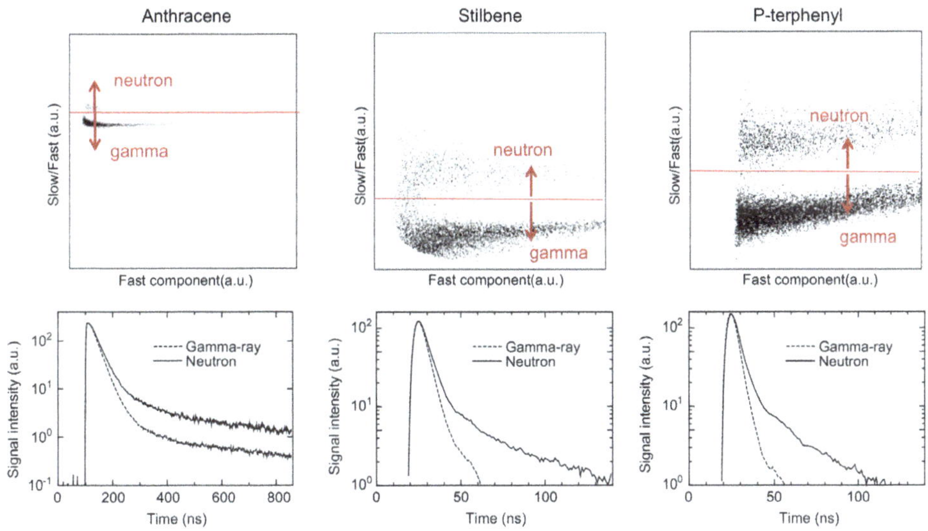

Figure 9.2. Neutron/gamma-ray discrimination using anthracene, stilbene and P-terphenyl. The neutrons are from a ^{252}Cf source and the gamma rays from a ^{60}Co source. (Bottom) the scintillation decay spectrum for neutrons and gamma-rays for each of the three materials. (Top) discrimination of neutrons from gamma rays using the ratio of the slow/fast components plotted against the fast component. In each case, the corresponding slow/fast windows were: 20–40 and 40–150 ns (anthracene), 20–40 and 40–150 ns (stilbene) and 20–30 and 30–150 ns (P-terphenyl). (Reprinted from [6] with the permission of Elsevier.)

windows may then be set on the different fast and slow components of the light output to discriminate neutrons from gamma rays (see figure 9.2 (top)).

9.1.1 Liquid scintillator detectors

Liquid scintillator detectors typically comprise a detector volume with a metal can containing an organic liquid scintillator. The can usually has a bellows arrangement to allow for expansion of the liquid under change of temperature. The cans have optical windows in them to allow coupling to a PMT. Typical liquid scintillators need to be handled with care; depending on the choice of liquid it may be harmful and carcinogenic, and it may also have a low flash point. The classic liquid scintillator material is BC501A. It has a decay time of 3.2 ns and a peak wavelength of 425 nm.

Liquid scintillator detectors are conventionally employed as fast neutron counters since they usually have no particular sensitivity to the incident neutron energy. They are often installed as arrays replacing some portion of a germanium detector array to provide fast neutron detection. An example of a contemporary liquid scintillator neutron array is the NEDA array (see figure 9.3) designed to be used at the SPIRAL2 facility in Normandy, and elsewhere. A related example is the DESCANT neutron array at TRIUMF (see figure 9.4) which is a variant of the standard approach since it makes use of deuterated liquid scintillator [7]. The n–d scattering reaction is asymmetric in the centre-of-mass as the neutron and deuteron have different masses. This means that the pulse-height information recorded from a deuterated scintillator

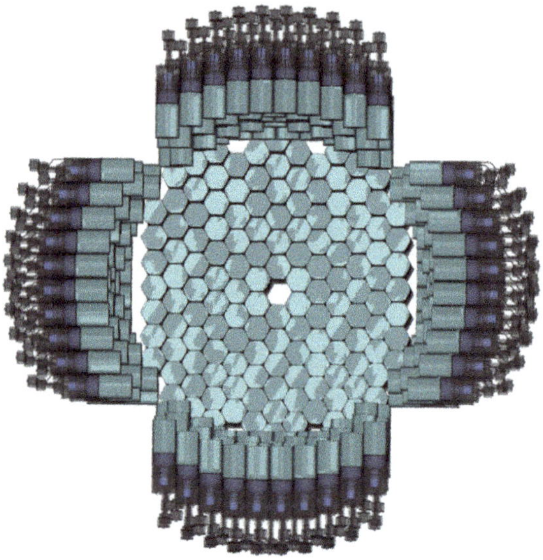

Figure 9.3. 3D CAD drawing of the NEDA array with a 2π solid angle coverage at a distance of 1 m. This full design comprises 331 individual identical detectors. (Reprinted from [8] with the permission of Elsevier.)

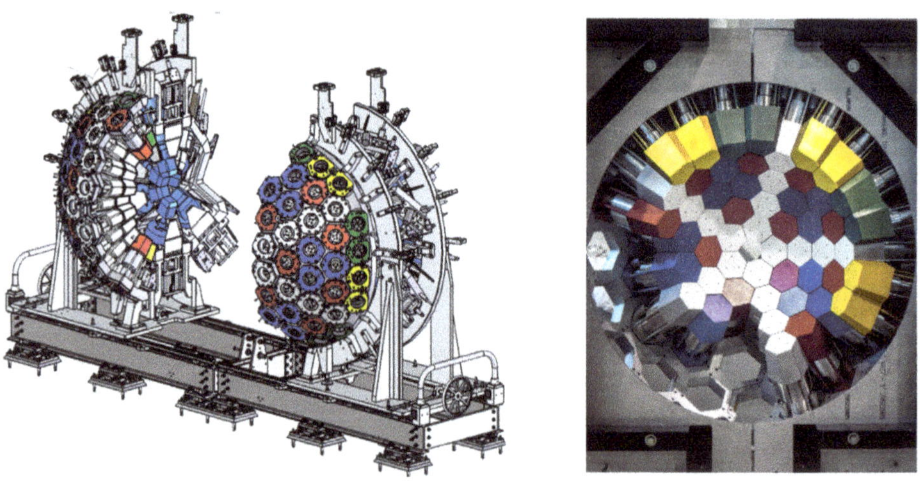

Figure 9.4. The DESCANT neutron array: (left) CAD drawing of DESCANT coupled to the GRIFFIN germanium detector array and (right) partially assembled DESCANT array. (Reprinted from [9] with the permission of Elsevier under Creative Commons License.)

detector, such as those making up DESCANT, is sensitive at some level to the incident neutron energy (see figure 9.5).

Principles of neutron discrimination with liquid scintillators

In typical in-beam nuclear physics experiments, neutrons and gamma rays may both be produced, which leads to ambiguity in the signal recorded in liquid scintillator

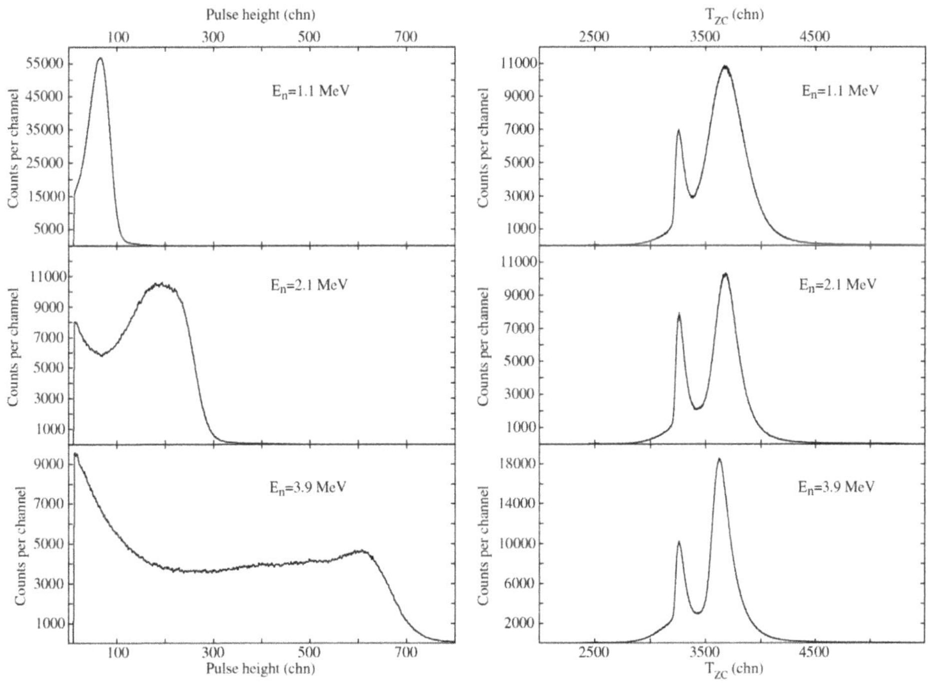

Figure 9.5. Pulse height (left) and neutron-gamma discrimination (right) for a DESCANT detector using monoenergetic neutrons of 1.1 MeV (top), 2.2 MeV (middle) and 3.9 MeV (bottom). (Reprinted with the permission of [10] under Creative Commons License.)

detectors as such detectors respond to both types of indirectly ionising radiation. However, neutrons can be discriminated in two ways. First, the time-of-flight for neutrons and gamma rays is different since gamma rays travel at the speed-of-light but typical fast neutrons (~MeV) travel only at a small fraction of the speed-of-light. If the production time can be well defined, e.g. an accelerated beam is pulsed with a pulse-width of order nanoseconds then the difference in time-of-flight between reaction target and the neutron detector can be used to discriminate. An example of neutron separation by time-of-flight is shown in figure 9.6 for ~2 MeV neutrons and gamma rays over a flight distance of around 5 m. The second technique is to discriminate neutrons from gamma rays using pulse shape analysis. This relies on neutrons generating a different response in the liquid scintillator leading to more amplitude in the tail of the pulse, as described in section 4.3.2. This can be exploited both using analogue electronics, splitting the signal into fast and slow components, as was achieved with the EUROBALL neutron wall [11] and with digital electronics as in the example of the more recent NEDA array [8]. Figure 9.7 shows an example of making use of this effect to discriminate neutrons from gamma rays. In practice, where both time-of-flight and pulse-shape analysis are available then they are applied in conjunction leading to a very high rejection of gamma rays and robust identification of fast neutron events.

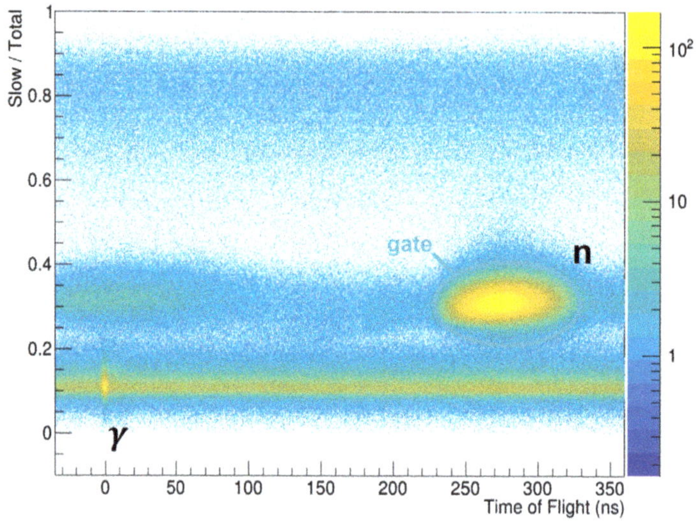

Figure 9.6. Discrimination of fast neutrons from gamma rays using a NEDA detector: the fraction of the slow component to total plotted against time of flight. The fast neutrons are well-localised.

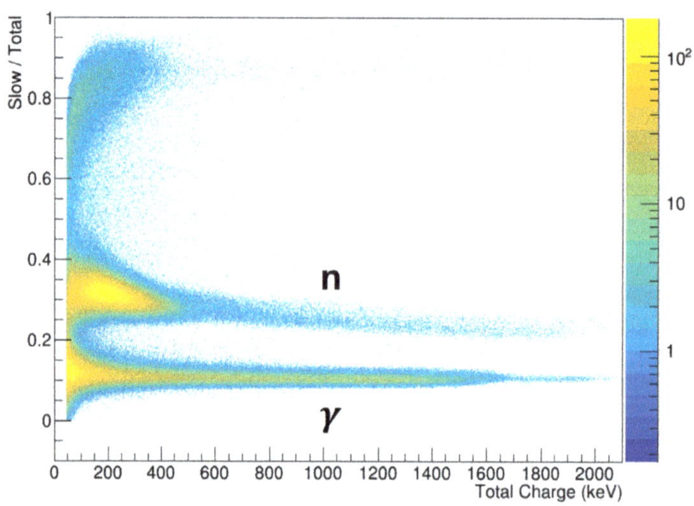

Figure 9.7. Neutron discrimination using NEDA: the fraction of the slow component to total plotted against total.

Complications enter the analysis of data from liquid scintillator arrays when it is required to identify two discrete neutron interactions. This is because neutrons interact predominantly through elastic and inelastic scattering and so may scatter out of one detector and into another leading to the detection of two 'neutron events' originating from the same neutron. A possible route to separation of such scattered events is to look at the time distribution of the interactions, however, this may not completely resolve the ambiguity. Figure 9.8 provides an example of Monte Carlo

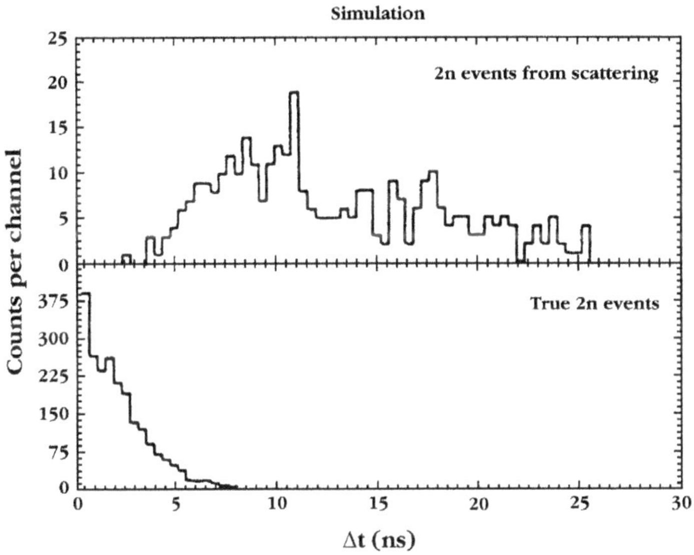

Figure 9.8. Monte Carlo simulation of time distributions from direct hits and scattered neutrons in the NORDBALL neutron wall. (Reprinted from [11] with the permission of Elsevier.)

simulation of the time distribution of direct neutron hits and such neutron scatters, which illustrates the experimental difficulty.

9.1.2 Plastic scintillator detectors

Plastic scintillators are a relatively inexpensive detector technology which is efficient for fast neutron detection. Here, the generation of scintillation light is initiated by the recoil of protons in the hydrogen-rich material. Plastic scintillators are characterised by relatively low light output but their chief advantage is their fast time response with rise times of order 1 ns and decay times of a few ns. This means that plastic scintillator detectors can count very fast and are also excellent for neutron/gamma discrimination based on time-of-flight. Due to its low effective Z, gamma ray interactions in plastic scintillator are predominantly associated with Compton scattering, which means they are not practical as spectroscopic gamma ray detectors but are useful for quantifying gamma flux or to provide a coincidence signal.

Plastic scintillators are usually based on a standard plastic matrix incorporating aromatic molecules such as polyvinyltoluene (PVT) or polystyrene (PS). Such plastics on their own are fluorescent but the light yield is extremely low. Moreover, the light emission is typically in the near-UV (300–350 nm) and such plastics re-absorb this light meaning the mean-free path of the scintillation light is very low (of order millimetres) and the decay time for light emission of, e.g. PS, is around 16 ns [12]. In order to achieve more useful properties such as a longer attenuation length and fast decay time, plastic scintillators typically have primary fluors incorporated into them at the level of about 1% by weight. Non-radiative energy transfer, known as *Fortster transfer* can occur between the plastic molecules and the primary fluor. This leads to a significantly higher light yield with emission of scintillation light at longer wavelengths than the base plastic (see figure 9.9). Moreover, the attenuation length for the light emitted by the fluors is of order 10 cm so scintillation light can be more readily extracted from a plastic scintillator with a useful volume.

In order to extend the attenuation length of plastic scintillators and shift the peak wavelength of light emission to match with different choices of photosensor, a wavelength shifter material is frequently added at the level of around 0.01% by weight. The inclusion of different fluors and wavelength shifter can be tuned to produce plastic scintillator with a variety of wavelength-emission characteristics (see figure 9.10). Standard plastic scintillators have peak emission around 400–430 nm which is well matched to the response of SiPMs, for example. Typical light emission from plastic scintillators is of the order of 10 000 photons MeV^{-1}. There are many suppliers of plastic scintillator but a good idea of what is available in terms of plastic scintillators with different wavelength emission, decay time etc, is available from ELJEN.

Plastic scintillator can be cast and machined like normal plastics allowing large and/or complex pieces to be created. It can also be produced in the form of scintillating fibres. Indeed, recently, there have been the first attempts to 3D print plastic scintillator [15], although the material produced has yet to reach the quality in terms of light output etc of conventionally produced materials. Nevertheless, this

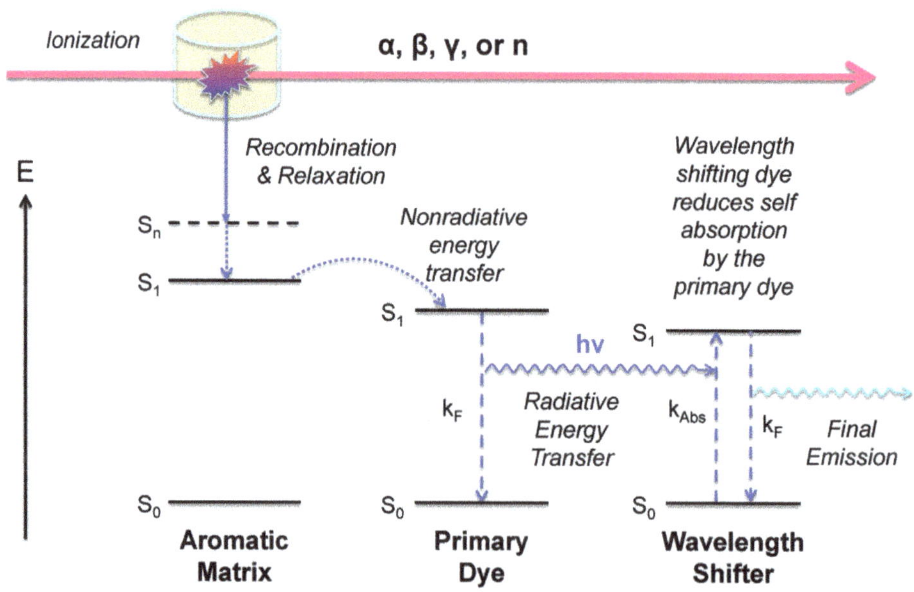

Figure 9.9. Physical processes in scintillation of organic materials. (Reprinted from [13] with permission from Wiley.)

Figure 9.10. Plastic scintillators with different wavelength emission when illuminated with ultraviolet light. (Reprinted from [14] with the permission of Elsevier under Creative Commons License.)

is clearly an area to watch as it should offer great flexibility and ease of design of prototypes.

Plastic scintillator can be formed into bars, rods, thin sheets or other planar geometries. It is readily connected to a variety of light sensors such as PMTs and SiPMs. An alternative to the regular approach of collecting light on a face is to machine grooves or channels into a slab of plastic scintillator and run optical fibres down these channels to collect the light and transport it to light sensors mounted at some distance away. Depending on the wavelength matching of the light emission of the plastic and the light sensor employed, sometimes wavelength-shifting optical fibres are employed to provide better wavelength matching to the light sensor.

Plastic scintillator can also be doped at the few % level with high-Z materials [13] such as tin or lead [16] to make them more efficient for gamma rays. Here, the improvement arises from increased probability for photoelectric absorption on the high-Z inclusions. Plastic scintillator may also be doped with gadolinium to turn it into a thermal neutron detector—an application more fully described in section 9.2.2.

Examples of neutron detector arrays based on plastic scintillator include VANDLE [17] constructed at Oak Ridge National Laboratory which is intended for studies of beta-delayed neutron emission (see figure 9.11) and nuclear reaction studies where the outgoing particle is a neutron. VANDLE comprises long bars of plastic scintillator coupled to PMTs on both ends. VANDLE is coupled to a digital data acquisition system based on Pixie-16 modules from XIA [17]. Other examples include NeuLAND [18] intended for the FAIR facility in Darmstadt, Germany.

Neutron-discriminating plastic scintillator

Plastic scintillators, e.g. EJ-276, are currently becoming available that offer pulse-shape discrimination between fast neutrons and gamma rays [19]. Figure 9.12 illustrates the capability to distinguish clearly between fast neutron and gamma ray interactions with such materials with a comparable performance to that of a standard liquid scintillator. Clearly the solid nature of the plastic offers significant advantages over liquid scintillators and none of the related hazards discussed above.

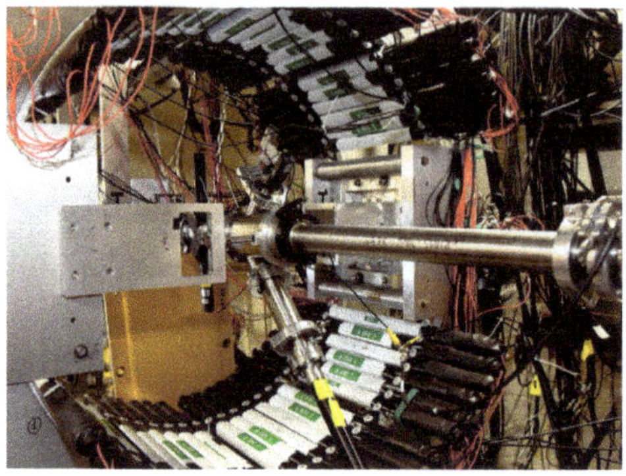

Figure 9.11. Photograph of the VANDLE neutron array arranged for study of beta-delayed neutrons comprising 48 individual modules. (Reprinted from [17] with the permission of Elsevier.)

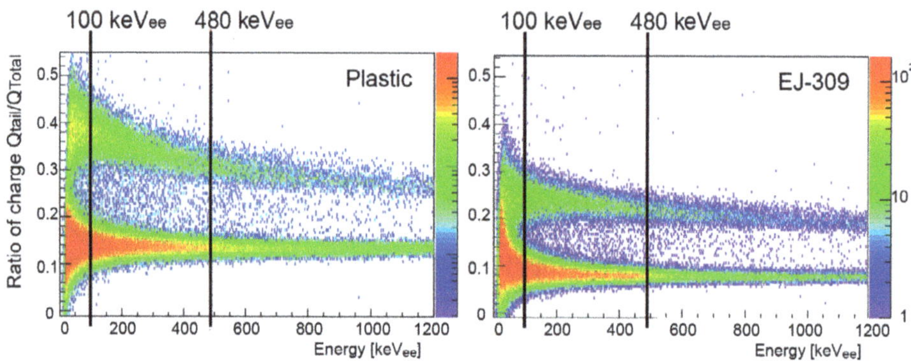

Figure 9.12. Neutron-gamma discrimination using plot of the ratio of charge in the tail of the pulse to total against electron-equivalent energy: (left) using 50 mm diameter, 76 mm long PSD plastic, and (right) using a liquid scintillator (EJ-309) of the same dimension. (Reprinted from [19] with the permission of Elsevier.)

In the longer term, this has significant potential for use in experimental nuclear physics and in application areas.

9.1.3 Emerging alternatives for fast neutron detection

Inorganic scintillator detectors

As noted in section 5.2.1, there are a number of inorganic scintillator detectors, e.g. CLYC and CLLBC, which can function as dual-purpose gamma-ray/fast neutron detectors; a critical review can be found here [20]. This functionality commonly depends on chlorine content in the crystal composition, where neutrons interact via the ^{35}Cl(n,p) reaction. Such detectors are relatively small, however, and expensive. The neutron detection efficiency is also rather low. For these reasons, they are yet to be widely adopted in experimental nuclear physics although they are undoubtedly valuable in nuclear security applications, chiefly because they can provide gamma-ray and neutron detection in a single compact unit.

^{4}He scintillation detectors

There is significant potential for liquid or gaseous ^{4}He-based scintillator detectors for fast neutron detection[3] [21, 22]. ^{4}He is a good scintillator emitting in the UV region, $\lambda\sim$ 80 nm. A typical implementation would be to have a gas-filled tube of ^{4}He with the walls of the tube covered with reflective material and a wavelength shifting phosphor that converts the scintillation photons to 430 nm so they can be more suitably recorded by PMTs at the two ends of the tube [22].

Since ^{4}He is a relatively light nucleus it can recoil in a neutron-scattering interaction although not with quite as high an efficiency in energy conversion as protons in an organic scintillator. However, it can offer significant advantages over organic scintillators in that the electron density of ^{4}He is low so gamma-ray interaction (i.e. Compton scattering) is much less probable than neutron interaction [22]. ^{4}He tubes are generally of small diameter limiting further the probability of gamma-ray interaction. In regular organic scintillators, gamma-ray interactions depositing the same amount of energy as a charged particle produce more scintillation light but in a noble gas such as ^{4}He, the light produced is essentially the same, i.e. linear to energy deposited irrespective of the origin. Taken together, these advantages greatly facilitate separation of fast neutrons from gamma rays using pulse-shape analysis. Figure 9.13 shows a comparison of the discrimination achievable with a ^{4}He scintillation detector and a comparable organic liquid scintillator detector.

9.2 Thermal neutron detectors

In experimental nuclear physics, fast neutron detection, as discussed above, is the most relevant technique. Thermal neutron detection is of lesser importance since while it can detect the presence of the neutron, by the time the neutron is detected all details on its initial energy and direction of travel are lost. Moreover, depending on the slowing medium, it takes some time for neutrons to thermalise, meaning that it would be difficult to establish prompt coincidences between detection of such neutrons and other radiation emitted when the initial fast neutron was produced. While relevance to nuclear physics is limited, thermal neutron detection is much more important in societal applications and a range of examples are reviewed in section 9.3.

9.2.1 ^{3}He gas-filled proportional counter

The classic thermal neutron detector is a ^{3}He gas-filled proportional counter (also known more commonly as a ^{3}He tube). The physics of gas-filled proportional counters was outlined in section 4.3.1. This is a regime where the electric field within the detector is sufficiently high that avalanche multiplication occurs. As discussed above for fast neutrons, neutrons are not themselves ionising but thermal neutrons

[3] Note that such detectors are based on scintillation and should not be confused with ^{3}He proportional counters for thermal neutron detection discussed in section 9.2.1.

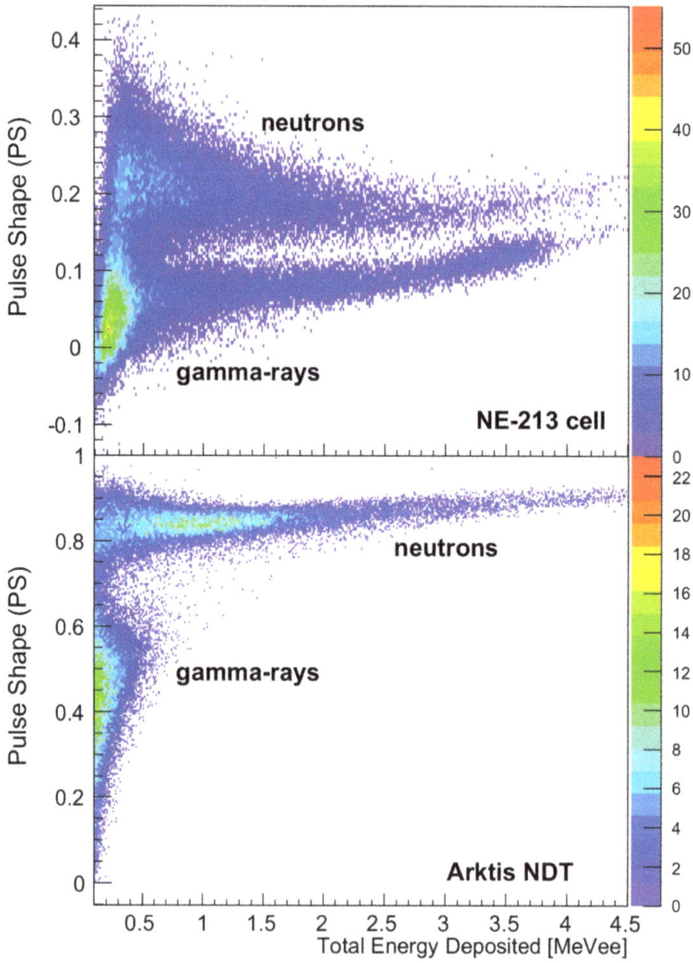

Figure 9.13. Pulse shape parameter as a function of the total energy deposited in an NE-213 liquid scintillator detector (top) and a ^{4}He scintillator detector (bottom). The pulse shape parameter is defined as the ratio of the scintillation-light yield in the tail of the signal to the total. (Reprinted from [22] with the permission of Elsevier under Creative Commons License.)

can interact with ^{3}He to produce light charged particles. The relevant nuclear reaction is

$$n_{\text{th}} + {}^3\text{He} \rightarrow t + p. \tag{9.3}$$

The outgoing proton and triton share the kinetic energy associated with the reaction Q-value (0.764 MeV); it is these light charged particles that ionise the gas and produce a large, proportional signal.

^{3}He detectors have the significant advantage of being largely insensitive to gamma rays. Gamma rays typically interact only at the detector walls and produce small amplitude signals which can be discriminated from the neutron signals by setting an appropriate electronic threshold. ^{3}He is, however, a scarce commodity and at times in the last decade there have been supply shortages and increased prices, leading to resurgence of interest in ^{3}He replacements, particularly for homeland security which is discussed in more detail in section 9.3.1.

9.2.2 ^{3}He replacements

There are a number of alternatives to ^{3}He leading to detectors with varying efficiencies and associated advantages/disadvantages. Most such alternative detectors are based on the employment of detector materials containing ^{6}Li or ^{10}B which have high thermal neutron cross sections and undergo breakup into energetic charged particles via the following reactions:

- $^{10}B + n_{th} \rightarrow {}^7Li + \alpha + 2.31$ MeV
- $^6Li + n_{th} \rightarrow {}^3H + \alpha + 4.78$ MeV.

The resulting charged particles can ionise a gas or induce scintillation in a solid material.

Boron trifluoride (BF_3)

Boron itself is a solid but boron trifluoride (BF_3) provides a useful alternative to ^{3}He since it makes boron available in a gaseous form. Moreover, it can be enriched with ^{10}B to produce a more efficient detector. BF_3 is used in gas-filled proportional counters where it serves both as the converter material and as the proportional gas. Figure 9.14 shows the typical response of a BF_3 detector; the *Q*-value of the neutron-converting reaction is evident from the peak around 2.3 MeV. The low-energy tail of the response arises from so-called *wall effects* where one of the reaction products

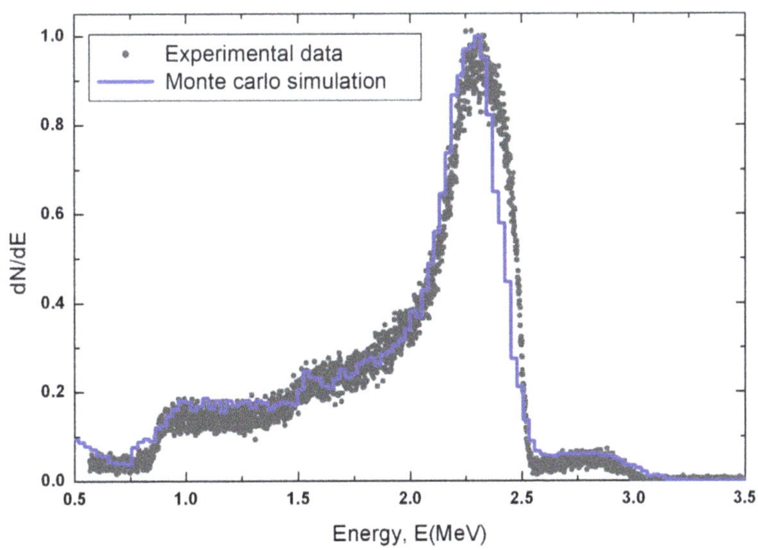

Figure 9.14. Comparison of the experimental response of a typical BF_3 detector with a Monte Carlo simulations. (Reprinted from [23] with the permission of Elsevier under Creative Commons License.)

may hit the detector wall without depositing all its energy in the gas. As with ^{3}He tubes, the gamma-ray component to the detector signal is of low amplitude relative to that from the charged particles and can be readily discriminated. The principal disadvantage of BF_3 is its high toxicity which poses challenges in its handling.

Boron straws

Boron straws are an interesting alternative to both ^{3}He and BF_3 tubes that employ boron converter material in a solid form [24–26]. A boron straw detector comprises one or more gas-filled proportional counters with their walls typically covered with a 1 μm thickness of B_4C, highly enriched in ^{10}B (see figure 9.15). The light charged particles, ^{7}Li and α from the neutron-conversion reaction enter the gas volume and ionise it. The proportional gas is typically a mixture (90/10 by volume) of Ar/CO_2 at low pressure (0.7 atm) [26]. Such a gas mixture is intrinsically safe compared to BF_3 discussed above. Since the particles are not produced directly in the gas volume, there is potential for them to be emitted in the wrong direction or to be stopped within the conversion layer itself due to their short range. This is what dictates the maximum thickness of the conversion layer that can be employed. To increase the effective surface area of the conversion layer and, hence, the thermal neutron detection efficiency, complex star-shaped tube geometries have been explored [24].

^{6}Li-doped scintillating glass

Glasses are an excellent detector medium because of their robustness and resistance to melting [27]. ^{6}Li-doped scintillating glasses are a well-established technology which can be competitive in some cases with ^{3}He tubes depending on the application [28]. They are available from manufacturers such as Scintacor. Typical lithium content of the scintillating glass is around 7% of which the ^{6}Li component can be enriched to 95%. The glasses are doped with cerium to improve the scintillation light output. Discrimination of thermal neutrons from gamma rays is carried out on the basis of the pulse amplitude since the energy deposited in the break-up of ^{6}Li

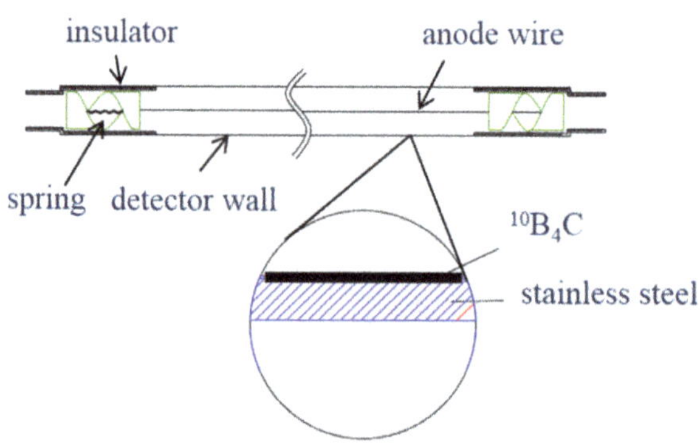

Figure 9.15. Design of a boron straw detector. Each tube is an 8.0 mm diameter proportional counter with 800 mm sensitive length. The detector is formed of stainless steel with 0.2 mm thickness. All tubes have been heated under high temperature for outgassing. An electrophoresis method is used to coat the inner walls with B_4C (90% ^{10}B enrichment). (Reprinted from [25] with the permission of IOP Publishing.)

following interaction with a thermal neutron is large in comparison to typical energy deposited in the glass through Compton scattering etc. A way in which gamma-ray discrimination can be improved is to operate a ^{6}Li-enriched detector in parallel with a ^{6}Li-depleted detector and compare the signal.

Sandwiched converter layers

The light charged particles produced in the ^{6}Li and ^{10}B thermal neutron-conversion reactions are emitted in a roughly back-to-back geometry in order to conserve momentum, and will have a short range in matter. These features have been exploited to produce thermal neutron detectors by sandwiching a ^{6}LiF film or ^{10}B layer between two silicon detectors [29] or two diamond detectors [30]. The efficiency in such applications is naturally limited by the dimensions, particularly the limited thickness of the converter layer.

Loaded plastic scintillator

Plastic scintillator (see section 9.1.2) can be converted into a high efficiency thermal neutron detector by loading it with suitable materials such as gadolinium, lithium or boron, which have each have isotopes high cross-sections for thermal neutrons. A gadolinium-doped plastic scintillator functions via neutron capture principally onto the component isotopes such as ^{155}Gd and ^{157}Gd, the latter of which has a thermal neutron capture cross section of 257 000 barns—the highest of any stable isotope [31]. The relevant reaction is

$$^{157}\text{Gd} + n_{\text{th}} \rightarrow {}^{158}\text{ Gd} + 7.94 \text{ MeV}. \tag{9.4}$$

The excess energy is released as a cascade of gamma rays for which the probability of at least one of them interacting in the plastic is high.

Boron-loaded [32] and lithium-loaded [33] plastic scintillators function by thermal neutron capture on to ^{10}B and ^{6}Li, respectively, which leads to energetic ionising charged-particles within the plastic. Figure 9.16 provides an example of pulse shape

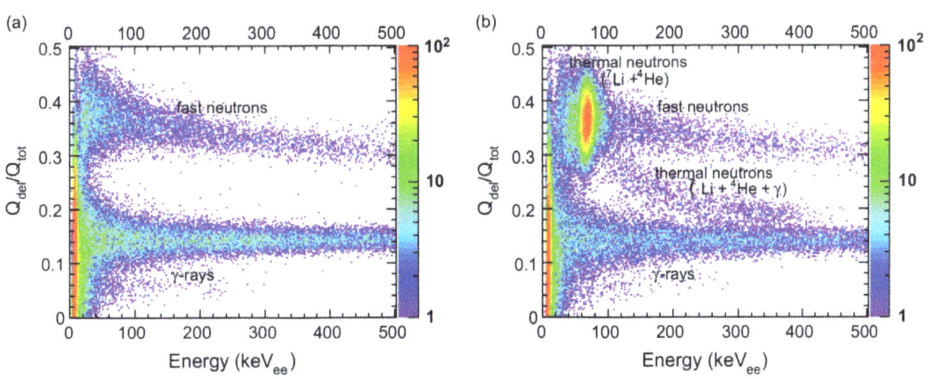

Figure 9.16. Pulse shape discrimination with different plastic scintillators: (left) regular pulse-shaping discriminating plastic (cf figure 9.12) and (right) borated plastic scintillator containing 5% by weight of m-carborane with natural boron abundance. The signature of the breakup of ^{10}B is clearly seen in the top left of the plot. (Reprinted from [32] with the permission of Elsevier.)

discrimination of thermal neutrons using a borated plastic where the thermal events are seen to be very well localised [32].

9.3 Industrial and security applications of neutron detection

There are a limited number of societal applications of neutron detection chiefly in the industrial and homeland security sectors.

9.3.1 Homeland security

There are very few natural sources of neutrons in our environment so detection of even a low counting rate of neutrons can be indicative of the presence in the locality of a fission neutron source. This is obviously of high interest in terms of nuclear security where special nuclear materials such as uranium and plutonium are under tight control.

Portal monitors

Radiation portal monitors (RPMs) are commonly found at ports and airports to identify illicit transport of special nuclear materials. Traditionally, such RPMs contain ^{3}He detectors, surrounded by moderating material such as polyethylene, to detect thermal neutrons; figure 9.17 shows an example of such a conventional radiation portal monitor (RPM). The choice of ^{3}He for this application arises from

Figure 9.17. Two ^{3}He tubes mounted within neutron-moderating material as a radiation portal monitor (RPM). (Reprinted from [34] with the permission of Elsevier.)

its high cross-section for thermal neutrons and from the fact that is largely insensitive to gamma rays as a background. There is high interest, as discussed above, in finding cost-effective ^{3}He replacements.

Handheld/wearable thermal neutron detectors

The Kromek D3S offers simultaneous thermal neutron detection and gamma-ray detection. This unit was already described in section 7.1.1 in terms of its ability to identify radioisotopes and map their location. The thermal neutron capability is provided by a layer of scintillator (zinc sulphide) mixed with ^{6}LiF converter material. The scintillation light is collected by a silicon photomultiplier. This sensor

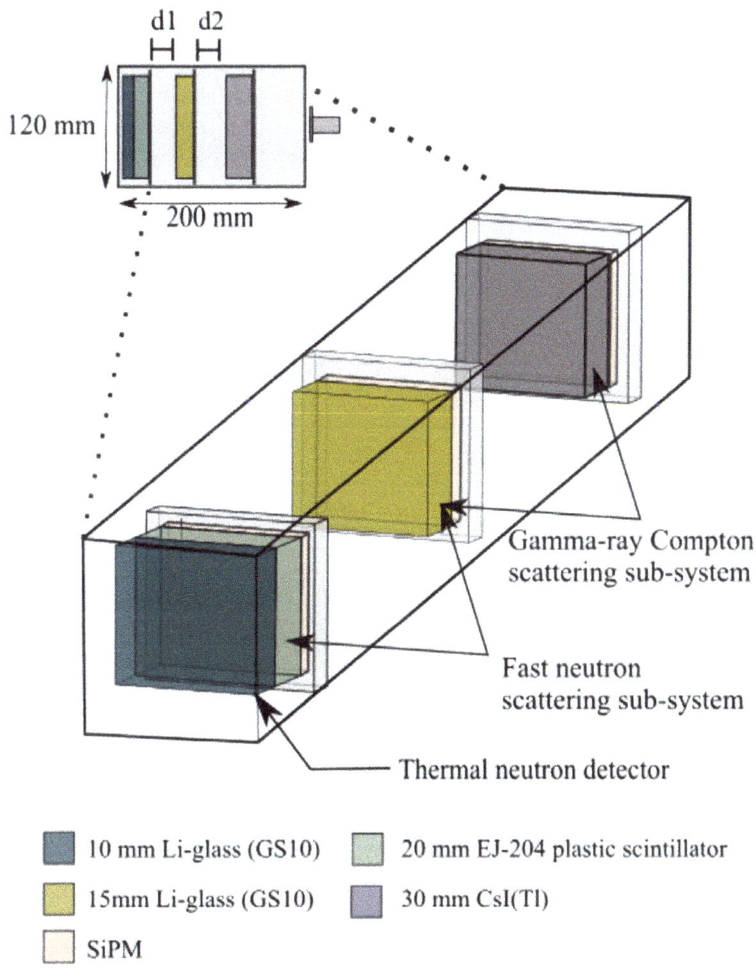

Figure 9.18. Diagram of a hand-held neutron/gamma-ray imager with three separate detector planes. The detector components are labelled on the diagram (see text for details). (Reprinted from [35] with the permission of Elsevier under Creative Commons License.)

is intrinsically thin and has high rejection of background from other sources such as gamma rays.

Al Hamrashdi *et al* [35] report on the development of a handheld imager which would measure both thermal and fast neutrons as well as gamma rays—see figure 9.18. It uses an elaborate series of three detector planes comprising different detector materials which draws together many of the concepts discussed in this chapter as well those in chapter 5. The first detector plane (closest to the neutron source) has two detector slabs optically coupled: a 10mm thickness of ^{6}Li-doped glass and a 20 mm thickness of plastic scintillator. The second detector plane comprises a further 15 mm thick slab of ^{6}Li-doped glass while the third and final detector plane is 30 mm thickness of CsI(Tl). Each detector plane is instrumented with an 8 × 8 array of SensL J-series SiPMs (see section 5.3.3). With these thin slabs of scintillator, the location where the scintillation light is detected is closely correlated with the position in the slab where the light is generated.

The detector shown in figure 9.18 is designed to identify thermal neutrons in the front of the first detector plane. The rear of the first detector plane serves as a fast neutron scatter and the scattered neutrons are detected in the second detector plane (see figure 9.19). This second plane doubles up as a Compton scattering plane for gamma rays where the scattered gamma rays are detected in the third detector plane, functioning as a Compton camera—a technique described in section 7.1.2.

9.3.2 Borehole logging

Thermal neutron detection is employed in borehole logging for porosity measurements. This reflects the fact that the scattering and thermalisation of neutrons is highly material dependent; moderation of neutrons will occur much more rapidly in hydrogen-rich material such as water-filled cavities than in solid rock. The porosity

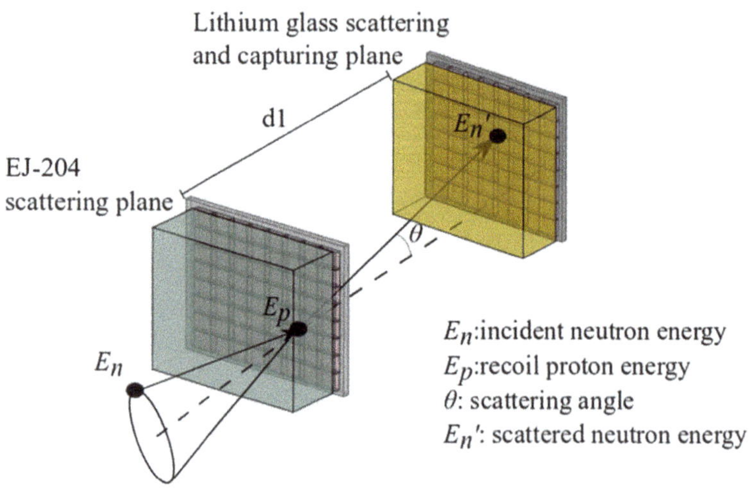

Figure 9.19. Concept of a plastic scintillator and ^{6}Li-doped detector plane as a neutron-scatter imaging detector. (Reprinted from [35] with the permission of Elsevier under Creative Commons License.)

measurements are implemented using a sealed ^{241}Am–Be neutron source in the neutron probe. Two ^{3}He proportional detectors are mounted sequentially along the borehole probe away from the source. The ratio of the thermal neutron flux reaching the near and far detectors depends on the hydrogen-content and porosity. Using two detectors with this ratio method provides a porosity measurement which is independent of borehole diameter.

References

[1] Birks J B 1964 *The Theory and Practice of Scintillation Counting* (International Series of Monographs on Electronics and Instrumentation vol 2) 1st edn (Oxford: Pergamon)

[2] Brooks F D 1979 Development of organic scintillators *Nucl. Instrum. Methods* **162** 477–505

[3] Schuster P 2016 Investigating the anisotropic scintillation response in anthracene through neutron, gamma-ray and muon measurements *IEEE Trans. Nucl. Sci.* **63** 1942

[4] Birks J B 1951 Scintillations from organic crystals: specific fluorescence and relative response to different radiations *Proc. R. Soc.* A **64** 874

[5] Ogawa T, Yamaki T and Sato T 2018 Analysis of scintillation light intensity by microscopic radiation transport calculation and Förster quenching model *PLoS One* **13** e0202011

[6] Yanagida T, Watanabe K and Fujimoto Y 2015 Comparative study of neutron and gamma-ray pulse shape discrimination of anthracene, stilbene and p-terphenyl *Nucl. Instrum. Methods* A **784** 111–4

[7] Garrett P E 2014 DESCANT—the deuterated scintillator array for neutron tagging *Hyperfine Interact* **225** 137–41

[8] Valiente-Dobon J J *et al* 2019 NEDA—NEutron Detector Array *Nucl. Instrum. Methods Phys. Res.* A **927** 81–6

[9] Wong J *et al* 2014 The DEuterated SCintillator Array for Neutron Tagging *EPJ Web Conf.* **66** 11040

[10] Garrett P E 2014 The evolving structure of the Cd isotopes *EPJ Web Conf.* **66** 02039

[11] Skeppstedt O *et al* The EUROBALL neutron wall–design and performance tests of neutron detectors *Nuclear Instrum. Methods Phys. Res.* A **421** 531–41

[12] Moser S W, Harder W F, Hurlbut C R and Kusner M R 1993 Principles and practice of plastic scintillator design *Radiat. Phys. Chem.* **41** 31–6

[13] Hajagos T J, Liu C, Cherepy N J and Pei Q 2018 High-Z sensitized plastic scintillators: a review *Adv. Mater.* **30** 1706956

[14] Giot M *et al* 2017 Nuclear instrumentation and measurement: a review based on the ANIMMA conferences *EPJ Nucl. Sci. Technol.* **3** 33

[15] Son J, Kim D G, Lee S, Park J, Kim Y, Schaarschmidt T and Kim Y K 2018 Improved 3D printing plastic scintillator fabrication *J. Korean Phys. Soc.* **73** 887–92

[16] Nusrat H, Pang G, Ahmad S B, Keller B and Sarefehnia A 2019 Quantifying the impact of lead doping on plastic scintillator response to radiation *Med. Phys.* **46** 4215–23

[17] Peters W A *et al* 2016 Performance of the versatile array of neutron detectors at low energy (VANDLE) *Nucl. Instrum. Methods* A **836** 122–33

[18] Douma C A *et al* 2017 Design studies for the NeuLAND VETO detector *J. Phys.: Conf. Ser.* **1024** 012027

[19] Zaitseva N P, Glenn A M, Mabe A N, Carman M L, Hurlbut C R, Inman J W and Payne S A 2018 Recent developments in plastic scintillators with pulse shape discrimination *Nucl. Instrum. Methods* A **889** 97–104

[20] Cieslak M J, Gamage K A A and Glover R 2019 Critical review of scintillating crystals for neutron detection *Crystals* **9** 480

[21] Chandra R, Davatz G, Friedrich H, Gendotti U and Murer D 2012 Fast neutron detection with pressurized ^{4}He scintillation detectors *J. Instrum.* **7** C03035

[22] Jebali R *et al* 2015 A first comparison of the responses of a ^{4}He-based fast-neutron detector and a NE-213 liquid-scintillator reference detector *Nucl. Instrum. Methods* A **794** 102–8

[23] Nasir R, Aziz F, Mirza S M and Mirza N M 2018 Experimental and theoretical study of BF_3 detector response for thermal neutrons in reflecting materials *Nucl. Eng. Technol.* **50** 439–45

[24] Lacy J L, Athanasiades A, Martin C S, Sun L and Vazquez-Flores G L 2013 The evolution of neutron straw detector applications in homeland security *IEEE Trans. Nucl. Sci.* **60** 1140–6

[25] Jiang Z G, Gong H, Li J M, Wang Y Q and Wang X W 2018 Experimental study of a boron coated straw detector module at CPHS *J. Instrum.* **13** P08004

[26] Klausz M *et al* 2019 Performance evaluation of the boron coated straws detector with Geant4 *Nucl. Instrum. Methods* A **943** 162463

[27] Bollinger L M, Thomas G E and Ginther R J 1962 Neutron detection with glass scintillators *Nucl. Instrum. Methods* **17** 97–116

[28] Xu M, Tang Z-C, Chen G-M and Tao J-Q 2013 Comparison of thermal neutron detection efficiency of ^{6}Li scintillation glass and ^{3}He gas proportional tube *Chinese Phys.* C **37** 106001

[29] Omar A *et al* 2019 Development of a handheld thermal neutron detector (GAMBE) using stacked silicon sensors coated with ^{6}LiF films *Radiat. Meas.* **129** 106180

[30] Almaviva S *et al* 2010 Improved performance in synthetic diamond neutron detectors: Application to boron neutron capture therapy *Nucl. Instrum. Methods* A **612** 580–2

[31] Ovechkina L, Riley K, Miller S, Bell Z and Nagarkar V 2009 Gadolinium loaded plastic scintillators for high efficiency neutron detection *Phys. Proc.* **2** 161–70

[32] Pawelczak I A, Glenn A M, Martinez H P, Carman M L, Zaitseva N P and Payne S A 2014 Boron-loaded plastic scintillator with neutron-gamma pulse shape discrimination capability *Nucl. Instrum. Methods* A **751** 62–9

[33] Mabe A N, Glenn A M, Carman M L, Zaitseva N P and Payne S A 2016 Transparent plastic scintillators for neutron detection based on lithium salicylate *Nucl. Instrum. Methods* A **806** 80–6

[34] Kouzes R T *et al* 2010 Neutron detection alternatives to ^{3}He for national security applications *Nucl. Instrum. Methods* A **623** 1035–45

[35] Al Hamrashdi H, Cheneler D and Monk S D 2020 A fast and portable imager for neutron and gamma emitting radionuclides *Nucl. Instrum. Methods* A **953** 163253

Part III

Electronics and data analysis

IOP Publishing

Radiation Detection for Nuclear Physics
Methods and industrial applications
David Jenkins

Chapter 10

Readout electronics and data analysis

This chapter provides an overview of the techniques employed in data acquisition and data analysis for applications involving radiation detectors. The initial focus is on data acquisition distinguishing between the traditional analogue acquisition systems and the more contemporary digital data acquisition systems. We will follow through the chain of analogue electronics to understand the conventional approach to produce an energy histogram from an initial detector pulse as well as how timing information for pulses is extracted. This will be complemented by a basic outline of digital data acquisition stressing the advantages of this approach over the traditional analogue systems. The chapter concludes with some remarks on data analysis, focussing particularly on the ROOT software package originally developed at CERN.

Concepts: electronics strategy; analogue electronics: energy chain and timing chain; preamplifiers; shaping amplifiers, analogue-to-digital converters; extraction of pulse timing characteristics: leading-edge timing and constant fraction discriminator; compactified readout: ASICs; digital data acquisition; pulse shape analysis; data analysis tools: ROOT, NPTool, Radware and GRAIN

10.1 Strategy for electronics readout of detectors

Assuming we have designed our detector setup and powered it on, what happens next? What we want nowadays is the data coming from the detector to be immediately available to us in digital form, displayed on a histogram on a screen, stored in a time-ordered array in our computer's memory, etc. The, often tiny, signal, e.g. the charge collected, in our detector needs to be transformed into an appropriate digital form while preserving as far as possible the fidelity of that signal. There are two features of the signal which are the most important: the amplitude of the signal which drives our goal of achieving spectroscopic performance, and the time-evolution of the signal. We have seen, many times in this e-book, areas where the time evolution of the signal can provide exquisite information on, for example,

doi:10.1088/978-0-7503-1428-2ch10

the position of interaction in a detector, e.g. see discussion of pulse shape information from germanium detectors in section 6.2.4, or for discriminating different types of radiation, e.g. the many examples given of neutron-gamma discrimination in chapter 9.

There are many examples of the challenge of ensuring signal fidelity in other areas of everyday life. Record players are a somewhat antique technology, although enjoying a trendy revival. They function through the tiny oscillations of a stylus on a grooved track on a record. This motion is transformed through electronics which can give a high fidelity rendition of the sound recorded on the record at a volume to fill an auditorium. Allied to the quality of the electronics in the system, which amplify while reducing noise as far as possible, the rotation of the turntable, the damping of the stylus arm etc, must all be exquisitely stabilised. Technology has moved on. We have already largely abandoned the compact disc which stores digital recordings of sound, retrieved by laser, to digital storage of our music on phones and computers downloaded from the cloud. Musical connoisseurs still argue passionately about whether the analogue or digital system provides the best fidelity of sound. What does this lengthy aside about the recording of music tell us about radiation detectors? The answer is that the developments in both fields are similar in the sense of transition from analogue to digital and, of course, this transition is by no means restricted to these two narrow fields—it has also influenced broad areas of science and engineering. The readout of radiation detectors in experimental nuclear physics and related applications has undergone a revolution in recent years. Historically speaking, *analogue* data acquisition systems were used but nowadays *digital* data acquisition systems are much more commonly found and analogue is rapidly becoming obsolete. Nevertheless, an analogue system is very good for learning, as all aspects are under manual control, i.e. one literally defines the settings of the electronics modules by hand by turning knobs and flicking switches. It can give the same satisfaction to the user as to the person who constructs their own music system from basic, or hand-built components.

In this chapter, we will first review aspects of analogue data acquisition systems before turning to the more contemporary digital systems.

10.2 Analogue electronics

There are some interesting parallels between the record player and the signal from a radiation detector. The stylus on the record player is a mechanical device—it might not be perfectly aligned, the stylus arm may wobble, the record may be scratched or dirty. In short, there are sources of noise, mechanical in nature, right at the beginning of the signal chain. We can be as careful as we like but we will never get rid of all these noise sources. The same is true of detectors. There are many sources of electronic noise, which we can try to minimise but will never eradicate and many are fundamental and have a basic physical origin. Even carrying the signal out of the detector will introduce further noise.

The overarching strategy behind an analogue electronics readout is to amplify and shape the signal from a detector prior to digitisation. There is no 'one size fits all' solution to this challenge and a whole book could be written on this topic alone. Indeed, the processing of the detector signal is not usually carried out in a single step and there are good reasons for that. Most detectors require a preamplifier stage before they enter a shaping amplifier although a few types of detector such as SiPMs produce large voltage pulses and, depending on the application, they may be input directly to a shaping amplifier. For many classes of detector, their output signal is small and preamplification is mandatory. The preamplifier has to be mounted as close as possible to the detector to limit signal degradation. Indeed, for some detectors such as HPGe detectors, the preamplifiers are even mounted within the detector casing itself. It cannot be stressed enough how important this issue is in terms of how close the preamplifier can be brought to the detector. Whatever noise enters at this stage when the signal is at its most fragile will be amplified and carried forward as the signal is processed. Minimising noise pickup can be a significant task for the physicist building up a sensitive system such as silicon DSSD detectors within a vacuum chamber. The laboratory is often a 'noisy' environment from an electronics point of view. Vacuum systems have mechanical pumps generating noise back on to their power supply. The power supply for detectors is therefore usually taken from a 'clean power' source, i.e. an electrical supply system isolated from the 'dirty power' system which powers the heavier infrastructure of the laboratory such as vacuum pumps. A further important point is the grounding of the detector system and its readout electronics. These should all be attached to a common ground to avoid ground loops. The signal cables used in the system should be of high quality. For the cables, often ribbon cables, originating directly from the detector to the preamplifier, this is especially important. You will often see such cables wrapped in aluminum foil in efforts to shield them from noise pick-up. It is hard to offer a detailed prescription around dealing with issues associated with pick-up of electronic noise, it is only something the physicist will lean through trial and error (and much frustration!). The reader who would like more guidance on this topic is directed to chapter 9 of the book *Semiconductor Detector Systems* by H Spieler (full reference details in appendix A) which has the rather splendid title 'Why things don't work'.

Analogue electronics components range from single modules which process the signals from a single detector, to modules which can process, e.g. 16 channels of signal readout in one unit. Modules can often comprise multiple components such as preamplifiers, shaping and timing filter amplifiers in a single unit. Historically, analogue systems would comprise a set of *NIM modules* (see figure 10.1); there are other parallel systems but we will focus on NIM for the purposes of simplicity. NIM is a long-established standard for nuclear electronics comprising rack-mounted modules of standard size which fit neatly into a powered bin. The modules have connectors on the back which slot into mating connectors in the bin to provide clean power of ± 12 and ± 24 V (and generally ± 6 V as well). Signal connections between NIM modules are carried on BNC cables while SHV cables are used for high voltage supplies.

Here, we simplify the discussion of analogue electronics by focussing on the readout of a single detector but the reader should bear in mind that this is readily scaleable, although when the number of detector channels enters the 100s, digital

Figure 10.1. A NIM bin containing a set of standard NIM modules. (Photo credit: the author.)

electronics is a preferred option. As discussed above, a preamplifier is a common, but not necessarily mandatory, first step in the signal processing chain. There are several types of preamplifier: *charge sensitive*, *current sensitive* and *voltage sensitive*. The former two are most commonly used in experimental nuclear physics and for most high-resolution spectroscopic applications, particularly those involving germanium detectors and silicon detectors, a charge-sensitive preamplifier is the appropriate choice. In other applications, where the input signal is relatively large and has a fast-rise time, such as that from a photomultiplier tube, the preamplification stage is less demanding and a current-sensitive preamplifier may be used. Here, in the interests of clarity and breadth of application, we will focus on charge-sensitive preamplifiers and mention current-sensitive preamplifiers briefly afterwards (in section 10.2.4).

10.2.1 Charge-sensitive preamplifier and signal chain

A charge-sensitive preamplifier effectively integrates the input charge pulse from the detector, whose duration is typically 1–10 μs, turning it into a voltage pulse. The basic components of such a preamplifier circuit are a low-noise amplifier with a feedback capacitor, C_f, and feedback resistor, R_f, in parallel (see figure 10.2). The function of the feedback resistor is to reset the signal to the baseline. The respective capacitor and resistor components are chosen so that the time constant (RC) of the feedback loop is longer than the duration of the input pulse which will ensure that the output voltage pulse is linearly proportional to the integral of the input charge pulse. Moreover, the resistor value, R is chosen to be as large as possible to minimise *Johnson–Nyquist noise*, i.e. electronic noise associated with thermal agitation of charge carriers. The output of a preamplifier is often given as mV MeV^{-1} for a given detector. For pulses with a very high dynamic range, a logarithmic preamplifier can be used in certain circumstances.

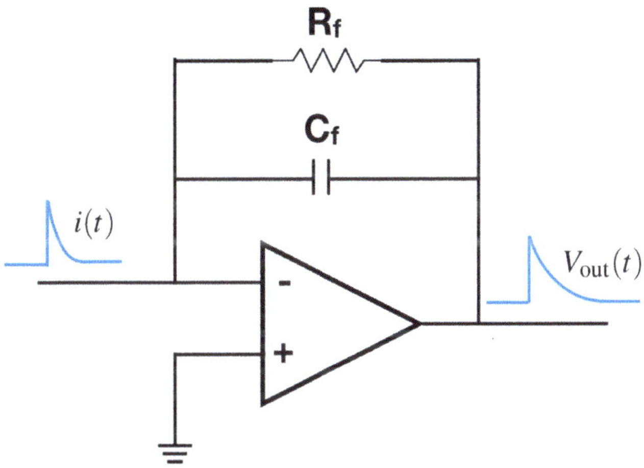

Figure 10.2. Basic circuit diagram for a charge-sensitive preamplifier.

The typical rise time of a preamplifier pulse is of order 100 ns, and the decay time is of order 10s of μs. The rising edge of the pulse encodes details of the charge collection process—a technique exploited, for example, in gamma-ray tracking, described in section 6.5.2. With a simple preamplifier operating at realistic counting rates, pulses will pile up on each other due to the slow fall time of the individual pulses. This is an acceptable situation because the next component in the signal chain—the pulse shaping amplifier, as described below—can resolve these individual pulses. Where this becomes problematic is when the counting rate is sufficiently high that the pile-up proceeds so far as to reach the maximum voltage of the preamplifier circuit—a status known as *lock up*. A preamplifier in lock up cannot record further pulses and so the information is lost. More sophisticated preamplifiers employ a transistor reset which drives the circuit back to the baseline. The preamplifier provides a logic signal while this reset is taking place so that it can be used to veto onward shaping of the pulses as the information in this reset time is of no use. This introduces some element of deadtime into the system (see section 10.2.8 for further elaboration on deadtime issues).

Most preamplifiers have a pulser input which provides a means of disentangling the noise contribution of the electronics to the achievable energy resolution from the noise contribution of the detector. This can be invaluable in optimising an electronics setup. Pulse generator modules such as the Ortec 419 can be used to generate a charge pulse to be injected into the preamplifier that mimics the typical pulse amplitude and rise time originating from the detector. Such a unit is highly stabilised and its output will be as close to 'perfect' as possible as opposed to the output of the radiation detector which will inevitably carry noise of different types in its signal.

The output of the preamplifier, depending on the application, can be split into either an 'energy' route and a 'timing' route or both; these are detailed in the flowchart presented in figure 10.3. First, we describe the electronic chain which leads from the preamplified pulse to an ADC spectrum.

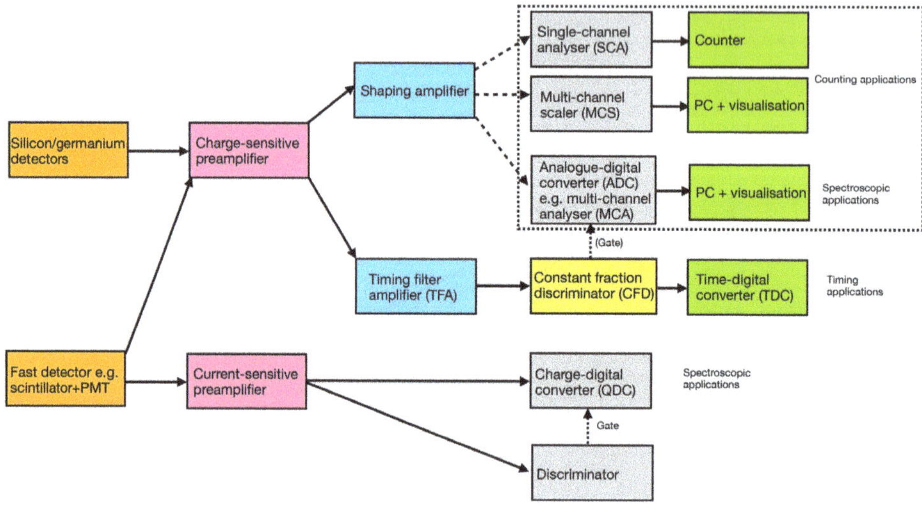

Figure 10.3. Flowchart of the different routes through an analogue signal processing system: (top) pathway for semiconductor detectors through a charge sensitive preamplifier and then split into an energy and timing branch; (bottom) alternative pathway for fast scintillators using a current-sensitive preamplifier and a QDC.

10.2.2 Pulse shaping amplifier

The output of a preamplifier is not suitable for a pulse height measurement since the preamplifier output has a fast rising edge and a slow exponential decay. Indeed, the ideal pulse shapes for making a pulse-height measurement would have a cusp shape or triangle shape. Moreover, the preamplifier output is small, typically of order 10s of mV and needs to be amplified into the 0.1–10 V range[1]. A shaping amplifier provides the necessary functionality, namely shaping the signal so that the output is a semi-Gaussian pulse shape as well as amplifying the signal by a variable gain from factors of 10s–1000s. Shaping of the pulse is achieved with two types of filters:

- a differentiator (or high-pass filter) which comprises a capacitor, C_d, with a resistor, R_d, in parallel;
- an integrator (or low-pass filter) which comprises a series resistor, R_i, with a capacitor, C_i, in parallel.

Figure 10.4 shows a basic schematic for a pulse shaping amplifier based on a differentiator on the input side and a differentiator on the output side. (Some amplifiers have one stage of differentiation followed by two or more stages of integration which leads to a *Semi-Gaussian* peak shape.) The components are chosen so that $C_dR_d = C_iR_i = \tau$, where τ is the shaping time of the circuit. The shaping time is the same for the differentiator and integrator elements since this minimises noise. The shaping time needs to be set appropriate to the detector being used and the characteristic of the corresponding pulse from the preamplifier. For example, for a

[1] The dynamic range should be tailored appropriately to the ADC employed for digitisation since the voltage range they accept varies.

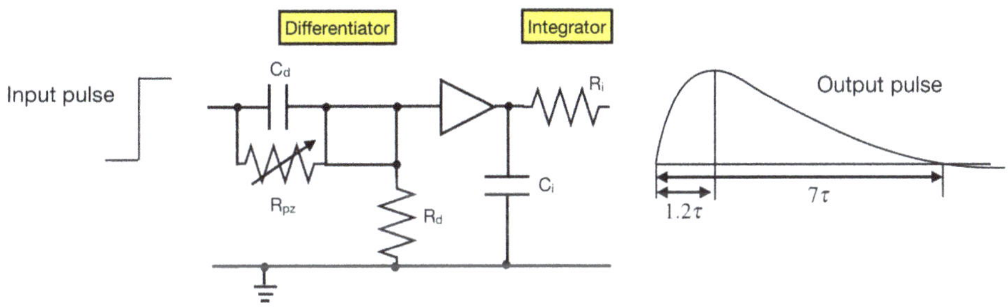

Figure 10.4. Basic circuit diagram for a pulse shaping preamplifier incorporating a differentiator on the input side and an integrator on the output side (a so-called *CR–RC circuit*). The various resistor and capacitor components are labelled. The shaping time of the system, $\tau = R_iC_i = R_dC_d$. The output of the circuit from an input step function is shown, leading to a semi-Gaussian pulse shape which peaks after 1.2 τ and begins to undershoot at 7 τ. The figure shows a pole-zero variable resistor in parallel with the differentiator capacitor which could be adjusted to remove undershoot.

scintillator detector, the shaping time, τ, should be around three times the decay constant of the scintillator. This implies that for very fast detectors like plastic scintillator (see section 9.1.2), the shortest shaping time, e.g. 0.5 μs, would be the best. For slower inorganic scintillators such as CsI(Tl) (see section 5.1.2), rather longer shaping times, e.g. 6 μs would be more appropriate. Similarly, for high-purity germanium detectors, typical shaping times are 4–10 μs. It is good practice to collect calibration spectra with different shaping times and explore the setting which leads to optimisation of the data collected, e.g. the energy resolution being optimised. Since the shaping time is finite, charge collection may be incomplete so that there is a discrepancy between the pulse height with a practical shaping time and an ideally very long shaping time; this is known as *ballistic deficit* [1]. This would not be a problem if the ballistic deficit was fixed and proportional to pulse height but with realistic detectors, there is commonly variation in pulse rise time.

There are other features of shaping amplifiers which need correction. For example, as can be seen in figure 10.4, the tail of the pulse may undershoot. This means that the next pulse arriving may ride on this undershoot and so appear to have a reduced pulse height. The effect of this in a spectroscopy application would be to broaden the peaks in the spectrum. Many traditional amplifiers have a variable resistor in parallel with the differentiator, as shown in figure 10.4. This is often controlled through an inset screw on the front panel of the amplifier which can be adjusted with a suitable jeweller's screwdriver while monitoring the output pulse from the amplifier on an oscilloscope. This is known as a *pole zero* adjustment. A further aspect requiring correction is *baseline shift*. This originates from the fact that the differentiator part of the amplifier circuit has a capacitor in series meaning that the circuit is AC coupled. Shifting of the baseline will occur and this will become more severe at higher rates. This will affect the analogue-to-digital conversion in the next stage of the signal chain since the pulse height is measured relative to a fixed reference. Accordingly, shaping amplifiers contain a baseline restorer which will drive the baseline back to zero between pulses.

An interesting question is: what happens if a second pulse from the preamplifier arrives as the amplifier is shaping the first one—a phenomenon known as pile-up and discussed in general terms in section 4.1.3? Clearly, the output pulse would be distorted and the pulse height information would be misleading as to the details of the first pulse. Accordingly, some amplifiers have in-built pile-up rejection. This is achieved by having a parallel fast amplifier within the unit which has a very short shaping time and a fast discriminator. This generates a logic pulse which can then be used to inhibit the collection of the pile-up event at the next stage, i.e. the digitisation via the analogue-to-digital converter (ADC).

10.2.3 Analogue-to-digital conversion

If the application is solely about counting then the output of the amplifier can go to a single-channel analyser (SCA). More commonly, in spectroscopic applications, the output of the amplifier would be processed in an analogue-to-digital converter (ADC).

Single channel analyser (SCA)

An SCA is a simple device to support the counting of events falling within a certain pulse height range. It comprises a lower-level discriminator (LLD) and an upper-level discriminator (ULD). If the pulse amplitude passes these two discriminator thresholds then a logic pulse is generated. This logic pulse could be taken to a further unit known as a counter/timer which counts pulses on a display. Settings allow the user to count for specific lengths of time.

Multi-channel analyser (MCA)

Most data analysis focuses on generating an *energy spectrum*, i.e. a histogram of the number of events as a function of the amplitude of the pulse height. In order to achieve this, the output of the amplifier can be fed into a multi-channel analyser (MCA) that can perform the necessary analogue-to-digital conversion. An MCA is logically equivalent to an array of SCAs and counters. The key parameter in this digitisation process is the number of bits (N) employed which will lead to a spectrum containing 2^N ADC channels. If the number of bits is too low, then the *binning* of the resulting spectrum will be too coarse. To take an extreme example, if a spectrum from a germanium detector has a dynamic range of 4 MeV, then if it were digitised with only 10 bits (1024 ADC channels), then each channel would correspond to a bin of approximately 4 keV width. Since the energy resolution of a typical germanium detector is 2 keV or better at 1.33 MeV then nearly all the events corresponding to a photopeak with an energy of 1 MeV would fall in a single bin. The value associated with the energy resolution of the detector has therefore been lost. There is danger in going to the opposite extreme of having too many bits as then the data is divided into twice as many bins each time the number of bits is increased by one leading to a loss of statistical significance in the observed peaks. As a rule-of-thumb, the number of bits should be chosen so that the FWHM corresponds to around four ADC channels.

An MCA needs to be chosen that has high linearity and a suitable dynamic range for the relevant application. The Ortec EASY-MCA is an example of a simple MCA which interacts directly to a PC using a USB cable and has accompanying software which histograms the data obtained on screen. It has two variants, one with 11 bit (2048 ADC channels) digitisation appropriate for, e.g. inorganic scintillator detectors, and one with 13 bit (8192 ADC channels) appropriate for germanium detectors. There are variants of MCAs available called *multi-channel scalers (MCS)* which come with software that allow the user to mark a region in a spectrum and count the number of events falling in that region per unit time which is a more flexible way of achieving the functionality of an SCA.

While an ADC is processing a pulse, it cannot accept a second one and this introduces dead time into the system. Dead time is discussed in more detail in section 10.2.8. Most MCAs have an in-built function to report system live time.

Modular ADCs

MCAs are appropriate to the readout of a single detector channel. To scale up to systems with many channels, the solution is to employ modular ADCs which have, e.g. 32 input channels. An example is the CAEN V785 ADC Such units are rack-mounted in a VME crate. Here, we begin to move away from the NIM platform to a platform more suitable for multichannel data acquisition systems.

10.2.4 Charge-to-digital-converter (QDC)

An alternative to the charge-sensitive preamplifier signal chain is available for fast detectors such as plastic scintillators coupled to PMTs. This employs a current-sensitive preamplifier which is a linear fast amplifier; the shape of the signal is unchanged through the amplification process. This is advantageous where timing is less important and the system is counting fast so pile-up is a concern. The output of a current-sensitive preamplifier is fed into a QDC (charge-to-digital converter) rather than a standard ADC. A QDC simply integrates the signal through an integration window defined by a gate. The signal for the gate can be generated from, e.g. the RF signal of the accelerator or it can be generated by splitting the detector signal and using some kind of threshold discriminator, as described below, to generate a gate.

10.2.5 Timing chain

As discussed in section 4.1.2, there are distinct advantages in achieving as good a timing resolution from a detector as possible in view of studying coincident events and in reducing experimental background. Determining the timing of an event corresponds to determining when a pulse rises above some pre-determined threshold. This is a fundamentally uncertain process and there are two main effects which can influence fixing the timing of an event—*timing walk* and *timing jitter*. Timing walk reflects the fact that pulses of different magnitude would cross the discriminator threshold at different times: small pulses later and large pulses earlier. Timing jitter is associated with the electronic noise inevitably superposed on the detector pulse. If the instantaneous noise should be adding to the pulse then the pulse would exceed the discriminator threshold earlier than it ought to have done. Timing jitter can, of course, be controlled to some extent by eliminating sources of electronic noise as far as possible. We will describe two basic approaches to timing measurements—*leading edge timing* and the *constant fraction discriminator*. Prior to such measurements, a split-off of the detector signal must first be amplified, conventionally using a *timing filter amplifier* (TFA).

Timing filter amplifier (TFA)
A TFA has different characteristics to shaping amplifiers described above. TFAs are designed to have fast output rise time in the region of 1 ns; the rise time needs to be set to a value shorter than the rise time of the preamplifier to preserve the characteristics of the signal rise time. This fast response can only be achieved by sacrificing some of the linearity and stability of the amplification stage.

Leading edge timing
The simplest approach to timing would be to determine when a pulse exceeds a fixed threshold—an approach known as *leading edge timing*—implemented with a simple voltage comparator. This approach may be acceptable for some applications where high precision timing resolution is not necessary since leading edge timing suffers severely from timing walk. The timing achieved would therefore be dependent on the pulse height which is undesirable unless all pulses are of very similar magnitude which is not a situation commonly encountered in an experiment.

Constant fraction discriminator
A *constant fraction discriminator* (CFD) [2] can provide improved timing resolution compared to leading edge timing through largely eliminating timing walk for pulses which have the same characteristic shape. This is achieved by splitting the input pulse into two copies and attenuating one copy by a fraction (typically 20%–40%) of the original pulse amplitude, while delaying and inverting the other copy. The two split signals are then recombined into a single bipolar pulse which has the property of crossing the zero-axis at a time which is independent of the initial pulse amplitude. The output of a CFD is a logic pulse. This could be fed into a *gate and delay generator* to generate an appropriate gate for the ADC used in the 'energy branch' described above. An example of a CFD is the Ortec 584.

Measuring time differences—time-to-amplitude converter (TAC)
A *time-to-amplitude converter* (TAC) is a device which produces an output pulse whose amplitude is proportional to the time difference between two input pulses—a 'start' and 'stop' pulse. Such a signal can then be fed into a multichannel analyser to produce a 'time spectrum' akin to the energy spectra described above. Such a device is useful for establishing that two events are in prompt coincidence or for measuring timing resolution, or for measuring time differences, for example, between two successive decay events. An example of a TAC is the Ortec 566 TAC which can cover time differences ranging from 10 ns to 2 ms.

Time to digital converter (TDC)
A *time-to-digital converter* (TDC) integrates the functionality of a TAC and an ADC providing a direct digital output.

10.2.6 Concept of a trigger

A common component of an analogue acquisition system is a *trigger*. In its simplest implementation for one detector with one acquisition channel, this could correspond to the timing discriminator that generates the gate for the readout of an ADC. In a slightly more complicated system, there might be a system comprising an alpha source, a silicon detector and a germanium detector. The timing signal from the signal chain of the silicon detector could generate a gate for the ADC readout of the germanium detector, in order to acquire only gamma-ray events in coincidence with detection of an alpha particle, i.e. corresponding to alpha decay to excited states in the daughter nucleus. Similarly, two germanium detectors could be positioned around a multi-line gamma-ray source and gate and logic units used to trigger readout of the their respective ADCs when two events are detected in coincidence. These are relatively simple triggers. In a more complex system with 10s or even 100s of individual detector channels, more elaborate triggers can be setup using logic units where events are only readout when, e.g. two or more individual detectors fire in coincidence—what is known as a *multiplicity trigger*. A trigger, therefore, defines which classes of events are recorded. This may be important because the system may not be able to process events quickly enough unless some kind of multiplicity trigger is applied. Inevitably, this imposes a bias in the data collected, which can impose complications in the data analysis and may lead to the conclusion, after the experiment, that important aspects of the process under study were not adequately captured. Unfortunately, there is no way to go backwards and disconnect this trigger from the data obtained and so careful design of a trigger system is important when an experiment is planned. We will revisit the issue of triggers when we discuss digital data acquisition systems in section 10.3.2.

10.2.7 Compactified read-out systems: ASICs

In nuclear physics experiments, it is becoming common to have significantly complex detector setups with hundreds or thousands of individual channels. The demands of high-energy collider experiments in particle physics can require orders of magnitude more channels. Clearly, rack-mounted analogue data acquisitions become unfeasible in these kind of regimes and a highly compactified readout system is preferred on grounds of cost, power consumption and close integration to the detectors. This is a regime where *application-specific integrated circuits* (ASICs) are preferred. An ASIC can combine, in a highly-integrated CMOS platform, all the elements required in signal acquisition such as charged sensitive preamplifier, shaping amplifier, ADC etc. Many tens of individual channels of electronics can be incorporated on a single chip of dimensions typically less than 10×10 mm, e.g. [3, 4]. This has been applied, for example, to the readout of the type of pixellated CZT detectors discussed in section 6.6. The example discussed in reference [5],

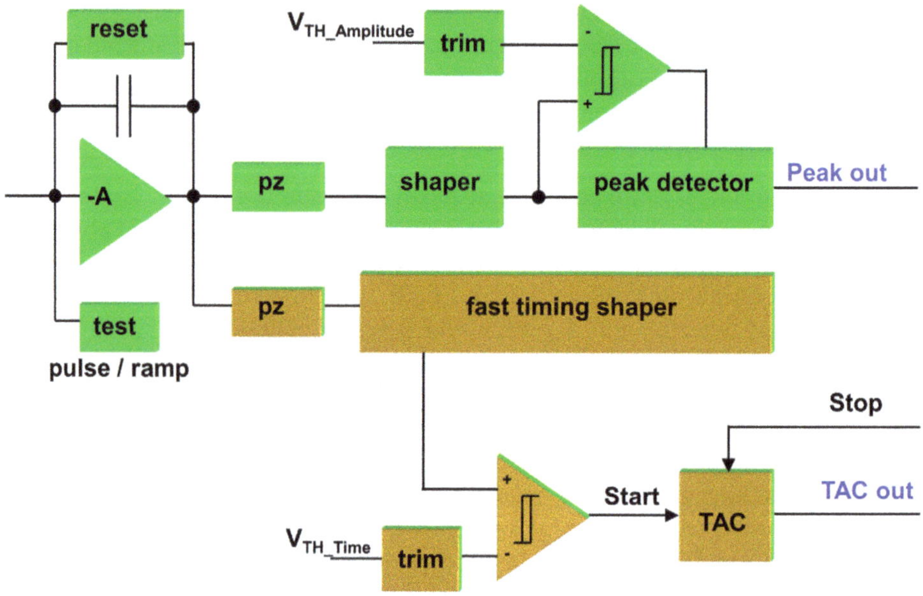

Figure 10.5. Channel block diagram for one of the four cathode channels on an ASIC developed to readout virtual Frisch-grid cadmium zinc telluride (VFG CZT) detectors for gamma ray spectroscopy. The signal path splits at the output of the preamplifier. The event amplitude (Peak out) is processed by the green signal path while the corresponding timing (TAC out) is processed by a fast third order shaper and a time-to-amplitude converter (TAC). (Reprinted from [5] with the permission of Elsevier.)

comprises an ASIC with 48 channels for anode readout and four channels for cathode readout. The block diagram for the cathode readout is shown in figure 10.5 which gives a sense of the complexity achievable in these circuits which displace whole racks of analogue electronics of the type discussed above.

Unfortunately, design and production of an ASIC is an industrial-scale operation with a development cost which could be several million dollars. Their take-up in nuclear physics is accordingly rather low as most detector systems are one-offs. By contrast, adoption in industrial/societal applications is growing all the time (see example in section 8.3.4). Indeed, a major application is in medical imaging such as PET; an example being the TOFPET2 ASIC which provides 64 channel fast readout of an 8 × 8 SiPM array which could be coupled to LYSO scintillator sticks. This system can achieve high rates and high timing resolution suitable for time-of-flight PET.

10.2.8 Dead time and pile-up

Losses in counting in data acquisition with a radiation detector can arise from many different aspects of the associated signal chain [6]:

- the finite time for an ADC to process a pulse during which it cannot accept another pulse;

- saturation of data transfer and storage for large, multi-channel data acquisition systems;
- time taken during reset of a preamplifier;
- summing of pulses due to them occurring at essentially the same time;
- pile-up rejection of pulses which occur sequentially with short time separation;
- very large pulses which paralyse a detector channel, e.g. randomly-occurring cosmic ray interactions with very high energy deposition.

It is almost inevitable that some events will be lost or misrecorded and the extent of this will depend heavily on the detector system concerned and the rates in the detector system. Traditionally, deadtime has been understood under two scenarios: type I (non-paralyzable model) and type II (paralyzable model); excellent reviews of this topic may be found in reference [6] and reference [7]. Under the paralyzable (or *extending deadtime (E DT)*) model, the system is unable to record a second output pulse until a resolving time of τ has elapsed. If a second event occurs in this time window, then the deadtime of the system extends by a further time, τ. The system is in a state of *paralysis* until at least a time, τ, elapses without a successive event. The observed counting rate, m, can be related to the true counting rate of the system, n, by:

$$m = ne^{-n\tau}. \tag{10.1}$$

Under the non-paralyzable (or *non-extending deadtime (NE DT)*) model, the system is unaffected by further events occurring in the deadtime window, τ. Such events are *lost* but the system will recover after the elapsed time, τ and resume counting. It is not *paralysed* for an indefinite time. In this case, the fraction of recorded events is given by:

$$\frac{m}{n} = 1 - m\tau. \tag{10.2}$$

The two deadtime models are plotted in figure 10.6. These should be considered two extreme scenarios—a realistic system will fall between these two extremes [7].

The picture of deadtime as presented above does not take account of the influence of pile-up rejection which is important in modern detector systems [6]. Figure 10.7 illustrates the combined effects of pile-up rejection and deadtime. Although the effective dead-time length is the same in all cases, there is a clear difference in which pulses are accepted and which are lost.

Deadtime is a complicated topic but a central one for analogue data acquisition. While a system can perform near perfectly and count linearly for very low counting rates, it will rapidly start to deviate as counting rates increase. How much of a problem this is depends on the application. If the application is to produce an energy spectrum and identify, e.g. gamma-ray peaks then nothing about the peak positions or their relative intensities might change but the absolute counting rate of the system and resulting spectrum would not reflect the true counting rate or efficiency of the detector. If this is important then considerable efforts have to be made to understand

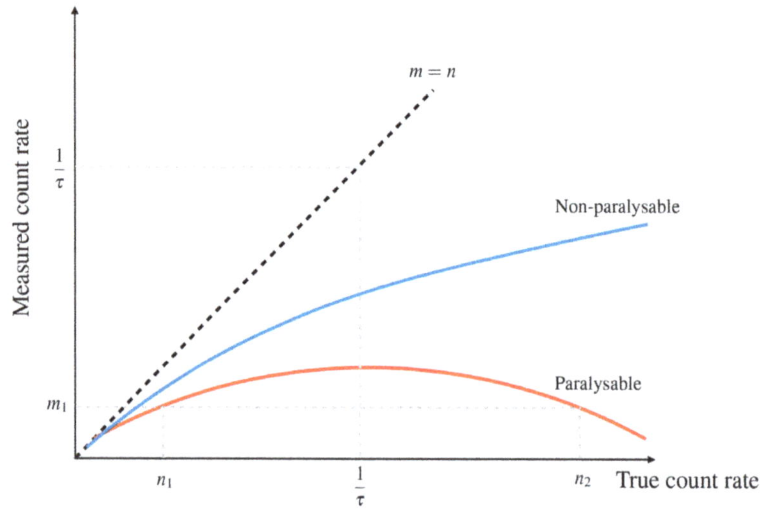

Figure 10.6. Illustration of the concept of dead-time plotting measured count rate for a system against the true count rate. An ideal system would correspond to the dashed line. The two extreme deadtime models of paralyzable (equation (10.1)) and non-paralyzable (equation (10.2)) are shown. (Redrawn after [7] with the permission of Korean Nuclear Society under Creative Commons License.)

the deadtime of the acquisition system and its conformity to the models described above. There are various techniques which address deadtime measurement such as introducing a pulser into the acquisition system with fixed frequency and noting the fraction of events recorded. Other techniques rely on real measurements. For example, two sources can be used individually and in combination. When the sources are combined, the counting rate will be reduced with respect to the true combined counting rate and in this manner, the deadtime can be extracted [7].

10.3 Digital data acquisition

Increasingly, analogue data acquisition systems, as described above, are being superseded by digital data acquisition systems at all levels, from the readout of single detectors to the readout of large complex detectors systems in the laboratory. Digitisers sample a preamplifier signal and convert it into digital form. There are generally two ways in which digital data can be handled. Firstly, raw digital traces can be captured which can be inspected and analysed off-line in software. This can be an extremely powerful method for understanding the performance of an individual detector but can rapidly become challenging in terms of data rate for large numbers of channels. The second mode is to analyse the pulse shapes as they come in using an appropriate algorithm which can extract parameters such as the time of the event, the amplitude and also details of the rise time/pulse shape which can be used to infer the interaction position for certain classes of detector, e.g. gamma-ray tracking detectors (see section 6.5.2). This is achieved using in-built *field-programmable gate arrays* (FPGAs); FPGAs are complex digital circuits which can be programmed to perform different signal processing tasks using a hardware

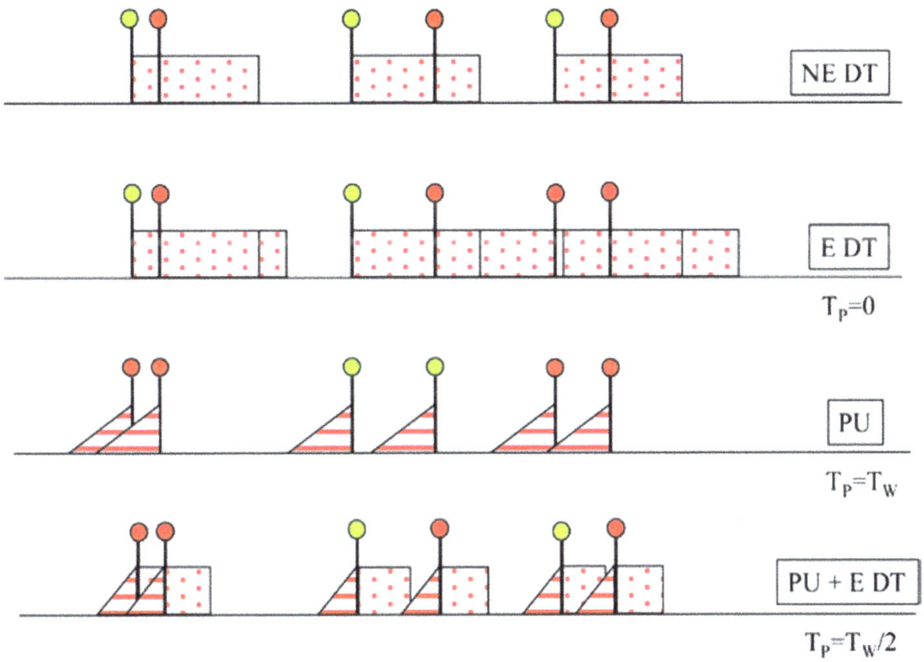

Figure 10.7. Event train passing through counters having different loss mechanisms; NEDT = non-extending dead time, EDT = extending dead time and PU = pile-up rejection. The case of combined PU and EDT corresponds to a spectrometer with Gaussian pulse shaping and fast ADC with negligible dead time. The dark (red) circles correspond to events which are not counted. T_W represents the pulse width and T_P the leading edge. (Reprinted from [6] with the permission of IOP Publishing under Creative Commons License.)

description language such as VHDL [8–10]. This offers tremendous flexibility since, not only is it possible to reproduce the functions of conventional analogue electronics-based signal processing as described above, e.g. CR–RC shaping [11], but it is also possible to explore and implement alternative processing of the raw pulse shape data which may be difficult to achieve with an analogue system.

Digital systems have a number of key advantages over conventional analogue systems:

- high density is possible—analogue systems become very cumbersome to scale up beyond of the order of 100 channels;
- a single digital channel can displace a whole rack of NIM analogue electronics;
- a digital data acquisition system is free running and events are timestamped. This means that coincident events can be reconstructed afterwards in software. This avoids the complexity of 'triggers' in analogue systems and coincident-logic electronics which necessarily bias the data obtained;
- once the data is in digital form, successive processing steps in FPGAs will not introduce electronic noise in the way conventional amplifiers do;
- pulse-specific corrections can be applied to correct for ballistic deficit;

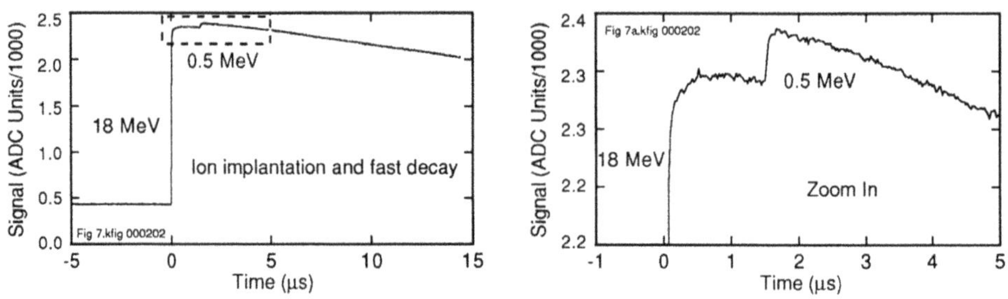

Figure 10.8. Pulse shape of a fast proton decay following ion implantation. The data were captured using a DGF-4C digitiser. The amplitude of the proton signal is very small compared to the ion implantation and follows only around 1 μs after the implantation. A standard analogue acquisition system would not resolve such events. (Reprinted from [12] with the permission of Elsevier.)

- pulse-shape analysis can be used to discriminate particle ID in suitable detectors without the need for a complicated analogue system;
- pulse-shape analysis can resolve real, successive, pulses with a small time separation, e.g. two successive alpha decays in the same pixel of a DSSD or proton emission following ion implantation (see figure 10.8).

A key parameter for any such digital system is the sampling frequency, i.e. over what time step should a preamplifier pulse be sampled to ensure signal fidelity? The Nyquist–Shannon sampling theorem states that 'If a function $x(t)$ contains no frequencies higher than B hertz, it is completely determined by giving its ordinates at a series of points spaced $\frac{1}{2B}$ s apart'. In other words, the sampling frequency should be at least double that of the highest frequency component in the input signal. If this condition is not met, then *aliasing* will occur. Figure 10.9 provides an example of the aliasing effect for two sine waves with different frequencies sampled with a period of $T = 1.0$. The sampling employed would be unable to distinguish these two sine waves. This is a rather extreme example as the preamplifier output of detectors is not in the form of a sine wave but clearly, in order to have access to the information encoded in the rising edge of the pulse then the sampling must be sufficiently fast to be able to make several samples during the time the pulse is rising. As we have seen elsewhere in this e-book, pulse rise times vary considerably from fast plastic scintillators to much slower pulses originating from germanium detectors. The sampling frequency of the digitiser therefore has to be chosen carefully to the specific detector application; typical digitzers range between around 100 MS s^{-1} (mega samples per second) to 4 GS s^{-1} (gigasamples per second). It is best to avoid overspecifying the sampling frequency as high sampling rate digitisers are usually significantly more expensive than low sampling rate systems and typically come in modules with fewer channels.

There are a wide range of digitiser systems commercially available which are suitable for working with one or a few detectors as well as being scaleable to more complex detector systems. Example of commercial products at the time of writing are the CAEN digitisers and Pixie-16 from XIA. In addition to the commercial systems, large laboratories have frequently developed their own platforms in

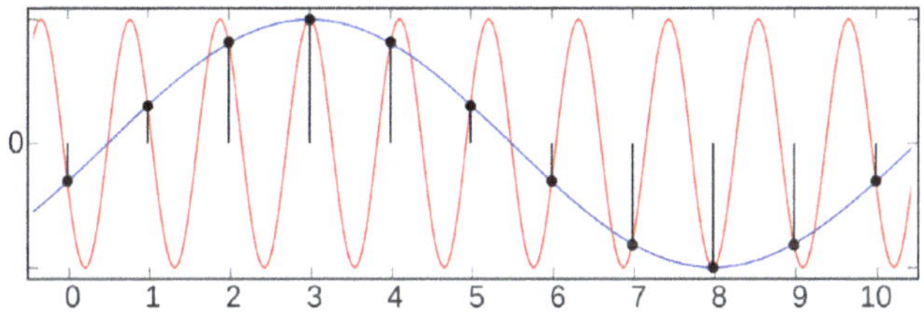

Figure 10.9. A graph showing aliasing of an $f = 0.9$ sine wave by an $f = 0.1$ sine wave by sampling at a period of $T = 1.0$. This image has been obtained by the author from the Wikimedia website where it was made available by Moxfyre under a CC BY-SA 3.0 licence. It is included within this article on that basis. It is attributed to Moxfyre.

response to the demands of their own specific apparatus; examples include the GRIFFIN data acquisition system [13] at TRIUMF in Canada, and the FEBEX system developed by GSI, Darmstadt.

10.3.1 Pulse processing

Let us discuss some aspects of pulse shaping in digital acquisition systems. We noted in the discussion of shaping amplifiers (section 10.2.2) that ideal pulse shapes for pulse-height extraction would be a cusp, or a triangle shape. These are incredibly difficult to realise or even approximate with analogue electronics. In the 1990s, the first attempts were made to develop digital filters that could achieve triangular or trapezoidal pulse shapes from preamplifier inputs [14]. Figure 10.10 shows the results of conversion of exponential pulses with different rises into trapezoidal pulses shapes (the details of the method are in reference [14]). Despite the differing rise times, all transformed pulses achieve the flat top of the trapezoid (albeit at slightly different times). This is a pathway to measurement of pulse height without the effect of ballistic deficit which is a strong challenge for analogue acquisition systems. These techniques have been further elaborated and implemented over the last twenty years, e.g. [15], finding particular application to readout of silicon and germanium detectors. The filter techniques have also been elaborated to achieve semi-Gaussian and cusp-shaped pulses, e.g. [16]. Neural networks [17] and machine learning have also been applied to pulse shape analysis.

A more recent implementation of these approaches to pulse shape analysis within FPGAs is found in the GRIFFIN data acquisition system [13]. Figure 10.11 shows how the initial digitised preamplifier signal can be processed via different digital filters to obtain a sharp pulse for timing measurements and a trapezoidal pulse for pulse-height evaluation.

10.3.2 Timestamping, rates and triggerless acquisition

As data enters a digital data acquisition system, it is timestamped using a clock time which typically extends over many weeks of realtime. This provides a pathway to reconstructing correlated events over very long time periods, which is difficult to achieve with analogue acquisition systems. Allied to this is the development of triggerless data readout pioneered at University of Jyväskylä [18] and developed

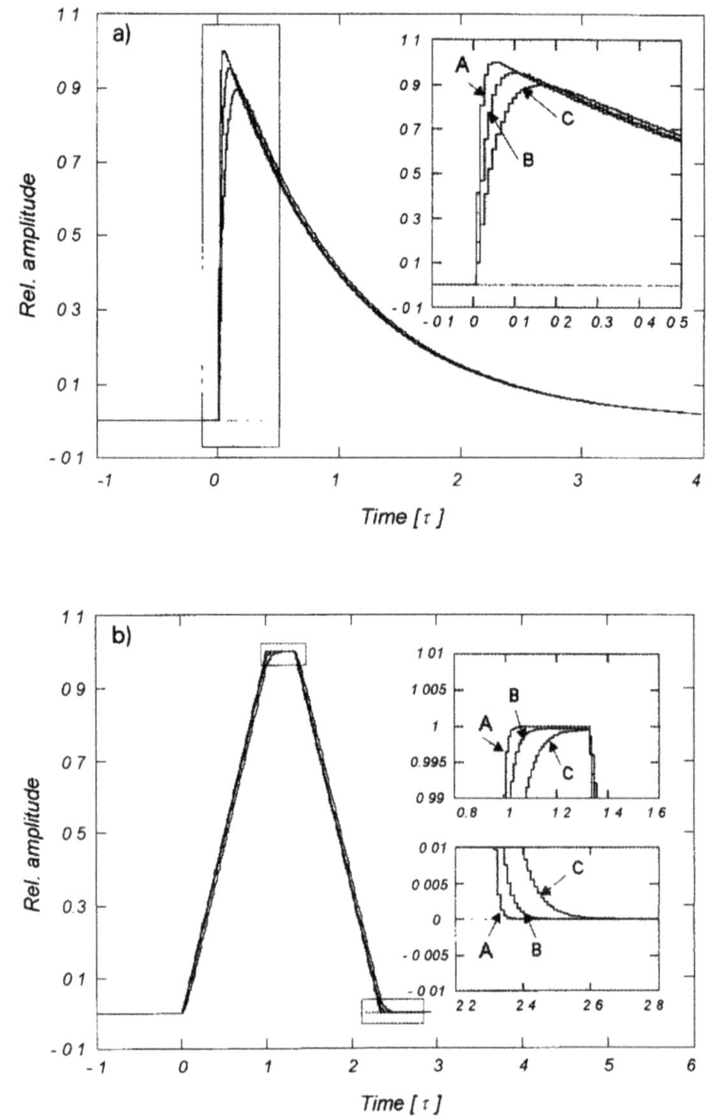

Figure 10.10. Exponential pulses with different rise time (a) and the corresponding trapezoidal pulse shapes (b). (Reprinted from [14] with the permission of Elsevier.)

further since, e.g. [19]. As discussed in section 10.2.6, analogue data acquisition systems conventionally use some kind of trigger which naturally biases the data obtained. A free-running, triggerless data acquisition system avoids this bias and coincidences between events in different detectors can be reconstructed later in software. A further advantage of a digital data acquisition system is its ability to support much higher counting rates in individual detectors through mitigation of dead-time effects in ADC readout. For example, the Gammasphere germanium array discussed in section 6.4.2 has been reinstrumented with a digital data acquisition system which has allowed individual germanium detectors to count at 40 000 cps with the same level of pile-up as they did with the previous analogue system running at 10 000 cps [20].

10.4 Data analysis

The earliest nuclear physics experiments such as the famous gold-foil scattering experiment of Rutherford actually relied on direct human observations of scintillations on a zinc sulfide scintillating screen. Later experiments made use, for example, of photographic plates which again had to be interpreted manually when the experiment was completed. As the sophistication and complexity of experimental setups grew and with advances in computer technology, data has been increasingly digitised and its analysis carried out by computer. Nowadays, a nuclear physics experiment can easily result in many terabytes of data. This is particularly the case where digital data acquisition systems are used, which may be *free-running* and pulse shapes are captured (see example of the nToF experiment in section 3.5.1). The demands of nuclear physics experiments are not on the scale of many high energy physics experiments and astronomy surveys where data can be in the petabytes and require significant dedicated infrastructure such as the GRID to permit its storage and manipulation. Nevertheless, handling terabytes of data is not a trivial undertaking.

Historically, there were as many *sort codes* as there were experimentalists, with each physicist writing their own code to sort, i.e. order and correlate the raw data and to visualise the data. The sorting was carried out in single passes through the data which might have been stored on magnetic tape and so took many hours or days to replay. Histograms would be defined in advance and filled and plotted as the data was retrieved. Examples might be to plot a gamma-ray spectrum for events recorded in germanium detectors in prompt time coincidence with an ancillary detector such as a charged particle detector. This entailed defining a number of *gates* on different parameters in the data, such as time differences, energy ranges, detector IDs, etc. This was a rather rigid approach where you got what you defined in advance and errors in the coding might not be spotted until sufficient statistics had accumulated in the histograms, forcing an edit of the code and a restarting of the sorting process. In recent years, the direction of travel has been away from this individualistic approach and towards more standardised methods of data manipulation.

10.4.1 ROOT

Nowadays, the most commonly found platform for data analysis in experimental nuclear physics is ROOT [21]. ROOT is an extremely powerful C++ based framework for data analysis which was originally developed at CERN for handling data from high energy physics experiments. In many ways, it works even better for nuclear physics where the datasets are of more manageable size. The philosophy is rather different to the old-style of working described above. The user will need to develop some basic subroutines to read the raw data (often these will have been developed in a standardised way in parallel with development of the detector apparatus) which identify events in the data and establish their associated energy

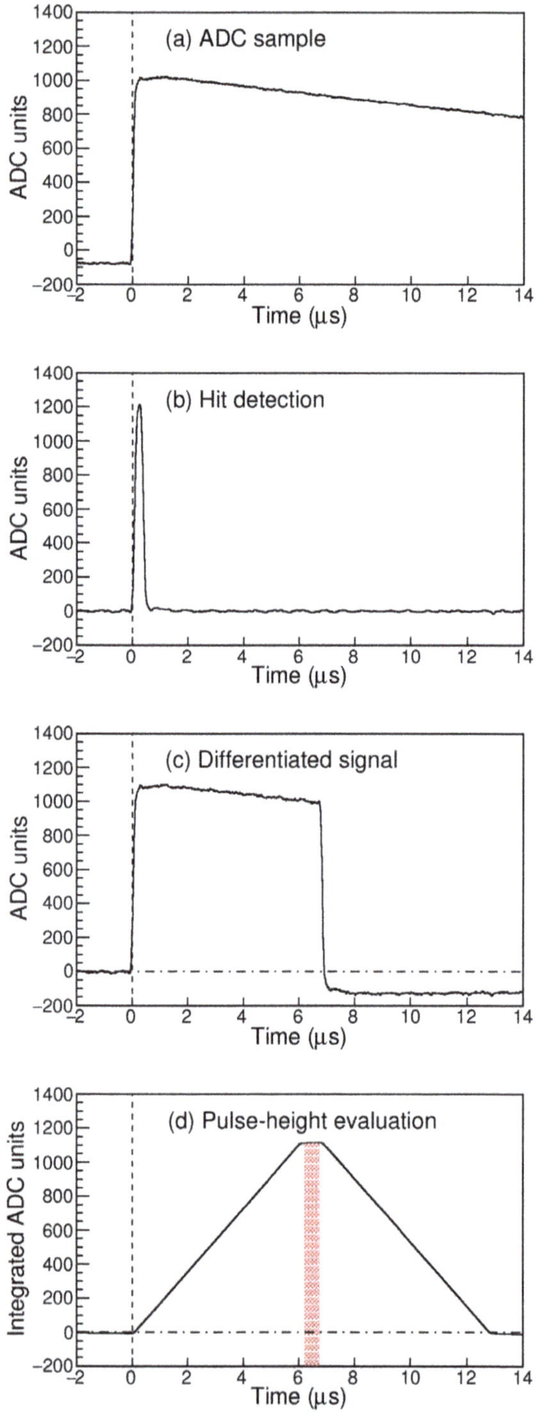

Figure 10.11. Details of GRIFFIN digital analysis: (a) digitised preamplifier signal from the detector; (b) application of the hit-detection algorithm to the signal in (a) using a set of three finite impulse response (FIR) digital filters; (c) differentiated signal; (d) trapezoidal signal for pulse-height evaluation. (Reprinted from [13] with the permission of Elsevier.)

(following appropriate calibration), time order, detector ID etc, which can be related to physical properties such as the detector angle, solid angle coverage etc. Depending on the format of the data, this may include stages of pulse-shape analysis on digital data to extract the relevant parameters; such a processing step greatly reduces the size of the dataset as raw pulse shapes occupy significant storage. The resulting data can be reordered into a TTree structure within ROOT which is held in memory (see figure 10.12). The TTree structure uses the concept of *branches* and *leaves* to separate the different parameters. Rather than using arrays of fixed size, e.g. an array of size 20 to hold energies of 20 different germanium detectors, the arrays (known as vectors) can float in size on an event-by-event basis since the number of germanium detectors which fire in a given event may be only two on average. This leads to a significant memory saving.

Histograms can be produced from the ROOT Tree pulling data directly from memory so the situation is significantly more flexible than the earlier principles of data analysis. Histograms could for example, plot one parameter, e.g. energy of particles/gamma rays measured in one detector conditioned with as many gates as desired on other parameters such as time difference, angular range of detectors, the energy range of particles detected in a different detector. Such histograms can be rendered near instantaneously depending on the size of the Tree and the power of the computer processor. Many different permutations of analysis can be explored in a

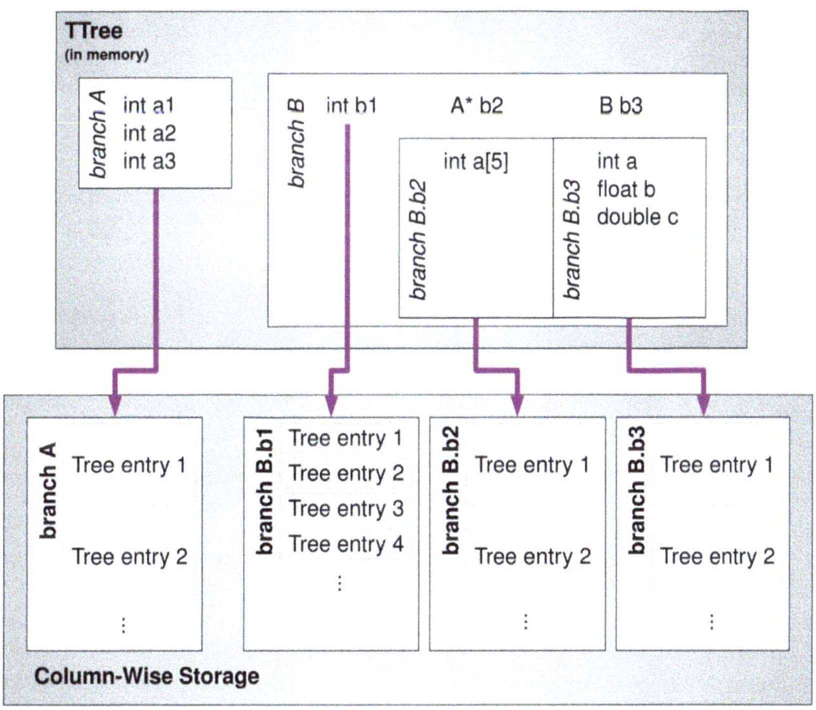

Figure 10.12. The TTree structure—part of the functionality of the ROOT data analysis software. (Reprinted from [21] with the permission of Elsevier.)

short time. This is extremely powerful and it encourages serendipity in discovery of new features in the data as the rigidity of the old data analysis model has been superseded. Moreover, the histograms produced by ROOT are in near-publication-ready format and have become a standard for data presentation.

10.4.2 Further selected examples of data analysis software

NPTool

NPTool [22] is a successful attempt to combine the functionality of ROOT and GEANT4 to assist in the design and analysis of nuclear physics experiments. Figure 10.13 shows an example of using NPTool to simulate the performance of the

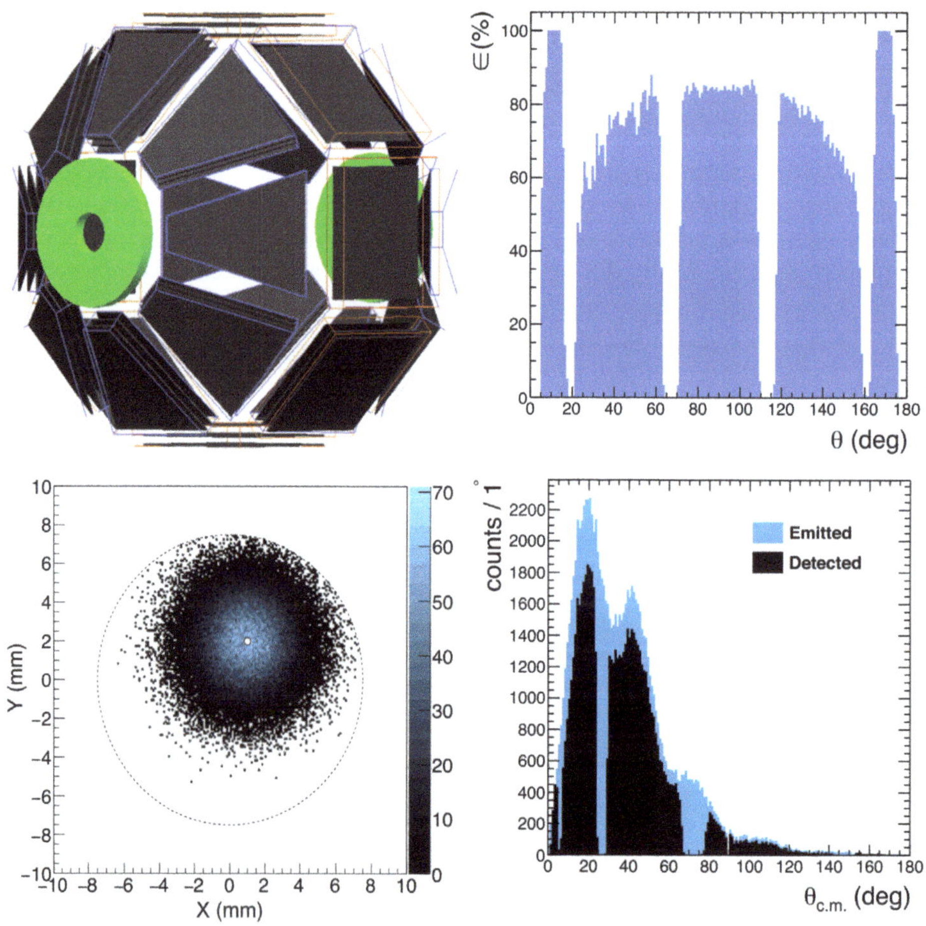

Figure 10.13. Simulation of the $^{132}Sn(d,p)$ reaction with the GASPARD silicon detector array using NPTool [22]. The radioactive ^{132}Sn beam is accelerated to 10MeV/u and is interacting with a 5 μm thick CD_2 target. (Top left) the geometry of GASPARD. (Top right) the particle detection efficiency of GASPARD as a function of angle with respect to the beam axis. (Bottom left) the spatial distribution of the ^{132}Sn radioactive beam on the target. (Bottom right) the emitted and detected protons from the $^{132}Sn(d,p)$ reaction. (Reprinted from [22] with the permission of IOP Publishing.)

future GASPARD silicon detector array for study of the ^{132}Sn(d,p) reaction using an accelerated radioactive beam. The geometry has been used to define the efficiency of the detector array in order to produce realistic particle spectra. Figure 10.14 provides a second example of the value of NPTool where simulated charged particle events for a silicon detector telescope arrangement are compared with real data as an aid to identify loci associated with different charged-particle species.

RadWare

RadWare is a suite of software focused on gamma-ray spectroscopy, particularly high fold, i.e. high multiplicity data collected with large germanium detector arrays. There are tools to assist in calibrating gamma-ray spectra and extracting efficiency curves from source data. There are also dedicated tools to manipulate γ–γ matrices and γ–γ–γ cubes. Information on gamma ray energies and intensities can be collated into level schemes using further tools.

GRAIN

GRAIN [23] is a JAVA-based software package for data analysis. While it is largely focussed on analysis of data from the University of Jyväskylä Accelerator Laboratory in Finland, it is worth describing here since it is designed to handle the challenge of free-running digital data acquisition systems with large numbers of individual channels. It can be used to correlate data over long time scales to search

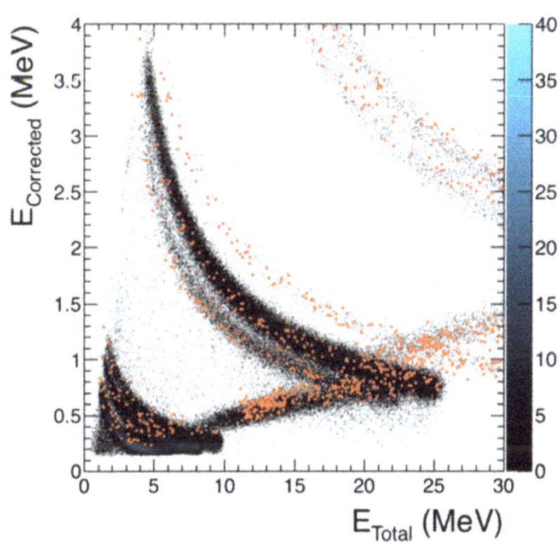

Figure 10.14. Comparison of experimental data with simulation for charged particles detected from the reaction of a 50 MeV/u ^{9}Li radioactive beam on a 1.8 mg cm^{-2} CD_2 target. The particles are detected using a telescope arrangement of silicon detectors where the first layer is 20 μm-thick and the second layer is thick in order to fully stop the charged particles. The data is plotted with energy loss in the front layer on the y-axis and the total energy deposited on the x-axis. The red points are randomly generated particles inserted into the simulation. These serve to identify the different loci for charged particles in the $\Delta E - E$ plot. (Reprinted from [22] with the permission of IOP Publishing.)

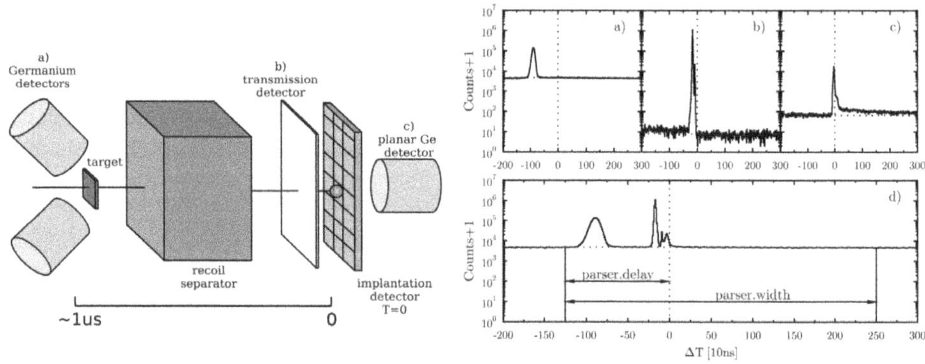

Figure 10.15. Schematic illustration of a typical RDT setup and the time structure of a typical RDT experiment data stream with respect to any signal in the implantation detector. Panels (a)–(c) represent the typical response of some individual detector groups. Panel (d) shows the summed structure and the main event builder timing parameters. (Reprinted from [23] with the permission of Elsevier.)

for decays following an implantation event in a DSSD pixel and correlate them. Figure 10.15 provides a schematic for how GRAIN handles a recoil-decay-tagging (RDT) experiment (see section 3.3.1) obtained using the Total Data Readout system [18].

References

[1] Foo B W and Goulding F S 1988 Ballistic deficits in shaping amplifiers *IEEE Trans. Nucl. Sci.* **35** 114

[2] Gedcke D A and McDonald W J 1968 Design of the constraint fraction of pulse height trigger for optimum time resolution *Nucl. Instrum. Methods* **58** 253–60

[3] Shimazoe K, Takahashi H, Yeom J Y, Furumiya T and Ohi J 2013 Multi-channel waveform sampling ASIC for radiation detection and measurement *Radiat. Meas.* **55** 87–9

[4] Ren J 2020 A high time resolution and low-power ASIC for MRPC applications *Radiat. Detect. Technol. Methods* **4** 63–9

[5] Vernon E *et al* 2019 Front-end ASIC for spectroscopic readout of virtual Frisch-grid CZT bar sensors *Nucl. Instrum. Methods* A **940** 1–11

[6] Pommé S, Fitzgerald R and Keightley J 2015 Uncertainty of nuclear counting *Metrologia* **52** S3–17

[7] Usman S and Patil A 2018 Radiation detector deadtime and pile up: A review of the status of science *Nucl. Eng. Technol.* **50** 1006–16

[8] Warburton W K and Grudberg P M 2006 Current trends in developing digital signal processing electronics for semiconductor detectors *Nucl. Instrum. Methods* A **568** 350–8

[9] Arriojas A, Barros H, Walter J and Sajo-Bohus L 2014 FPGA based digital signal processing for high resolution low energy gamma ray spectrometry *AIP Conf. Proc.* **1607** 65

[10] Yang Y *et al* 2020 A real-time peak-detection approach for nuclear detection and its implementation on an FPGA *Radiat. Detect. Technol. Methods* **4** 161–73

[11] Liu Y, Zhang J, Liu L, Li S and Zhou R 2018 Implementation of real-time digital CR-RC^m shaping filter on FPGA for gamma-ray spectroscopy *Nucl. Instrum. Methods* A **906** 1–9

[12] Warburton W K, Momayezi M, Hubbard-Nelson B and Skulski W 2000 Digital pulse processing: new possibilities in nuclear spectroscopy *Appl. Radiat. Isot.* **53** 913–20

[13] Garnsworthy A B *et al* 2017 The GRIFFIN data acquisition system *Nucl. Instrum. Methods* A **853** 85–104

[14] Jordanov V T and Knoll G F 1994 Digital synthesis of pulse shapes in real time for high resolution radiation spectroscopy *Nucl. Instrum. Methods* A **345** 337–45

[15] Grzywacz R 2003 Applications of digital pulse processing in nuclear spectroscopy *Nucl. Instrum. Methods* B **204** 649–59

[16] Salathe M and Kihm T 2016 Optimized digital filtering techniques for radiation detection with HPGe detectors *Nucl. Instrum. Methods* A **808** 150–5

[17] Flores J L, Martel I, Jiménez R, Galán J and Salmerón P 2016 Application of neural networks to digital pulse shape analysis for an array of silicon strip detectors *Nucl. Instrum. Methods* A **830** 287–93

[18] Lazarus I H *et al* 2001 The GREAT triggerless total data readout method *IEEE Trans. Nucl. Sci.* **48** 567

[19] Agramunt J *et al* 2013 A triggerless digital data acquistion system for nuclear decay experiments *AIP Conf. Proc.* **1541** 165

[20] Anderson J T *et al* 2012 A digital data acquisition system for the detectors at Gammasphere *IEEE Nuclear Science Symp. and Medical Imaging Conf. Record (NSS/MIC) (Anaheim, CA)* 1536–40

[21] Antcheva I *et al* 2009 ROOT—A C++ framework for petabyte data storage, statistical analysis and visualization C++ framework for petabyte data storage, statistical analysis and visualization *Comput. Phys. Commun.* **180** 2499–512

[22] Matta A 2016 NPTool: a simulation and analysis framework for low-energy nuclear physics experiments *J. Phys.* G **43** 045113

[23] Rahkila P 2008 Grain–a Java analysis framework for total data readout *Nucl. Instrum. Methods* A **595** 637–42

Chapter 11

Closing remarks

In drawing this volume to a close[1], I would like to give some suggestions for how to follow up on your interest in detectors for nuclear physics and applications. This e-book has been very much a personal view on the topic—indeed, this whole e-book series is intended to focus on personal views. I have unashamedly placed a heavy focus on what is contemporary and what is eye-catching. My motivation has been to think about what I would want to read were I starting out again as PhD student in nuclear physics but this book is by no means the only book on this topic. There are several listed in the further reading section listed at the end. Many are, admittedly, much longer than this e-book and have the space to go into much more detail on aspects such as electronics. They often give a more historical perspective.

Journals are a considerable aid in expanding knowledge of this topic. The key journal as far as nuclear physics detectors and instrumentation is concerned is *Nuclear Instruments and Methods A* but there are other journals such as *Journal of Instrumentation* where more and more such material has been appearing in recent years. On the applications side, but not exclusively so, the *IEEE Transactions on Nuclear Science* are a goldmine of useful articles. Many IEEE articles have more of an engineering bent than a physics one but the physicist cannot ignore the engineering challenges in this field. If you were attentive to the references provided in each chapter, you will see that the literature on detectors is rather broad and that discussion of such detectors can appear in a diverse range of journals.

Attending conferences is also a valuable way to learn more about the field. There are usually several conferences per year held worldwide on this topic. The premier such conference is the annual IEEE Nuclear Science Symposium (NSS) and Medical Imaging Conference (MIC). This is a conference of spectacular scale with keynote talks by world-leading researchers, a dense programme of parallel talks on specialist

[1] If you have reached this chapter by working through all the preceding chapters, then well done. If you have got here by skipping ahead to the end then well done for making the most of the e-book format!

doi:10.1088/978-0-7503-1428-2ch11

topics and poster exhibitions running to many hundreds of posters. The full gamut of applications is covered from homeland security to medical imaging—the latter being a particular focus. There are also smaller scale conferences at, e.g. the European level.

Let me close by offering some particular advice on navigating through this complex field:

- Nuclear physics is still a big field internationally and it is one driven by advances in detector capability. It is not the only player in the field who makes use of radiation detection. Radiation detection is crucial to other big science areas such as particle physics, astronomy and space science. Their demands can be different in terms of energy range of particles they want to detect and the scale of their instrumentation but it should not be neglected that the technology they develop can be repurposed. Physicists, like many, are apt to sit in silos. There is a real danger of reinventing the wheel to solve problems that others have already addressed. Similarly, industry and medicine make wide use of radiation detection. Some industries can be heavily conservative, such as the nuclear industry where detector development is often more evolution than revolution. Other fields like medical imaging are investing very significant money in detector development. The rapid advance of the silicon photomultiplier, driven by the demands of PET imaging, is a very good example of this. Silicon photomultiplier technology in terms of nuclear physics is still in its infancy but could be transformational.
- In chapter 4, we emphasised the need to match the detector to the measurement task in hand. It is easy to be wowed by reports of a prototype detector with breakthrough performance but if this detector is presently highly experimental, small in scale and significantly expensive, you need to think carefully if this is the path to follow. Many such technologies are quite fragile. They may perform well in isolated tests in the laboratory but be difficult to scale up to large experimental set-ups. For applications, they will need to be suitable to be used in the field and are almost certain to be rough-handled by some potential users.
- While this book has aimed for a contemporary picture, I believe it is a mistake not to try to get an acquaintance with some of the historical aspects of the field. Indeed, I often come across ideas and concepts for experiments in older textbooks, even from the 1950s and 1960s, where a technique has great promise but the detector technology was not right: too expensive, too complicated, too large and too cumbersome. It then gets bypassed by history and becomes a dead end. These ideas are ripe for re-invention. Maybe some of the discussion of newer detector technology in this e-book will inspire you to reinvent a technique that others dropped because it was simply too hard at the time.
- At the risk of this sounding too much like a self-help book, I would encourage people to get out of their comfort zone. Why not cold call a potential industrial partner or strike up a conversation with a medical physicist about whether the detector and apparatus you are working on could solve their real-world problems? Working with industry can be immensely satisfying when one can

provide a solution or expertise. Just remember the other advice above that your technology needs to be robust, not just something that works well in the lab on a good day with a lot of care and attention. In the same spirit, why not speak with your astrophysics and high energy physics colleagues? Maybe they have already solved detection problems that you need answers to? Maybe your condensed matter colleagues have a wonder material that can be repurposed as a ground-breaking radiation detector?

Appendix A

Further reading

A.1 Nuclear structure models

- *Quantum Mechanics for Nuclear structure: a primer*, J L Wood and K Heyde, IOP Publishing, 2020, ISBN: 978-0-7503-2179-2 (e-book).
- *Quantum Mechanics for Nuclear structure: intermediate*, J L Wood and K Heyde, IOP Publishing, 2020, ISBN: 978-0-7503-2171-6 (e-book).
- *Introductory Nuclear Physics*, K S Krane, Wiley, 2008, ISBN: 978-8126517855.
- *Fundamentals of Nuclear Models: Foundational Models*, D J Rowe and J L Wood, World Scientific, 2010, ISBN: 978-981-3101-72-2 (e-book).
- *The Nuclear Shell Model*, K L G Heyde, Springer, 1990, ISBN: 978-3-642-97203-4 (e-book).
- *Nuclear Structure from a Simple Perspective (Second Edition)*, R F Casten, Oxford Studies in Nuclear Physics, 2001, ISBN: 9780198507246.
- *From Nucleons to Nucleus*, J Suhonen, Springer-Verlag Berlin Heidelberg, 2007, ISBN: 978-3-540-48861-3 (e-book).
- *Nuclear Stucture: Volume I: Single Particle Motion*, Aa Bohr and B Mottelson, World Scientific, 1998: ISBN: 981-0-239-793.
- *Nuclear Stucture: Volume II: Nuclear Deformations*, Aa Bohr and B Mottelson, World Scientific, 1998: ISBN: 981-0-239-807.
- *Simple Models of Complex Nuclei, The Shell Model and the Interacting Boson Model*, I Talmi, Harwood Academic Publishers, New York, 1993.

A.2 Nuclear astrophysics

- *Cauldrons in the Cosmos*, C E Rolfs and W S Rodney, University of Chicago Press, 2005, ISBN: 978-0226724577.
- *Nuclear Physics of Stars*, C Iliadis, Wiley, 2015, ISBN: 978-3527336487.

doi:10.1088/978-0-7503-1428-2ch12

A.3 Radiation detectors and applications

- *Radiation Detection and Measurement*, G F Knoll, Wiley, 2010, ISBN: 978-0-470-13148-0.
- *Techniques in Nuclear Structure Physics*, J B A England, Palgrave, *1974, ISBN: 978-0333174753.*
- *Techniques for Nuclear and Particle Physics Experiments: A How-to Approach*, W R Leo, Springer-Verlag, 1994, ISBN: 0-387-57280-5.
- *Practical Applications of Radioactivity and Nuclear Radiations*, G C Lowenthal and P L Airey, Cambridge University Press, 2001, ISBN: 978-0-521-01980-4.
- *Physics and Engineering of Radiation Detection*, S Ahmed, Elsevier, 2014, ISB: 9780128016442 (e-book).
- *Practical Gamma-ray Spectrometry*, G Gilmore, Wiley, 2011, ISBN: 978-1-119-96469-8 (e-book).
- *Radiation and Detectors*, L Cerrito, Springer, 2011, ISBN: 978-3-319-53181-6 (e-book).
- *Alpha, beta and gamma-ray spectroscopy: Vol 1 and 2*, K Siegbahn, North Holland, 1966.
- *Inorganic Scintillators for Detector Systems*, P Lecoq, A Getkin and M Korzhik, Springer, 2017, ISBN: 978-3-319-45522-8 (e-book)
- *Handbook of Nuclear Spectrometry*, J Kantele, Elsevier Science Publishing, 1995, ISBN: 978-0-123-96440-3.
- *Nuclear Physics: Principles and Applications*, J Lilley, Wiley, 2004, ISBN: 978-0471979364.
- *Nuclear Radiation Detectors*, S S Kapoor and V S Ramamurthy, Wiley, 1986, ISBN: 978-0852264966.
- *Semiconductor Detector Systems*, H Spieler, Oxford University Press, 2005, ISBN: 978-0198527848.
- *Signal Processing for Radiation Detectors*, M Nakhostin, Wiley, 2017, ISBN: 978-1119410225.

A.4 Statistics

- *Statistics for nuclear and particle physicists*, L Lyons, Cambridge University Press, 2018, ISBN: 978-1139167710 (e-book).

www.ingramcontent.com/pod-product-compliance
Ingram Content Group UK Ltd.
Pitfield, Milton Keynes, MK11 3LW, UK
UKHW051902100926
8556IPUK00005B/732